Introductory Statistics for
Engineering Experimentation

Introductory Statistics for Engineering Experimentation

Peter R. Nelson
Clemson University

Marie Coffin
Paradigm Genetics Inc

Karen A. F. Copeland
Statistical Consultant

ELSEVIER
ACADEMIC
PRESS

Amsterdam Boston Heidelberg London New York Oxford
Paris San Diego San Francisco Singapore Sydney Tokyo

Academic Press
An Imprint of Elsevier
525 B Street, Suite 1900, San Diego, California 92101-4495, USA
http://www.academicpress.com

Academic Press
An Imprint of Elsevier
84 Theobald's Road, London WC1X 8RR, UK
http://www.academicpress.com

Library of Congress Control Number:

0-12-515423-2

PRINTED IN THE UNITED STATES OF AMERICA
04   05   06   07   08   9   8   7   6   5   4   3   2

CONTENTS

PREFACE

The Accreditation Board for Engineering and Technology (ABET) introduced a criterion starting with their 1992–1993 site visits that "Students must demonstrate a knowledge of the application of statistics to engineering problems." Since most engineering curricula are quite full with requirements in their own discipline, they generally do not have the time for a traditional two semesters of probability and statistics. Attempts to condense that material into a single semester often result in so much time being spent on probability that the statistics useful for designing and analyzing engineering/scientific experiments is never covered.

Therefore, we began to develop and teach a one-semester course whose purpose was to introduce engineering/science students to some useful statistical methods. We tried a number of different texts, but never found one we felt covered the right material in a one-semester format. Our initial approach to covering the desired material was to skip around in the text. That, however, was not well received by many of the students and proved to be difficult for the instructor. In order to alleviate some of the problems caused by skipping around, we started to supplement the text with a set of class notes. This book is the outgrowth of those class notes. We have used it as the text in our introductory statistics course for engineering and science students at Clemson University since 1998. Continuous improvement since then has resulted in the current book.

We have tried to use only as much theory as is necessary to motivate the techniques. While calculus is a prerequisite for our course, it is actually only used when obtaining probabilities, expected values, and variances from some simple density functions. In line with the more recent ABET requirement that students must be able to demonstrate "an ability to design and conduct experiments as well as analyze and interpret data" (Engineering Accreditation Commission, 1997) our main emphasis is on the design and analysis of experiments. We introduce the idea of modeling early (Chapter 3) by first modeling the deterministic aspects of an experiment. Then (in Chapter 4) we model the random fluctuations (the random error) around the deterministic part of the model. The remaining chapters put the two modeling aspects together in order to do inference on the complete model, starting with the simplest deterministic models and building up to the models most useful in engineering experimentation. We emphasize the use of graphics in both the analysis and presentation of results. For example, normal probability plots are used to assess normality. In this spirit we emphasize using the analysis of means (ANOM) to analyze fixed-effect designed experiments when it is appropriate. In our experience engineers and scientists find the ANOM much easier to understand and apply than the analysis of variance (ANOVA). We also discuss the ANOVA and point out that it is applicable to a broader range of problems than the ANOM.

All our students are required to do a project. Either individually, or in groups of up to three, students must choose a topic of interest to them, design an experiment to study that topic (we require at least two factors with one factor at more than two levels), conduct the experiment, analyze the results, write a project report, and give a presentation. The material is laid out so they are introduced to the design aspects of experimentation early enough for them to design their experiments and collect data for their projects while we are discussing exactly how this data should be analyzed. Examples of some project reports based on actual projects are given in one of the appendices.

We envision a course of this type being taught with some kind of statistical software; however, the text is not oriented toward any particular software package. All the data sets referred to in the text are available in a plain text format that can be easily imported into virtually any statistics package. The examples and problems reflect actual industrial applications of statistics.

Engineering Accreditation Commission (1997). "ABET Engineering Criteria 2000". Accreditation Board for Engineering and Technology Inc., Baltimore, MD.

1

INTRODUCTION

Most scientific disciplines are concerned with measuring items and collecting data. From this book you will learn how to analyze data in order to draw more powerful conclusions from your experiments. With the help of statistics, data can be used to:

- Summarize a situation
- Model experimental outcomes
- Quantify uncertainty
- Make decisions.

Because the quality of our conclusions depends on the quality of the original data, some attention will be given to efficient ways of collecting data. Generally, we obtain data by taking a **sample** from some larger **population**. The population is a conceptual group of all the items of interest to the scientist. We take samples because the entire population may be infinite, or at least too large to be examined in a timely and economic manner.

Example 1.0.1 A production process makes seat posts for mountain bikes that must fit into the seat post tube of a bike frame. The seat posts must be of a certain diameter or they will not fit. The design specifications are 26 mm ± 0.5 mm. To monitor the production process a technician takes 10 posts off the production line every hour and measures them. If more than 1 of the 10 are outside the specification limits, the process is halted, and the controls that determine diameter are investigated.

Frequently, questions of what to measure and how to measure it are more difficult than in the above example. When the quantity of interest is simple (like diameter) it is obvious what needs to be measured and how.

Example 1.0.2 Air pollution is a serious problem in many cities and industrial areas. In deciding what to measure one needs to consider the time, cost, and accuracy of different measurement systems. One good indicator of pollution is ozone concentration. This can be measured fairly easily by taking an air sample and counting the ozone in it (measured in parts per hundred million).

Variability

Clearly, if the answer to a statistical study is important, we should make our measurements as precisely as possible. In fact, we often think of measured quantities as completely "correct". No matter how much care is taken, however, the measurement process will almost inevitably introduce some **measurement error** into our calculations. That is, even if a careful person measures the same item repeatedly, he will not arrive at the same answer every time. Furthermore, we are usually measuring a new item each time, and the items (even if they appear to be identical) will have their own inherent **random variability**. Because our calculations involve both of these two sources of variability, scientific conclusions always contain some uncertainty. One of the functions of statistics is to quantify that uncertainty. For example, if we say a component will fail at an ambient temperature of 40°C, it matters if the estimate is $40° \pm 1°$ or $40° \pm 10°$.

Experimental Design

In order to minimize the variability so that it is possible to more precisely determine what treatments in an experiment have significant effects, one must carefully plan how the data should be collected. Particularly when one is interested in studying more than one factor at a time, such as studying the effects of different temperatures and different times on a chemical reaction, much greater efficiency can be obtained if the proper data are collected.

Example 1.0.3 Suppose one is interested in studying the effects of time and temperature on the yield of a chemical process. The experimenter believes that whatever effect time has will be the same for the different temperatures. A common (but inefficient) approach is to study the effects of time and temperature separately (this is referred to as a one-factor-at-a-time design). Such a design might call for studying three times (for a fixed temperature) and taking three observations for each time. Then one might study two temperatures (for a fixed time) using three observations for each temperature. With these 15 total observations we can estimate both the average yield at a particular temperature and the average yield at a particular time. The uncertainty in the average yield for a fixed temperature would be less than that for a fixed time because more observations were taken (nine versus six). (Exactly how to determine this uncertainty is discussed in Chapter 5.)

A much more efficient way to study the two factor (discussed in Chapter 8) is to study them together. Taking two observations for each Time/Temperature combination would require only 12 total observations, but would provide estimates for both the average yield at a particular temperature and the average yield at a particular time with uncertainties of only 87% ($= 100\%\sqrt{9/12}$) of the uncertainty for the average temperature from the previous design. Thus, by studying the two factors together one can obtain better information with fewer observations.

Random Sampling

Even when studying only a single population, care must be taken as to how the data are collected. In order to quantify the uncertainty in an estimate of a population characteristic (e.g., the diameter of a piston), one has to estimate the variability associated with the particular characteristic. As already mentioned, this variability consists of two parts: variability due to the characteristic itself (e.g., differences in the pistons) and variability due to the measurement process. In order to correctly assess the variability due to the characteristic of interest, one must collect samples that are representative of the entire population. This is accomplished by taking samples at random (referred to as **random sampling**). The idea is that samples should be chosen such that every possible sample is equally likely to be obtained. While it is not always possible to perform exact random sampling, it is important that the sample be representative. For example, if one is interested in comparing the lifetimes of two brands of batteries, one would want to obtain samples of the batteries from different locations to help ensure that they were not all manufactured at the same time in the same plant. Random sampling ensures that (on the average) the samples are independent and representative. That is, they are not related to one another other than coming from the same population.

Randomization

In addition to random sampling the concept of **randomization** is also important. The basic idea is that treatments should be assigned to experimental units at random and, when applicable, experimental trials should be performed in a random order so that any biases due to the experimental units not behaving independently because of some unknown or uncontrollable factor are on average distributed over all the experimental trials.

Example 1.0.4   Consider an experiment to compare rates of tire wear, and suppose one is interested in only two brands. The tires are to be used on cars for say 10,000 miles of driving, and then the tread wear is to be measured and compared. Since it is desirable to subject the two brands of tires to the same conditions affecting wear, it would be reasonable to conduct a paired experiment (see the discussion on pairing in Chapter 6) and assign one of each brand of tire to each car used. Also, since it is known that front and rear tires do not necessarily wear in the same way, the two tires should both be mounted on either the front or the rear. Having chosen say the rear, all that remains is to decide which brand goes on which side, and this is where randomization comes into the picture. In order to average over both brands any effect on wear due to the side on which a tire is mounted, the brands should be assigned to the sides at random. This could be done, for example, by using a table of random numbers and associating odd numbers with the assignment of brand A to the driver's side.

Replication

Multiple observations under the same experimental conditions are referred to as **replicates**. Replicate observations are used to determine the variability under the specific experimental conditions, and therefore, a true replicate must be obtained from a complete repetition of the experiment. A replicate measure is not to be confused with a repeated measure, where multiple measurements are taken on the same experimental unit.

Example 1.0.5 In Example 1.0.3 (p. 2) the two observations for each Time/ Temperature combination are two replicates. A true replicate in this case would require the process being completely re-run under the same experimental conditions. Simply running two assays on the product for a particular Time/Temperature combination would be a repeated measure rather than a true replicate since it would only encompass the variability due to the assay (measurement error), and would not reflect the variability in the process from batch to batch.

Problems

1. A scientist working with new fibers for a tennis ball covering wants to compare the "bounce" of the balls with different coverings. One test will consist of dropping balls with each covering from a specified height and counting the number of times it bounces. Explain why it is better to use several different balls with each covering, rather than dropping the same ball repeatedly.

2. A student wants to measure the volume of liquid in cans of soft drink to see how it compares to the nominal volume of 12 oz. printed on the can. Suggest some sources of measurement error in this experiment, and discuss how the measurement errors could be reduced or avoided.

3. An industrial engineer specializing in ergonomics performs an experiment to see if the position (left, center, or right) within a work station affects the speed achieved by a production line worker who packages a product from a conveyor belt into boxes. Carefully explain what random error (i.e., random variability) signifies in this experiment. What should be randomized in this experiment and why?

4. Suppose you need to measure the size of some candies (Gummi Bears and Peanut M&M's), where size is defined to be the maximum diameter of the candies.
   (a) Which kind of candy will probably be subject to more measurement error? Explain.
   (b) Suggest how this measurement error could be reduced.
   (c) Which kind of candy will probably be subject to more random variability? Explain.

5. Suppose a food scientist wanted to compare the "tastiness" of different variations of a cookie recipe.
   (a) Suggest several ways to quantify (measure) "tastiness", and discuss the advantages and disadvantages of each.
   (b) Discuss the sources of variability (e.g., measurement error) in the context of this problem, and suggest ways to minimize the effects of variability on the results of the study.

6. R&D chemists conducted an extensive study of a new paint formulation. Four batches of the paint were made and ten test panels from each batch were sprayed for testing. One characteristic of interest was the gloss of the paint (an appearance measure). Discuss repeated measures versus replication in the context of this study.

2

SUMMARIZING DATA

In this chapter, we will consider several graphical and numerical ways to summarize sets of data. A data summary should provide a quick overview of the important features of the data. The goal is to highlight the important features and eliminate the irrelevant details.

The first (and often most important) thing to do with a set of data is to find a way to display it graphically. Graphs can summarize data sets, show typical and atypical values, highlight relationships between variables, and/or show how the data are spread out (what one would call the shape, or the distribution, of the data). Even for small sets of data, important features may be more obvious from a graph than from a list of numbers. We will discuss several ways to graph data, depending on how the data were collected.

2.1 SIMPLE GRAPHICAL TECHNIQUES

Univariate Data

Histograms

When data are **univariate** (one characteristic is measured on each experimental unit), a histogram is often a good choice of graph.

Example 2.1.1 (*Rail Car Data*)   A company ships many products via rail cars. These rail cars are either owned or leased by the company, and keeping track of their whereabouts is critical. The company is interested in minimizing the amount of time that a rail car is held by a customer. A histogram of the number of days rail cars from a particular fleet were held by customers during a 4-month period is given in Figure 2.1 and shows several interesting features of the data (which are found in `railcar.txt`).

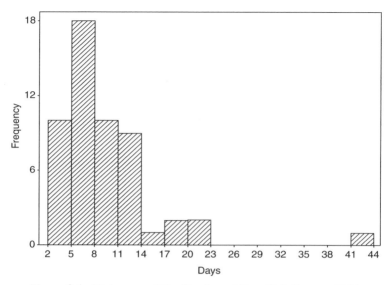

Figure 2.1. Histogram of the Number of Days Rail Cars are Held.

- The amount of time a customer keeps a rail car varies greatly. This makes scheduling of the rail cars difficult.
- There are a few observations in the right "tail" that are far away from the majority of the observations. Unusually low values are not of a concern in this situation, however, unusually high values indicate a possible customer issue that needs to be resolved.
- The number of days a rail car is held by a customer is centered around 7 or 8.
- Samples with long hold times are more spread out than the samples with short hold times. Thus, the distribution of hold times is skewed toward the higher values, or **right skewed**.

Several steps are involved in constructing a histogram. You must decide on the class boundaries, count the number of observations in each class, and draw the histogram. Generally, the class boundaries are chosen so that somewhere between 10 and 20 classes are obtained. For larger data sets, more classes are appropriate than for smaller data sets. There is no single correct way to construct the class boundaries. All that is required is an appropriate number of equal width classes that encompass all the data. One way to do this is to divide the range of the data (i.e., largest value − smallest value) by the desired number of classes to obtain an approximate class width.

Example 2.1.2 (*Silica Surface-Area Data*)   Samples of a particular type of silica (a chemical product with many applications such as a filler in rubber products) were tested for their surface area (a key property). The resulting 32 measurements are listed in Table 2.1.

Table 2.1. Surface Areas for a Particular Type of Silica

101.8	100.5	100.8	102.8	103.8	102.5	102.3	96.9
100.0	99.2	100.0	101.5	98.5	101.5	100.0	98.5
100.0	96.9	100.7	101.6	101.3	98.7	101.0	101.2
102.3	103.1	100.5	101.2	101.7	103.1	101.5	104.6

The range of values is $104.6 - 96.9 = 7.7$, and therefore to obtain approximately 10 classes, the class width should be $7.7/10 = 0.77 \doteq 0.8$. Starting with 96.9 as the lower boundary for the first class, one obtains classes of 96.9–97.7, 97.7–98.5, and so forth. The complete **frequency distribution** (a count of the number of observations in each class) is given in Table 2.2. Note that since some of the data values fall exactly on a class boundary, one has to decide which class to put them in. It was arbitrarily decided to put values that fell on the boundary in the upper class. For example, the two 98.5 values were both put in the 98.5–99.3 class. The important issue is not which class to put the value in, but to be consistent for all of the data. The corresponding histogram is given in Figure 2.2.

Table 2.2. Frequency Distribution for the Data in Table 2.1

Class	Frequency	Relative Frequency
96.9–97.7	2	0.0625
97.7–98.5	0	0.0
98.5–99.3	4	0.125
99.3–100.1	4	0.125
100.1–100.9	4	0.125
100.9–101.7	8	0.25
101.7–102.5	4	0.125
102.5–103.3	4	0.125
103.3–104.1	1	0.03125
104.1–104.9	1	0.03125
Total	32	1.000

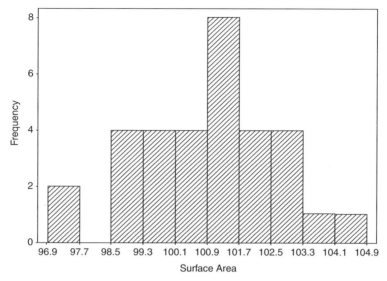

Figure 2.2. Histogram for the Frequency Distribution in Table 2.2.

The histogram shows that the surface area distribution is relatively balanced about a middle value of around 101. Typically with data such as this, measurements on multiple samples from the same source, one would expect more of a bell-shaped distribution than what is seen here. However, 32 data points is a small enough sample that the choice of class size or starting point can affect the look of the histogram (see Problem 2.1.2, p. 15).

There are several things to note about histograms.

- All of the classes have the same width.
- The classes are adjoining, but not overlapping. This is so that every observation will be in one and only one class (i.e., it will get counted exactly once).
- The number of classes is somewhat arbitrary. Choosing too few classes will cause your graph to lose detail, while choosing too many may obscure the main features.
- Histograms for small data sets are more sensitive to the number of classes than histograms for large data sets.

Runs Charts

Time ordered plots or **runs charts** provide a view of the data based on the order in which the data were collected or generated. Runs charts are of particular interest when working with production data, however, they are not limited to such data. A runs chart is simply a chart on which the data are plotted in the order in which they were collected. By plotting the data with respect to time, one can check for time dependencies.

Example 2.1.3 (*Historical Production Data*)   The runs charts in this example were generated from **historical data**, that is, data that were collected and saved, but analyzed after the fact; as opposed to data that were generated in a controlled experiment with the type of analysis to be performed on the data known in advance. The runs charts were used to look at the process in the past to better understand the process for the future.

Data from a chemical process were collected at different time intervals depending on the particular variable being measured (some were collected every minute, some every 5 minutes, etc.). The values associated with a particular variable were then averaged over the shift. This resulted in one data point per shift and three data points per day for each variable. Three plots are shown below, two for process variables, and one for a quality variable. In reality this process had well over 100 process variables with recorded values and about a dozen quality variables. The runs charts helped the plant engineers separate the variables that had high variability, or were unusual, from all the others. For Process Variable 1 the runs chart (Figure 2.3) shows a drift down over the first 30 time periods (i.e., 10 days), then the values appear to stabilize. Such a drift is an event that the engineers might want to research for causes, especially as it precedes the jump in the quality variable. The runs chart for Process Variable 2 (Figure 2.4) does not exhibit any highly "non-random" behavior, and thus, is of little interest. The chart for the Quality Variable (Figure 2.5) shows a jump in the values at about time 35, but by time 70 it appears to have been "corrected" (the target for this variable is 22).

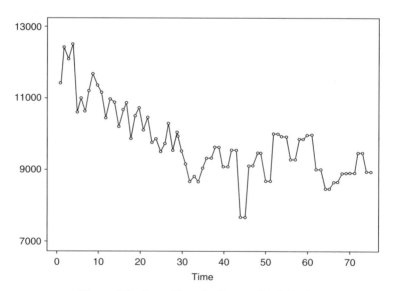

Figure 2.3. Runs Chart for Process Variable 1.

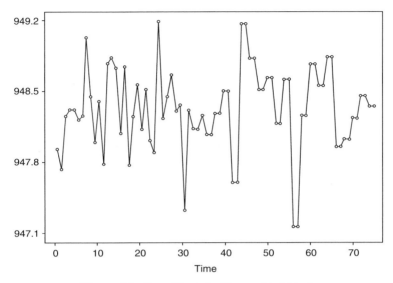

Figure 2.4. Runs Chart for Process Variable 2.

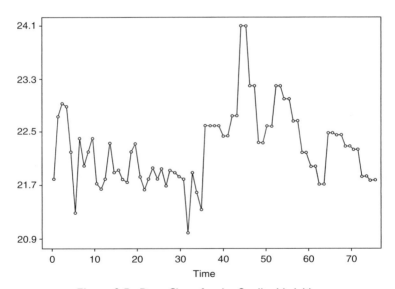

Figure 2.5. Runs Chart for the Quality Variable.

Bivariate Data

Scatter Plots

If data are **bivariate**, that is, two variables are recorded for each experimental unit (say x_i and y_i for the ith unit), a scatter plot (a plot of the (x_i, y_i) pairs) can be used to show the relationship between the variables.

Example 2.1.4 (*Test Temperature Data*)   Developers of automotive paint must test not only the appearance of the paint but also the durability. Some durability tests are run on test panels that have been put in a freezer to simulate exposure to cold temperatures. One such test was to be run at $-10°C$. To set up the test protocol a study was done to determine how long it took (in seconds) for a test panel to reach $-10°C$ after being removed from the freezer. This would allow the lab personnel to decide on a temperature for the freezer such that they would have sufficient time to remove the panel and take it to the test equipment. The study was conducted with both original equipment manufacture (OEM) and repaired panels (repaired panels have additional coats of paint on them to simulate touch up work done on a car after the paint job but before it leaves the paint area). The data are shown in Table 2.3 and stored in `timetemp.txt`.

Table 2.3. Durability Test Data

	Repair						
Temperature	−22.7	−25.5	−29.0	−24.8	−27.2	−23.5	−23.9
Time	5	11	16	9	14	6	6
Temperature	−28.7	−26.8	−25.9	−26.9	−24.2	−28.5	
Time	16	13	11	11	8	16	
	OEM						
Temperature	−24.1	−23.4	−26.6	−25.3	−22.9	−25.0	
Time	3	3	8	6	3	4	
Temperature	−23.0	−25.9	−25.7	−28.7	−28.5		
Time	4	7	7	11	12		

Three scatter plots of these data are given in Figures 2.6–2.8. In the individual plots for both OEM (Figure 2.6) and the repair panels (Figure 2.7), there are generally negative linear trends, indicating that the colder the starting temperature the longer it takes for the panel to reach $-10°C$ (as would be expected). However, the trends for OEM and repair panels differ, as is more apparent when both groups of data are plotted together (Figure 2.8) using two different symbols to differentiate between data sets.

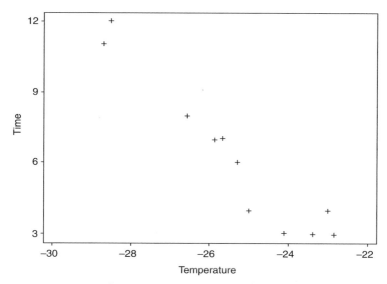

Figure 2.6. Scatter Plot for the OEM Test Data.

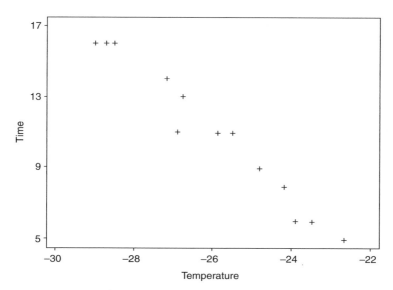

Figure 2.7. Scatter Plot for the Repair Test Data.

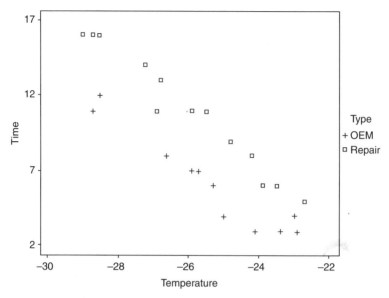

Figure 2.8. Scatter Plot Comparing OEM and Repair Data.

From Figure 2.8 it appears that, in general, the rate of warming is similar for the two types of panels, but the repair panels take longer.

Problems

1. The following data (`webvisit.txt`) are the number of visits to a web site over a 3-week period.

15321 15382 15337 15338 14320 15366 15251 15319 15405 14594
14453 14376 14249 15377 14489 15285 15425 15333 15319 15273
14290

   (a) Construct a histogram of these data.

   (b) Write a short report describing the distribution of the visits to this web site. Include some discussion of why this distribution might have that particular shape.

2. Construct a histogram for the data from Example 2.1.2 (p. 7) using a different class width than was used in the example. The data are stored in `surfarea.txt`.

3. A company wants to give one of its raw material suppliers a measure of how well a process is running with a new raw material. On a scale of 1–10 they rate a number of key areas of the process and average those ratings. This average is then given to the supplier as a measure of the success of the raw material. The data for the ratings are stored in `ratings.txt`.

   (a). Construct a histogram for these data.

   (b). Write a description of the distribution for the ratings.

4. The data set `ph.txt` contains measurements from three batches of material that were made at a chemical plant. For each batch numerous pH readings were taken during the course of production. For this particular material the pH should have been consistent throughout production.

   (a) Construct a histogram for these data.

   (b) Discuss the sources of measurement error in this study. What could have been done to minimize the measurement errors?

   (c) Construct a histogram for each individual batch.

   (d) Write a short report based on the histograms for these data.

5. A company was developing a new vitamin for those at risk from ailments associated with low calcium levels. In order to test the effectiveness of three new formulations a group of test subjects had their calcium levels tested before and after taking one of the three vitamins. The data are stored in `vitamin.txt`.

   (a) Suggest a way to measure the effectiveness of treatment, and calculate this measure for each subject in the study.

   (b) For each vitamin group construct a histogram for your measurement.

(c) Describe the shapes of the three histograms. Compare and contrast the three treatments.

(d) Suggest another way the scientists could have tested the vitamin formulations, especially if the number of test subjects was limited.

6. The data set `drums.txt` contains weights that were made on 30 drums of a chemical product. Weights were taken before the drums were filled (empty drum weight) and after the drums were filled (full drum weight).

(a) Construct a runs chart for the empty weight. Include a line at the average empty weight for reference and comment on the chart.

(b) Construct a runs chart for the full weight. Include a line at the average full weight for reference and comment on the chart.

(c) Construct a runs chart for the net weight. Include a line at the average and a line at the target of 411 and comment on the chart.

7. The data set `absorb.txt` contains data from a silica processing plant. One key quality parameter for silica is the amount of oil it can absorb since silica is often mixed with rubber and oil in various rubber applications (battery separators, tires, shoe soles, etc.). The measurements here represent the average oil absorption of the silica produced on one shift.

(a) Construct a histogram for the data and describe its properties.

(b) Construct a runs chart for the data.

(c) Write a brief summary of the two data displays contrasting the information gained by each plot.

8. The ratings discussed in Problem 3 were constructed because the company felt that the moisture levels in the raw material were causing them to have problems in their plant. The data set `ratings2.txt` contains both the ratings and the corresponding moisture levels for the raw material.

(a) Make a scatter plot of the data.

(b) Write a paragraph describing the relationship between the two variables. Is there evidence of a relationship between the ratings and the moisture levels?

(c) Discuss the potential merits or downfall of the rating measurement.

9. The data set `odor.txt` contains three responses gathered from a factorial design (described in Chapter 9). The purpose of the experiment was to study the effect of formulation changes for a lens coating on three primary responses: odor, hardness, and yellowing. The odor of each formulation was ranked on a scale of 1–10 (with 10 being the smelliest) by 10 people and then all 10 rankings were averaged to obtain the value of odor for the analysis. The purpose of this portion of the analysis is to look for relationships between the formulation odor and the two other responses.

(a) Make scatter plots of the data.

(b) Do the responses appear to be related to the odor?

(c) Describe the relationships between the variables.

10. In the development of coatings (such as for paper, optical lenses, computer screens, etc.) a researcher will often study a "thin film"; that is, the coating cured on some substrate such as glass; rather than studying the coating cured on the actual product. The data set `thinfilm.txt` contains data from a study of thin films made on two different substrates: glass and foil.

   (a) What rationale do you think is behind studying a thin film as opposed to the entire product?

   (b) Make scatter plots for both the glass and foil portions of the data to study the relationship between the formulation component and the max load (a measurement of the strength of the coating).

   (c) Describe how the max load is affected by the amount of the formulation component and how the substrate affects that relationship.

11. The data set `vitamin.txt` contains the results of a study on vitamins. Three groups of test subjects each received a different treatment (three different vitamin formulations). Each subject's calcium levels before and after treatment were recorded.

   (a) To reduce bias in the experiment, the subjects should have been assigned a treatment regimen at random.

      i. Explain what this means.

      ii. Using just the "before" calcium levels, how could we check if any obvious non-random assignment was used?

      iii. Perform this check, and comment on the results.

   (b) Make separate scatter plots for the three sets of data.

   (c) Write a report summarizing the experiment and its results.

2.2 NUMERICAL SUMMARIES AND BOX PLOTS

In addition to graphical methods, there are also certain numbers that are useful for summarizing a set of data. These numbers fall into two categories: measures of location and measures of spread.

Notation: For the formulas that follow we will need the following notation. The number of observations in our data set will be n, and the individual observations will be denoted as $y_1, y_2, \ldots, y_n$.

Measures of Location

A location statistic tells us where the data are centered. We will look at two such measures: the sample mean and the sample median.

Sample Mean

The sample mean is defined as

$$\bar{y} = \frac{1}{n} \sum_{i=1}^{n} y_i. \tag{2.2.1}$$

This is the usual average. We are simply adding up all the observations and dividing by the number of observations. Physically, it represents the balance point if each data point has the same weight (see Figure 2.9).

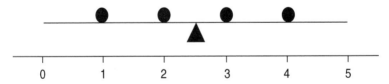

Figure 2.9. Physical Representation of the Sample Mean.

Sample Median

The sample median is the middle observation when the observations are ordered from smallest to largest. If n is even, there are two "middle" observations, and the median is as the average of these two observations.

Notation:
- We will use $y_{(i)}$ to denote the ith smallest observation. Therefore, $y_{(1)} \leq y_{(2)} \leq \cdots \leq y_{(n)}$.
- We will use $\tilde{y}$ to denote the sample median.

The sample median is defined as

$$\widetilde{y} = \begin{cases} y_{\left(\frac{n+1}{2}\right)} & \text{if } n \text{ is odd} \\ \frac{1}{2}\left[y_{\left(\frac{n}{2}\right)} + y_{\left(\frac{n}{2}+1\right)}\right] & \text{if } n \text{ is even.} \end{cases} \tag{2.2.2}$$

Definition (2.2.2) can be rephrased as

$$\widetilde{y} = \begin{cases} y_{\left(\frac{n+1}{2}\right)} & \text{if } \frac{n+1}{2} \text{ is an integer} \\ \text{average of the two observations} & \\ \text{closest to position } \frac{n+1}{2} & \text{if } \frac{n+1}{2} \text{ is not an integer.} \end{cases} \tag{2.2.3}$$

Example 2.2.1 (*Test Temperature Data, p. 12*)   Consider the time it takes for OEM test panels to reach $-10°C$. The sample mean of theses values is

$$\overline{y} = \frac{1}{11}(3 + 3 + 8 + 6 + 3 + 4 + 4 + 7 + 7 + 11 + 12) = \frac{1}{11}(68) = 6.18.$$

So 6.18 seconds is the average time it takes for OEM test panels to reach $-10°C$. The sample median is found by first ordering the data from smallest to largest:

$$3 \quad 3 \quad 3 \quad 4 \quad 4 \quad 6 \quad 7 \quad 7 \quad 8 \quad 11 \quad 12.$$

Since n is odd, the sample median is

$$\widetilde{y} = y_{\left(\frac{11+1}{2}\right)} = y_{(6)} = 6.$$

Thus, we can think of either 6.18 seconds or 6 seconds as measuring the center for this group. Neither one is "more correct" than the other. In this case, the mean and median were about the same (in practical terms they are the same), but in some cases they may be quite different.

Example 2.2.2 (*Rail Car Data, p. 6*)   In this example, $\overline{y} = 8.89$ days, and the median is 7 days. The mean and median are different because the distribution of the data is skewed. A few large observations (such as the customer that held the rail car for 42 days) will tend to make the mean, but not the median, larger. For skewed data or data with a few extremely large or extremely small values (called **outliers**), the sample median may be more representative of the center of the data set than the sample mean.

Example 2.2.3 (*Silica Surface-Area Data, p. 7*)   For this data set there are $n = 32$ measurements, so the sample median is the average of the two observations closest to position $(32 + 1)/2 = 16.5$.

$$\widetilde{y} = \left[y_{(16)} + y_{(17)}\right]/2 = [101.2 + 101.2]/2 = 101.2.$$

The sample mean is $\bar{y} = 100.94$. In this case the sample mean and median are quite close, as would be expected from the histogram of the data (p. 9), which is fairly symmetric.

Measures of Spread

These statistics quantify how spread out, or variable, or dispersed, the data set is. We will consider four such sample measures: the range, the interquartile range, the variance, and the standard deviation.

Sample Range

The sample range is the difference between the largest and the smallest data values

$$R = y_{(n)} - y_{(1)}.$$

The range is very sensitive to outliers in either direction.

Example 2.2.4 (*Test Temperature Data, p. 12*)   For the OEM panel data, the largest observation is $y_{(11)} = 12$ and the smallest observation is $y_{(1)} = 3$. Therefore, the range is $R = 12 - 3 = 9$.

Interquartile Range

The interquartile range (IQR) avoids sensitivity to outliers by measuring the spread of the middle half (50%) of the data. It is defined as

$$\text{IQR} = q(0.75) - q(0.25)$$

where $q(p)$ is the pth **sample quantile** and is defined as

$$q(p) = \begin{cases} y_{(np+\frac{1}{2})} & \text{if } np + \frac{1}{2} \text{ is an integer} \\ \text{the average of the two} \\ \text{observations closest to} & \text{if } np + \frac{1}{2} \text{ is not an integer.} \\ \text{position } np + \frac{1}{2} \end{cases} \qquad (2.2.4)$$

This definition of sample quantiles is a generalization of definition (2.2.3) for the sample median, and $\tilde{y} = q(0.5)$.

Note: Sample quantiles can be defined in slightly different ways, so one should not be concerned if, for example, sample quantiles obtained from a computer program are slightly different than the ones computed using equation (2.2.4).

Note that if $q(0.25)$ is the observation in position k (or the average of the two closest), then $q(0.75)$ is the observation in position $n - k + 1$ (or the average of the two closest).

Example 2.2.5 (*Test Temperature Data, p. 12*)   Consider the time to reach $-10°C$ for the 11 OEM test panels, which were ranked (i.e., ordered) in Example 2.2.1 (p. 19). Since $np + 0.5 = 11(0.25) + 0.5 = 3.25$,

$$q(0.25) = \left[y_{(3)} + y_{(4)}\right]/2 = (3 + 4)/2 = 3.5$$
$$q(0.75) = \left[y_{(11-3+1)} + y_{(11-4+1)}\right]/2 = \left[y_{(9)} + y_{(8)}\right]/2 = (8 + 7)/2 = 7.5.$$

Therefore, IQR $= 7.5 - 3.5 = 4$, and the times for the middle six test panels are within 4 seconds of one another.

Example 2.2.6 (*Silica Surface-Area Data, p. 7*)   For the silica surface-area data, $q(0.25) = 100$ units and $q(0.75) = 102.05$ units. Therefore, IQR $= 102.05 - 100 = 2.05$ units, so the middle 50% of the data fall within 2.05 units of one another.

Sample Variance

The sample variance is defined as

$$s^2 = \frac{1}{n-1} \sum_{i=1}^{n} (y_i - \overline{y})^2. \tag{2.2.5}$$

It measures the average (squared) distance of the observations from the sample mean. If the observations are all close to each other, then each $y_i - \overline{y}$ will be small, and hence s^2 will be small. On the other hand, if the observations are far apart, some of the $(y_i - \overline{y})$'s will be large, and s^2 will be large. One disadvantage of the sample variance is that it measures spread in units that are the square of the original units. Therefore, it is harder to interpret.

The sample variance (equation (2.2.5)) is not exactly an average since we have divided by $n - 1$ rather than n. Ideally, we would like to measure the distance between each observation and the center of the population from which our sample was obtained. Using the sample mean as an estimate of the population center results in distances that tend to be too small. This is compensated for by dividing by $n-1$ rather than n.

Sample Standard Deviation

The sample standard deviation is defined as

$$s = \sqrt{s^2} = \sqrt{\frac{1}{n-1}\sum_{i=1}^{n}(y_i - \overline{y})^2}.$$

The sample standard deviation is simply the square root of the sample variance, resulting in a measure of spread that is in the same units as the original observations. The sample variance and standard deviation are both influenced by outliers.

All of the measurements of spread have one property in common: they will be large if the data are spread out and small if the data are tightly packed together. Thus, they are mainly useful for comparing one data set to another.

We have referred to all of the measures computed from a sample as "sample" statistics (e.g., the sample mean) to emphasize the point that they are numbers calculated from a sample. It is important to distinguish between our sample statistics (easily calculated quantities) and the corresponding population parameters (something we can probably never know for sure). The population consists of all the items or individuals of interest, and the population parameters (e.g., the population mean and variance) are not directly observable because generally we cannot measure the whole population. In Chapter 5 we will discuss using sample statistics to make **inferences** about the corresponding population parameters.

Notation: We will use the Greek letters μ and σ to denote the population mean and standard deviation, respectively. We will use σ^2 to denote the population variance.

Box-and-Whisker Plots

The IQR introduced earlier as a measure of spread is the key to a graphical display known as a box-and-whisker plot, or simply a box plot. This compact plot provides a quick way to display a summary of a set of data. It is especially useful for comparing several sets of data. A box plot uses the following five sample statistics:

$$y_{(1)} \quad q(0.25) \quad \widetilde{y} \quad q(0.75) \quad y_{(n)}.$$

The **box** of the box plot runs from $q(0.25)$ to $q(0.75)$, with the sample median marked in the box. The **whiskers** are lines that extend from $q(0.25)$ to $y_{(1)}$ and from $q(0.75)$ to $y_{(n)}$.

A box plot for the 32 silica surface-area measurements is given in Figure 2.10. Notice that the two whiskers are approximately the same length and the sample median is close to the center of the box. This indicates the data are close to being symmetric.

Note: Some statisticians recommend a modified box plot where the whiskers extend only to the furthest data points within (1.5)IQR above q(0.75) and below q(0.25). If there

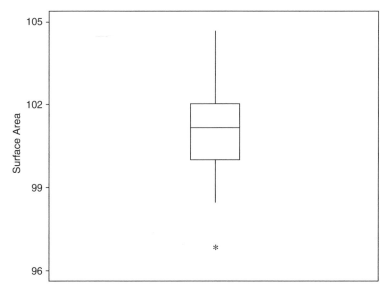

Figure 2.10. A Box Plot for the Silica Surface-Area Data.

are any data points outside this range they are plotted as single points to indicate that they may be outliers. This modified box plot is more difficult to draw but is common in computer packages (such as the one used to produce the plot in Figure 2.10).

When placed side-by-side, box plots can be used to help compare groups of data.

Example 2.2.7 (*Tablet Data*) A process engineer suspected a problem with a batch of chlorine used to make chlorine tablets for home pool chlorinators. The material had already been made into tablets. In order to check the batch the engineer selected 10 tablets at random from the suspect batch (Batch 1) and 10 tablets from each of two other batches. The lab tested all 30 tablets and recorded the lifetime of each tablet in `tablets.txt`. The data were plotted using side-by-side box plots (see Figure 2.11), and it was immediately clear that there had indeed been a problem with Batch 1.

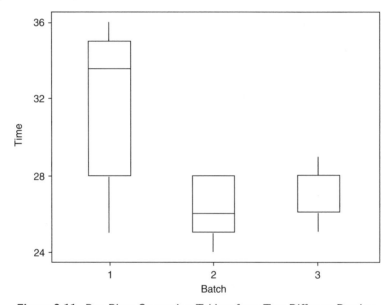

Figure 2.11. Box Plots Comparing Tablets from Two Different Batches.

Problems

1. For the data in `webvisit.txt` calculate the sample median and interquartile range and make a box plot.

2. For the data set `odor.txt` (see Problem 2.1.9, p. 16) do the following.

   (a) Calculate the sample mean and standard deviation of hardness.

   (b) Calculate the sample median and interquartile range for hardness.

   (c) Which set of summary statistics do you prefer, and why?

   (d) Construct a histogram for the hardness values.

   (e) Use the previous parts of the problem to write a report summarizing the distribution of hardness.

3. Calculate the sample mean and standard deviation for the time to reach $-10°$C for the Repair test panels discussed in Example 2.1.4 (p. 12).

4. Calcium levels were measured for three groups of test subjects after taking a vitamin (see Problem 2.1.11, p. 17). The data are recorded in `vitamin.txt`.

   (a) Calculate the sample median and interquartile range for each set of test subjects.

   (b) Make side-by-side box plots for the three sets of subjects.

   (c) Summarize the differences in calcium levels (if any) between the three sets of subjects.

5. A large industrial corporation produces, among other things, balls used in computer mice. These plastic balls have a nominal diameter of 2 cm. Samples of 10 balls were taken from each of two different production lines, and the diameters of the sampled balls were measured. The results are given below.

   Line 1:   2.18   2.12   2.24   2.31   2.02   2.09   2.23   2.02   2.19   2.32
   Line 2:   1.62   2.52   1.69   1.79   2.49   1.67   2.04   1.98   2.66   1.99

   (a) Calculate summary statistics and make side-by-side box plots for the two production lines.

   (b) Two important issues in production are being on target and having minimal variation. Write a short discussion of the two production lines describing which line seems to be doing a better job, and why?

6. The data set `safety.txt` contains the number of recorded safety violations at a company's six plants over a 5-year period. Use graphs and summary statistics to explore this data set, and then write a brief report summarizing your findings. Be sure to include supporting data and graphs in your summary.

2.3 GRAPHICAL TOOLS FOR DESIGNED EXPERIMENTS

A designed experiment is one in which design variables (or **factors**) are systematically changed and the outcome (or **response**) is measured for each change. The purpose of conducting designed experiments is to study how the factors affect the response.

Example 2.3.1 (*Optical Lens Data*) In the development of coatings (such as for UV protection or anti-reflexiveness) for optical lenses it is important to study the durability of the coating. One such durability test is an abrasion test, simulating day-to-day treatment of lenses (consider how one cleans lenses: on the corner of a T-shirt, etc.). In this study the controlled variable is the surface treatment of the lens, and the response Y is the increase in haze after 150 cycles of abrasion. Minimal increase in haze is desired. The data are given in Table 2.4 as well as in dhaze.txt. There are various ways we could summarize and display these data. For example,

Table 2.4. Optical Lens Data

	Treatment		
1	2	3	4
8.52	12.5	8.45	10.73
9.21	11.84	10.89	8.00
10.45	12.69	11.49	9.75
10.23	12.43	12.87	8.71
8.75	12.78	14.52	10.45
9.32	13.15	13.94	11.38
9.65	12.89	13.16	11.35

box plots for the four treatment groups are given in Figure 2.12. Also, summary statistics are given in Table 2.5.

Table 2.5. Summary Statistics for the Optical Lens Data

Treatment 1	Treatment 2	Treatment 3	Treatment 4
$\bar{y} = 9.45$	$\bar{y} = 12.61$	$\bar{y} = 12.19$	$\bar{y} = 10.05$
$s = 0.716$	$s = 0.417$	$s = 2.084$	$s = 1.302$
$n = 7$	$n = 7$	$n = 7$	$n = 7$

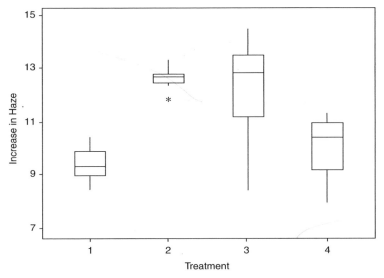

Figure 2.12. Box Plots of the Optical Lens Data.

This information can also be displayed in an error-bar chart as shown in Figure 2.13. An error-bar chart has a circle (or some other symbol) for each group's mean, and

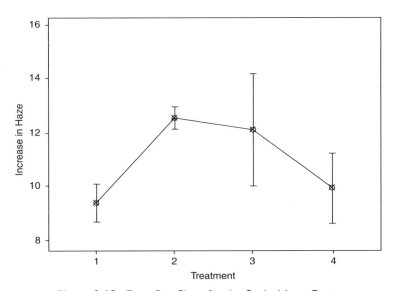

Figure 2.13. Error-Bar Chart for the Optical Lens Data.

lines on each side of the mean that are some measure of variability. The lines on the above chart are one standard deviation in length. From the chart, one can see that the mean haze varies from one treatment group to another (Treatment 1 seems to be the best), and also that the random error in Treatment 3 is greater than that of the other three groups.

In Example 2.3.1 there is just one **factor**. That is, the different treatments are different versions of one thing: a lens coating. We can easily imagine experiments with more than one factor. Perhaps we would want to change the density of the plastic (i.e., lens), as well as the lens coating, since there may be some combination of density and coating that provides the best wear. In such a case, we call the density one factor, and the lens coating another factor. One possible experimental design for such a study is a **factorial design**, where there is at least one observation for each possible combination of density and lens coating levels.

Note: The terminology "error-bar chart" is somewhat confusing since the "bar" refers to the lines used to indicate variability (i.e., error) and not to the type of chart. For a two-factor experiment an error-bar chart with group means for one factor calculated at each level of the second factor is an effective way to display the results. Such a graph is easier to read without the error bars and is usually referred to as an **interaction plot**.

Example 2.3.2 (*Paint Formulation Data*) In the development of an automotive coating (i.e., paint) an experiment was run that varied the types of two components (chemicals) in the coating formulation. Each test formulation was sprayed onto two metal panels and cured as if on a car. Responses were measured on the appearance and performance of the coating. In this example an appearance measurement, long wave (LW), is considered. Low values of LW are desired. Each treatment combination had two replicates, and the data are given in Table 2.6 and an interaction plot is shown in Figure 2.14. The data are also stored in `lw.txt`.

Table 2.6. The Appearance Measure Long Wave for the Paint Formulation Data

		Component 1				
		1	2	3	4	Average
	1	10.4	5.3	10.5	6.5	7.98
		8.7	5.9	8.5	8.0	
Component 2	2	6.5	4.2	7.0	3.4	5.44
		8.3	5.4	5.1	3.6	
	3	11.5	7.4	11.4	10.1	10.55
		11.2	8.1	12.9	11.8	
	Average	9.43	6.05	9.23	7.23	7.99

From the interaction plot we can see that regardless of which chemical is used for Component 1, chemical 3 is the worst choice for Component 2. Notice that the graph has nearly "parallel" lines. This means that the effect of Component 2 is the same, no matter what chemical is used for Component 1. One would say in

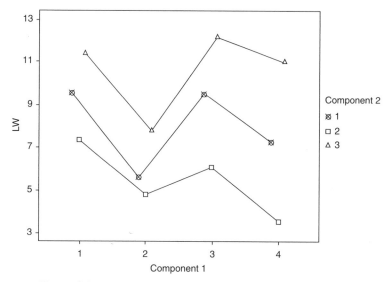

Figure 2.14. Interaction Plot for the Paint Formulation Data.

this case that the two factors (Components 1 and 2) do not **interact**. The overall best combination of factors appears to be chemical 2 for Component 2 and either chemical 2 or 4 for Component 1.

Example 2.3.3 (*Coating Formulation Data*)   A similar experiment to the one described in Example 2.3.2 was run in which there were two factors, the level of Chemical A and the order in which Chemical B was added to the formulation (either before or after Chemical A). Again, an appearance measurement, distinctness of image (DOI), is considered. Higher values correspond to better appearance of the coating, and the data are given in Table 2.7, and an interaction plot is shown in Figure 2.15. From the interaction plot, we can see the strong interaction between the amount of Chemical A used and the time that Chemical B was added to the

Table 2.7. Coating Formulation Data

Amount of A	Order of B	DOI
7.7	1	92
7.7	2	87
7.7	1	90
7.7	2	85
3.86	1	88
3.86	2	91
3.86	1	87
3.86	2	91

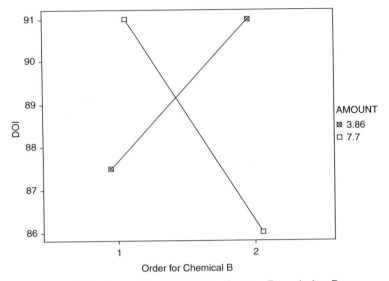

Figure 2.15. Interaction Plot for the Coating Formulation Data.

formulation. The effect of increasing Chemical A depends on when Chemical B is added to the formulation. When Chemical B is added at time 1, DOI increases as the amount of Chemical A is increased, whereas, when Chemical B is added at time 2, DOI decreases as the amount of Chemical A is increased. The nature of this interaction provides the experimenter with options for a final formulation. There are two different combinations that yield high DOI values. Which of the two combinations (Time 1 with high Chemical A or Time 2 with low Chemical A) is best would depend on other responses and/or cost issues.

Problems

1. An experiment was performed to compare the effect of pH on the adhesive qualities of a lens coating. The test procedure for measuring adhesion is "noisy", that is, there is a large amount of test error in the measurement. For that reason, 10 samples from each formulation were made and tested. The data are given below as well as in `adhesion.txt`.

pH										
4.3	15.17	14.27	13.3	17.85	15.17	13.81	17.97	13.25	13.79	13.62
5.3	15.72	15.78	15.72	15.48	15.13	18.61	14.04	13.84	14.22	16.17
6.3	15.87	14.50	15.82	14.39	17.52	15.62	16.78	15.81	15.24	16.53

(a) Make side-by-side box plots for these data.

(b) Make an error-bar chart for these data.

(c) Which graph do you think provides better information, and why?

(d) How does pH affect the adhesion of the coating?

2. The data set `labcomp.txt` is the result of an "inter-lab" study where multiple labs, in this case each located at a different plant, participate in a study to compare measurements within and between the labs. For this study, the surface area of a silica was measured by each lab seven times. All labs had the capability of measuring using a 1-point CTAB procedure, and in addition two labs also measured using a 5-point CTAB procedure.

(a) Calculate summary statistics for each lab.

(b) Make side-by-side box plots comparing the labs (one for the 1-point data and one for the 5-point data)

(c) Make an error-bar chart of the data.

(d) Describe the distributions, and note any differences among the labs.

3. A second experiment on the lens coating discussed in Problem 1 was run to study the effects of both the pH level and the catalyst level on the adhesion. Three levels of pH were studied along with two levels of catalyst. Five measurements were taken for each of the treatment combinations. The data for the experiment are stored in `adhesion2.txt`.

(a) Make plots of the data, including an interaction plot.

(b) Write a paragraph summarizing the results of this experiment.

4. A major automotive paint manufacturer has a "panel farm" on the coast of Florida. The purpose of this "farm" is to expose paints to different environmental conditions (sun, rain, salt, and sand) to test durability. A set of panels from three test formulations was sent to the farm and left out in the elements for 6 months. After that time they were rated on their appearance using a scale of 1–10 (1 being good). The data is in `exposure.txt`.

(a) Calculate summary statistics for each formulation.

(b) Graph the data.

(c) Summarize your findings.

2.4 CHAPTER PROBLEMS

1. Consider the following small sample of five SAT verbal scores.

$$580 \quad 475 \quad 602 \quad 350 \quad 713$$

   (a) Find the mean SAT score.
   (b) Find the median SAT score.
   (c) Find the variance of the SAT scores.
   (d) Find the IQR of SAT scores.

2. Suppose that the data given below is the length of time in months before the first major repair is required on samples of Brand X and Brand Y computers.

$$\text{Brand X: } 8 \quad 9 \quad 11 \quad 12 \quad 18 \quad 32$$
$$\text{Brand Y: } 0 \quad 0 \quad 9 \quad 11 \quad 20 \quad 32$$

   (a) Compute the sample mean for each brand.
   (b) Compute the sample standard deviation for Brand X.
   (c) Using the same scale, draw box-and-whisker plots for the two data sets.
   (d) Based on the above information, which brand would you prefer, and why?

3. Consider the following data on the diameter of ball bearings from a particular production line.

$$1.18 \quad 1.42 \quad 0.69 \quad 0.88 \quad 1.62 \quad 1.09 \quad 1.53 \quad 1.02 \quad 1.19 \quad 1.32$$

   (a) Find the sample mean and the sample variance.
   (b) Find the IQR.

4. A chemical process was run using two different alcohols (Factor A) and three different bases (Factor B). The process was run twice with each alcohol/base combination, and the resulting yields are given below.

	Base		
	93	76	65
Alcohol	95	78	61
	60	75	95
	64	79	93

Draw an interaction plot for these data. What would you conclude?

5. A manufacturer of cheese has a number of production facilities across the country. One product that is produced is used by a major pizza chain, and it is critical that there is consistency in the cheese made by individual plants. The data set `cheese.txt` contains data from six batches at each of three plants. The factor of interest is the fat content. Summarize these data for the quality manager of the company using numerical and graphical summaries.

6. Time and temperature are key properties in many chemical curing processes. Process engineers are often concerned with determining a "bake window" for a product, that is, finding a time/temperature combination that achieves the best results in the smallest amount of time. In one process the engineers studied three temperature levels on their oven at two different time intervals and measured the yield of parts (if the parts were not cured correctly they could not be used). Using the data set `cure.txt` summarize the data that the engineers collected, and based on the data make a recommendation for the best time/temperature combination.

7. As a safety precaution employees who work in a particular area of a chemical plant are monitored monthly for the level of mercury in their urine. Data in `urine.txt` are results from four employees over a 1-year time period. Summarize these data for the safety officer. Be sure to look at the data over time and note any unusual occurrences in the data.

3

MODELS FOR EXPERIMENT OUTCOMES

In Chapter 2, we discussed graphical and numerical summaries of data. Another way to summarize data is with a **mathematical model**. A mathematical model is simply a mathematical expression for the relationship among several variables. For example,

$$d = \frac{gt^2}{2} \qquad (3.0.1)$$

is a mathematical model for the distance an object will travel due to gravitational acceleration. In model (3.0.1), the quantities $d =$ distance and $t =$ time are variables. The quantity g is *not* a variable. It is the gravitational acceleration constant: $g = 9.8\,\mathrm{m/sec.}^2$. A number like g (one with a fixed value) is called a **parameter** of the mathematical model. Unlike g, many parameters have unknown values. A large part of statistics is concerned with estimating unknown parameter values.

Consider the problem of estimating the shelf life of a drug in order to obtain an expiration date from the Food and Drug Administration (FDA). An expiration date must be granted by the FDA before a drug is allowed onto the market. Initial estimates are often obtained using accelerated testing methods. Accelerated testing consists of exposing the product to an extreme condition such as high heat to increase its rate of degradation. From the data collected under extreme conditions inferences are made as to how the product will behave under normal conditions.

In order to infer the rate of degradation at normal temperatures from the rate at high temperatures, a mathematical model is needed. In this case that mathematical model is based on two other mathematical models. One needs a model for (i) the relationship between the amount of chemical product, known as the reactant, and the length of time it is stored at a particular temperature, and for (ii) the relationship between the degradation rate and the temperature.

The relationship between the amount of reactant and the length of time stored at a particular temperature can be described by

$$\frac{dz}{dt} = -K(T)z^c \tag{3.0.2}$$

where z is the amount of reactant, t is time, c is the order of the reaction, and $K(T)$ is the rate constant, which is dependent on temperature T. There are various models for the rate constant that may be appropriate. However, in the pharmaceutical industry the Arrhenius equation is frequently employed. The rate constant $K(T)$ is said to follow the Arrhenius equation if

$$K(T) = a \exp\left[-\frac{E}{RT}\right] \tag{3.0.3}$$

where T is temperature in Kelvin, a is the frequency factor, E is the activation energy, and R is the gas constant. Integrating equation (3.0.2) when $c = 1$ (a common value for drugs) and using the Arrhenius equation (3.0.3), the degradation of a drug can be described by

$$\ln(z) = \ln(z_0) - a \exp\left(-b/T\right) t \tag{3.0.4}$$

where

$$z = \text{the amount of the drug at time } t$$
$$z_0 = \text{the amount of the drug at time zero}$$
$$b = E/R.$$

The parameters a and b in model (3.0.4) would need to be estimated. This would involve storing the product at a fixed (high) temperature, and at specified time points assaying a quantity of the product to determine how much drug remained. Values of a and b would then be chosen to give the best prediction of the measured values.

Notice that if we estimate a and b and then use equation (3.0.4) to predict the rate at which the drug will degrade when stored at room temperature, our prediction will contain some uncertainty due to small errors in the assayed values (measurement error) and due to the fact that every batch of the product is slightly different from every other batch (random variability). We will refer to this uncertainty as **random error**. However, because we do not know how well our model will predict reality, there is still another source of error in our prediction. This is called **modeling error**. Despite all these errors, model (3.0.4) can do a very good job of predicting shelf life.

As we have already seen, most interesting experiments involve data that are obtained from different treatments (i.e., taken under a variety of experimental conditions) for the purpose of comparing the treatments to one another. Therefore, we need to consider models that facilitate comparing the responses from different treatments. In this chapter, we will be concerned with estimating the parameters in models for different experimental designs and studying the modeling error. In Chapter 4, we will study models for the random error, and in later chapters we will combine the two types of models.

3.1 MODELS FOR SINGLE-FACTOR EXPERIMENTS

We will first consider single-factor experiments, that is, experiments in which only the levels of a single factor are changed. It is reasonable to suppose that the observations may be centered at different locations for different levels of the factor. A mathematical model for this situation can be written as

$$Y_{ij} = \mu_i + \epsilon_{ij}$$

(3.1.1)

where

Y_{ij} = the jth replicate observation at the ith level of the factor

μ_i = the mean response at the ith level

ϵ_{ij} = the individual random error associated with observation Y_{ij}.

Note that the Y_{ij} in equation (3.1.1) represents a random quantity since it contains the random error ϵ_{ij}.

Once we have actually obtained a value for the jth replicate observation at the ith level of the factor, we would refer to it as y_{ij}. Thus, y_{ij} is a specific realization of the random quantity Y_{ij}.

Example 3.1.1 (*Optical Lens Data, p. 26*)   Using model (3.1.1) to describe this set of data would correspond to saying we think the different lens coatings may result in different durabilities. The value 8.52 would be denoted as y_{11} since it is the first observation at the first level of the factor lens coating. Similarly, $y_{12} = 9.21$ and $y_{47} = 11.35$.

Notation: We will use I to represent the number of factor levels, and n to represent the number of observations at each level (i.e., the number of replicates).

Parameter Estimation

For model (3.1.1), the parameters to be estimated are $\mu_1, \ldots, \mu_I$. The mean response at the ith level (i.e., μ_i) is estimated using the sample mean (equation (2.2.1)) at the ith level. That is,

$$\boxed{\widehat{\mu}_i = \overline{y}_{i\bullet}.}$$

(3.1.2)

Notation:

- We use the notation $\widehat{\mu}_i$ (read "mu sub i hat") to represent an estimate of μ_i. In general, any parameter with a "hat" over it will represent an estimate of that parameter.

- We use the notation $\overline{y}_{i\bullet}$ to represent the average of all the y_{ij} values for a specific i. That is, $\overline{y}_{i\bullet} = \frac{1}{n}\sum_{j=1}^{n} y_{ij}$. In general, any subscript replaced by a dot will indicate that the observations have been averaged over all the values for that particular subscript.

Estimating the Magnitude of the Error

The magnitude of the random error is also of interest. It is usually measured in terms of the variance of the random error. Model (3.1.1) says that the variance of the observations (the Y_{ij}'s) is due to the variance of the random errors (the ϵ_{ij}'s). Therefore, the variance of the random error is the same as the variance of the Y_{ij}'s. For any specific level of the factor (i.e., for a specific i), all the observations amount to a single sample, and the variance at that level (call it σ_i^2) can be estimated using equation (2.2.5). Thus, an estimate of the variance of the random error for level i is

$$\widehat{\sigma}_i^2 = s_i^2 = \frac{1}{n-1}\sum_{j=1}^{n}(y_{ij} - \overline{y}_{i\bullet})^2.$$

(3.1.3)

Often it is reasonable to suppose that the variance of the random error will be the same for each factor level, and we will refer to this common variance as σ^2. In that case the estimates (3.1.3) can be combined to obtain an estimate of σ^2 as

$$\widehat{\sigma}^2 = \frac{s_1^2 + s_2^2 + \cdots + s_I^2}{I}.$$

(3.1.4)

In Chapter 7, we will see how to modify this estimate to account for different sample sizes at different levels.

The Observed Errors

The random errors ϵ_{ij} are not parameters; they are random variables. However, once we obtain specific values y_{ij}, the corresponding specific errors can be obtained. Rearranging model (3.1.1), it follows that

$$\epsilon_{ij} = Y_{ij} - \mu_i.$$

Therefore, for a specific value y_{ij} the corresponding error is the difference between the observed value y_{ij} and the predicted value $\widehat{\mu}_i$,

$$\hat{\epsilon}_{ij} = y_{ij} - \hat{\mu}_i = y_{ij} - \overline{y}_{i\bullet}. \qquad (3.1.5)$$

The $\hat{\epsilon}$'s are referred to as **residuals**. Note that one can obtain the estimate of σ^2 (equation (3.1.4)) using (see Problem 3.1.4)

$$\hat{\sigma}^2 = \frac{1}{I(n-1)} \sum_{i=1}^{I} \sum_{j=1}^{n} (\hat{\epsilon}_{ij})^2. \qquad (3.1.6)$$

Example 3.1.2 (*Optical Lens Data, p. 26*)   Consider again the optical lens coating example. The data are reproduced in Table 3.1 with their summary statistics.

Table 3.1. Optical Lens Data Together with Their Summary Statistics

	Treatment			
	1	2	3	4
	8.52	12.5	8.45	10.73
	9.21	11.84	10.89	8.00
	10.45	12.69	11.49	9.75
	10.23	12.43	12.87	8.71
	8.75	12.78	14.52	10.45
	9.32	13.15	13.94	11.38
	9.65	12.89	13.16	11.35
$\overline{y}_{i\bullet}$	9.45	12.61	12.19	10.05
s_i^2	0.513	0.174	4.344	1.695

Estimates of the μ_i's, obtained using equation (3.1.2), are

$$\hat{\mu}_1 = 9.45 \quad \hat{\mu}_2 = 12.61 \quad \hat{\mu}_3 = 12.19 \quad \hat{\mu}_4 = 10.05.$$

Using equation (3.1.4), the common variance is estimated as

$$\hat{\sigma}^2 = \frac{0.513 + 0.174 + 4.344 + 1.695}{4} = 1.68.$$

The residuals, obtained using equation (3.1.5), are

$$\hat{\epsilon}_{11} = 8.52 - 9.45 = -0.93$$

$$\hat{\epsilon}_{12} = 9.21 - 9.45 = -0.24$$

$$\vdots$$

$$\hat{\epsilon}_{43} = 9.75 - 10.05 = -0.3$$

$$\vdots$$

$$\hat{\epsilon}_{47} = 11.35 - 10.05 = 1.3.$$

Problems

1. Consider again the data from the data set `adhesion.txt`.
   (a) Suggest an appropriate model for this data set.
   (b) Make an error-bar chart for the data. Does it seem reasonable to assume that all three treatment groups have the same variability? Explain.
   (c) Estimate the parameters of the model you chose in part (a).
   (d) Estimate σ^2 and explain what it stands for.
   (e) Calculate the residuals.

2. Consider the data set `labcomp.txt` discussed in Problem 2.3.2 (p. 31).
   (a) Suggest an appropriate model for the 1-point data.
   (b) Estimate the parameters of the model you chose in part (a).
   (c) Calculate the residuals.
   (d) Calculate the sample variance for each lab and use these to estimate σ^2.

3. Re-do Problem 2 using the 5-point data.

4. Show that equation (3.1.6) reduces to equation (3.1.4).

3.2 MODELS FOR TWO-FACTOR FACTORIAL EXPERIMENTS

Recall that a **factorial** experiment is one where each level of every factor appears with each level of every other factor in at least one treatment combination. In that case there are several mathematical models to choose from. We will discuss two here, and others are discussed in Chapter 8.

Example 3.2.1 (*Paint Formulation Data, p. 28*) The purpose of this experiment was to understand how two components in an automotive paint formulation affect the paint's appearance. One factor in the experiment was a chemical denoted as Component 1, which occurred at four levels. The other factor was a second chemical denoted as Component 2, which occurred at three levels. There were two replicate observations for each treatment combination.

Notation: Let y_{ijk} be the kth replicate observation for the ij^{th} treatment combination, that is, the group where Factor A is at the ith level and Factor B is at the jth level. Let I denote the number of levels of Factor A and J denote the number of levels of Factor B. Then

$$\bar{y}_{ij\bullet} = \text{the mean for the } ij^{\text{th}} \text{ treatment combination}$$
$$s^2_{ij} = \text{the variance for the } ij^{\text{th}} \text{ treatment combination}$$
$$\bar{y}_{i\bullet\bullet} = \text{the mean for the } i^{\text{th}} \text{ level of A}$$
$$\bar{y}_{\bullet j\bullet} = \text{the mean for the } j^{\text{th}} \text{ level of B}$$
$$\bar{y}_{\bullet\bullet\bullet} = \text{the overall mean}$$
$$n = \text{the number of replicates}$$
$$\text{(i.e., the number of observations for each treatment combination)}$$
$$N = \text{the total number of observations.}$$

Example 3.2.2 (*Paint Formulation Data, p. 28*) To illustrate the use of this notation, the summary statistics for this data set are given in Tables 3.2 and 3.3 with each statistic labeled.

Table 3.2. Summary Statistic Notation for the Means of the Data in Example 2.3.2, p. 28

		Component 1				
		1	2	3	4	
Component 2	1	$\bar{y}_{11\bullet} = 9.55$	$\bar{y}_{12\bullet} = 5.60$	$\bar{y}_{13\bullet} = 9.50$	$\bar{y}_{14\bullet} = 7.25$	$\bar{y}_{1\bullet\bullet} = 7.98$
	2	$\bar{y}_{21\bullet} = 7.40$	$\bar{y}_{22\bullet} = 4.80$	$\bar{y}_{23\bullet} = 6.05$	$\bar{y}_{24\bullet} = 3.50$	$\bar{y}_{2\bullet\bullet} = 5.44$
	3	$\bar{y}_{31\bullet} = 11.35$	$\bar{y}_{32\bullet} = 7.75$	$\bar{y}_{33\bullet} = 12.15$	$\bar{y}_{34\bullet} = 10.95$	$\bar{y}_{3\bullet\bullet} = 10.55$
		$\bar{y}_{\bullet1\bullet} = 9.43$	$\bar{y}_{\bullet2\bullet} = 6.05$	$\bar{y}_{\bullet3\bullet} = 9.23$	$\bar{y}_{\bullet4\bullet} = 7.23$	$\bar{y}_{\bullet\bullet\bullet} = 7.99$

Table 3.3. Summary Statistic Notation for the Variances of the Data in Example 2.3.2, p. 28

		Component 1			
		1	2	3	4
Component 2	1	$s_{11}^2 = 1.45$	$s_{12}^2 = 0.18$	$s_{13}^2 = 2.0$	$s_{14}^2 = 1.13$
	2	$s_{21}^2 = 1.62$	$s_{22}^2 = 0.72$	$s_{23}^2 = 1.81$	$s_{24}^2 = 0.02$
	3	$s_{31}^2 = 0.05$	$s_{32}^2 = 0.25$	$s_{33}^2 = 1.13$	$s_{34}^2 = 1.45$

A Model with No Interaction

If there is no interaction between the two factors, then one can fit the model

$$Y_{ijk} = \mu + \alpha_i + \beta_j + \epsilon_{ijk} \tag{3.2.1}$$

where

$$\mu = \text{the overall mean}$$
$$\alpha_i = \text{the main effect of the } i^{\text{th}} \text{ level of Factor A}$$
$$\beta_j = \text{the main effect of the } j^{\text{th}} \text{ level of Factor B}$$
$$\epsilon_{ijk} = \text{the random error associated with } Y_{ijk}.$$

Estimates for the parameters in model (3.2.1) are

$$\begin{aligned} \widehat{\mu} &= \overline{y}_{\bullet\bullet\bullet} \\ \widehat{\alpha}_i &= \overline{y}_{i\bullet\bullet} - \overline{y}_{\bullet\bullet\bullet} \\ \widehat{\beta}_j &= \overline{y}_{\bullet j\bullet} - \overline{y}_{\bullet\bullet\bullet}. \end{aligned} \tag{3.2.2}$$

Rearranging equation (3.2.1), replacing Y_{ijk} with the observed value y_{ijk}, and using the estimates (3.2.2), the residuals are (see Problem 3.2.6)

$$\widehat{\epsilon}_{ijk} = y_{ijk} - \overline{y}_{i\bullet\bullet} - \overline{y}_{\bullet j\bullet} + \overline{y}_{\bullet\bullet\bullet}. \tag{3.2.3}$$

The appropriate estimate for σ^2, the variance of the individual observations, depends on the model. For model (3.2.1), the estimate of σ^2 is

$$\widehat{\sigma}^2 = \frac{1}{N - I - J + 1} \sum_{i=1}^{I} \sum_{j=1}^{J} \sum_{k=1}^{n} \left(\widehat{\epsilon}_{ijk}\right)^2. \tag{3.2.4}$$

Example 3.2.3 (*Paint Formulation Data, p. 28*)   The interaction plot for this data set (see p. 29) suggests that the two factors do not interact, and therefore, model (3.2.1) is appropriate. The parameter estimates for that model are (equations (3.2.2))

$$\widehat{\mu} = 7.99$$
$$\widehat{\alpha}_1 = 7.98 - 7.99 = -0.01$$
$$\widehat{\alpha}_2 = 5.44 - 7.99 = -2.55$$
$$\widehat{\alpha}_3 = 10.55 - 7.99 = 2.56$$
$$\widehat{\beta}_1 = 9.43 - 7.99 = 1.44$$
$$\widehat{\beta}_2 = 6.05 - 7.99 = -1.94$$
$$\widehat{\beta}_3 = 9.23 - 7.99 = 1.24$$
$$\widehat{\beta}_4 = 7.23 - 7.99 = -0.76.$$

These estimates can be used to help ascertain if one of the components has more of an effect than the other. One way to do this would be by comparing the maximum magnitudes of the two effects. Since

$$\max_i |\widehat{\alpha}_i| = 2.56 > \max_j |\widehat{\beta}_j| = |-1.94| = 1.94$$

this would indicate that Component 2 has a bigger effect. Alternatively, one could compare the averages of the squared effects. Since

$$\frac{1}{I} \sum_{i=1}^{I} \widehat{\alpha}_i^2 = 4.35 > \frac{1}{J} \sum_{j=1}^{J} \widehat{\beta}_j^2 = 1.99$$

this also indicates that Component 2 has a bigger effect.

Using equation (3.2.3), one can compute (for example)

$$\widehat{\epsilon}_{111} = 10.4 - 7.98 - 9.43 + 7.99 = 0.98.$$

The remaining residuals are computed in a similar fashion and are given in Table 3.4.

Table 3.4. Residuals for the Paint Formulation Data in Example 2.3.2, p. 28

		Component 1			
		1	2	3	4
Component 2	1	0.98	−0.74	1.28	−0.72
		−0.72	−0.14	−0.72	0.78
	2	−0.38	0.70	0.32	−1.28
		1.42	1.90	−1.58	−1.08
	3	−0.50	−1.21	−0.40	0.30
		−0.80	−0.51	1.10	2.00

Note: There is one residual for each data point, and the rows and columns of residuals each sum to zero (at least to the number of significant digits used).

Using equation (3.2.4), one obtains

$$\widehat{\sigma}^2 = \frac{(0.98)^2 + (-0.72)^2 + \cdots + (0.30)^2 + (2.00)^2}{18} = \frac{25.1}{18} = 1.39.$$

A Model Accounting for Interaction

What about two factors that interact? If there is interaction, then the result of changing one factor depends on the value of the other factor. In that case it does not make sense to talk about the "main effect" of a factor, and model (3.2.1) is not appropriate. There are several possible models that can be used in this situation, but we will only discuss the simplest one here. Additional models are discussed in Chapter 8.

The simplest mathematical model when two factors interact is

$$Y_{ijk} = \mu_{ij} + \epsilon_{ijk}. \tag{3.2.5}$$

This model simply says that each treatment combination has a different mean. Similarly to single-factor experiments, the μ_{ij}'s are estimated as

$$\widehat{\mu}_{ij} = \overline{y}_{ij\bullet} \tag{3.2.6}$$

and the residuals are

$$\widehat{\epsilon}_{ijk} = y_{ijk} - \overline{y}_{ij\bullet}. \tag{3.2.7}$$

For model (3.2.5), the estimate of σ^2 is

$$\widehat{\sigma}^2 = \frac{1}{IJ(n-1)} \sum_{i=1}^{I} \sum_{j=1}^{J} \sum_{k=1}^{n} (\widehat{\epsilon}_{ijk})^2 \qquad (3.2.8)$$

which, after a little algebra, reduces to (see Problem 3.2.7)

$$\widehat{\sigma}^2 = \frac{s_{11}^2 + s_{12}^2 + \cdots + s_{IJ}^2}{IJ} . \qquad (3.2.9)$$

Example 3.2.4 (*Oven Data*) The drying ability of several brands of gas and electric lab ovens was measured. For each trial the experimenter recorded the percentage of moisture remaining in a sample of silica after drying for 15 minutes. The data are given in Table 3.5 and in oven.txt.

Table 3.5. Moisture Remaining in Silica Samples after 15 minutes of Drying

Type	Oven Brand A	B	C
Gas	72	66	67
	68	69	64
	76	65	68
	74	64	66
	70	66	70
Electric	66	67	70
	63	71	66
	70	64	73
	65	66	67
	66	67	74

Let

$$i = 1 \text{ for Gas}$$
$$i = 2 \text{ for Electric}$$
$$j = 1 \text{ for Brand A}$$
$$j = 2 \text{ for Brand B}$$
$$j = 3 \text{ for Brand C}.$$

The interaction plot for this set of data is given in Figure 3.1. Since it shows strong evidence of interaction, model (3.2.5) is appropriate.

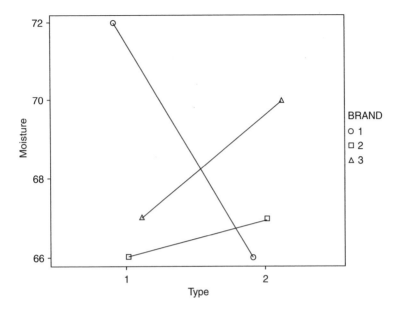

Figure 3.1. Interaction Plot for the Silica Moisture Data.

Using equation (3.2.6), the parameter estimates for model (3.2.5) are given in Table 3.6.

Table 3.6. Table of $\widehat{\mu}_{ij}$'s for the Data in Table 3.5

Type	Brand of Oven		
	A	B	C
Gas	72	66	67
Electric	66	67	70

Using equation (3.2.7), one can calculate

$$\widehat{\epsilon}_{111} = 72 - 72 = 0$$
$$\widehat{\epsilon}_{112} = 68 - 72 = -4$$
$$\vdots$$
$$\widehat{\epsilon}_{235} = 74 - 70 = 4.$$

The sample variances are given in Table 3.7, from which one can compute (equation (3.2.9))

$$\widehat{\sigma}^2 = \frac{10.0 + 3.5 + \cdots + 12.5}{6} = 7.33.$$

Table 3.7. Table of s_{ij}'s for the Data in Table 3.5

| Type | Brand of Oven | | |
	A	B	C
Gas	10.0	3.5	5.0
Electric	6.5	6.5	12.5

Problems

1. Using the data in the data set `temprate.txt`, and without using a computer, do the following.
   (a) Make an interaction plot and use it to determine an appropriate model for the data.
   (b) Find estimates for the parameters of the model you found in part (a).
   (c) Compute the residuals for the model you found in part (a).
   (d) Which factor seems to have a bigger effect?

2. Repeat Problem 1 using the data set `deink.txt` and a computer.

3. Repeat Problem 1 using the data set `dry.txt` and a computer.

4. For the data `cure.txt` discussed in Problem 3.6 (p. 34), do the following.
   (a) Make an interaction plot and use it to determine an appropriate model for the data.
   (b) Find estimates for the parameters of the model you found in part (a).
   (c) Use a residual plot to check the validity of your model.

5. Using the adhesion data set `adhesion2.txt` (see Problem 2.3.3, p. 31), answer the following questions.
   (a) In the terminology we have used to describe experiments, what kind of experiment is this? Be as specific as possible.
   (b) Consider the interaction plot you already made for this set of data. Do you think there is significant interaction between the two factors? Explain.
   (c) Does it seem that all the treatment groups have about the same variance? If not, how does the variance change from one group to another?
   (d) Based on your answer to part (b), choose a model for this situation. Explain why you chose this model.
   (e) Estimate the parameters of the model you chose in part (d).

6. Using model (3.2.1) and the estimates in equations (3.2.2), verify equation (3.2.3).

7. Using equation (3.2.7), show that the estimate of σ^2 for model (3.2.5) given in equation (3.2.8) reduces to equation (3.2.9).

8. A process that fills cylinders with small beads is under study. During the process the cylinders are periodically "tamped". Two control factors related to tamping are the distance the cylinder is lifted from the surface (in centimeters) before being tamped and the number of times it is tamped. The file `fill.txt` contains the results of an experiment that varied the tamp distance and the number of tamps and measured the final fill amount (in grams) of material that was used to fill the cylinder. The end use of the product is as a filter and a tight pack of the filling material is desired. Analyze these data and write a report for the product developers working on this project.

3.3 MODELS FOR BIVARIATE DATA

When two pieces of information (e.g., height and weight) are obtained at each experimental trial (e.g., for each person sampled), we have what is referred to as **bivariate data**. The two pieces of information are generically labeled as the **independent variable** (denoted with x) and the **dependent** or **response variable** (denoted with Y). This reflects the idea that Y (e.g., weight) depends on x (e.g., height), so that knowing x should tell us something about Y.

Notation:

$$x_i = \text{the value of the independent variable at the } i^{\text{th}} \text{ trial}$$
$$Y_i = \text{the response at the } i^{\text{th}} \text{ trial}$$
$$n = \text{the sample size.}$$

Fitting Lines

In many situations there is a linear relationship between x and Y and the model

$$Y_i = \beta_0 + \beta_1 x_i + \epsilon_i \qquad (3.3.1)$$

would be appropriate. Here, as before, ϵ_i represents the random variability of the ith experimental trial and y_i is a specific realization of Y_i.

Before proceeding further with this model one would want to draw a scatter plot of the data to be sure a line was a reasonable model. For example, the scatter plot of the Test Temperature Data for the OEM panels (see p. 13) indicates model (3.3.1) would be appropriate for that data set.

Once it has been determined that model (3.3.1) is appropriate, one would like to be able to use it to obtain information on the Y's for specific x values. For that we need estimates of β_0 and β_1. We would like these estimates to be as good as possible in the sense that the resulting line will fit the data points as closely as possible. To quantify this, suppose we had some estimates b_1 and b_0. For any x_i, these would give us a *predicted* value

$$\widehat{y}_i = b_0 + b_1 x_i.$$

The prediction errors (i.e., the residuals) are the distances between the actual observed y values and the predicted values $\widehat{y}$, that is

$$\widehat{\epsilon}_i = y_i - \widehat{y}_i . \qquad (3.3.2)$$

A good-fitting line would have generally small prediction errors, whereas a poor-fitting line would have generally larger prediction errors (even though it might go exactly through one or two data points).

To further quantify this, define

$$\mathcal{E} = \sum_{i=1}^{n} \widehat{\epsilon}_i^2$$

$$= \sum_{i=1}^{n} (y_i - \widehat{y}_i)^2$$

$$= \sum_{i=1}^{n} [y_i - (b_0 + b_1 x_i)]^2.$$

Our "best" estimates are the values of b_0 and b_1 (call them $\widehat{\beta}_0$ and $\widehat{\beta}_1$) that minimize $\mathcal{E}$. We can minimize $\mathcal{E}$ by

- Finding $\partial \mathcal{E}/\partial b_0$ and $\partial \mathcal{E}/\partial b_1$
- Setting both partial derivatives equal to zero
- Replacing b_0 and b_1 with $\widehat{\beta}_0$ and $\widehat{\beta}_1$, respectively
- Solving simultaneously for $\widehat{\beta}_0$ and $\widehat{\beta}_1$.

This process produces what are called **least-squares estimates**, which in this case are

$$\widehat{\beta}_1 = \frac{\sum_{i=1}^{n} (x_i - \overline{x})(y_i - \overline{y})}{\sum_{i=1}^{n} (x_i - \overline{x})^2} = \frac{S_{xy}}{S_{xx}}$$

$$\widehat{\beta}_0 = \overline{y} - \widehat{\beta}_1 (\overline{x})$$

$$(3.3.3)$$

where

$$S_{xx} = \sum_{i=1}^{n} (x_i - \overline{x})^2 \qquad (3.3.4)$$

and

$$S_{xy} = \sum_{i=1}^{n} (x_i - \overline{x})(y_i - \overline{y}). \qquad (3.3.5)$$

The fitted line

$$\widehat{y} = \widehat{\beta}_0 + \widehat{\beta}_1 x \qquad (3.3.6)$$

is called the **least-squares line**. (For historical reasons, this line is also called the **regression line**.) For a given set of data, the least-squares line is the best-fitting line, in the sense that it minimizes $\mathcal{E}$. The values $\widehat{y}$ are referred to as **predicted values** or **fitted values**.

Note: With a little algebra, it can be shown that

$$\sum (x - \bar{x})^2 = \sum x^2 - \frac{1}{n} \left(\sum x \right)^2$$

$$\sum (y - \bar{y})^2 = \sum y^2 - \frac{1}{n} \left(\sum y \right)^2$$

and

$$\sum (x - \bar{x})(y - \bar{y}) = \sum xy - \frac{1}{n} \left(\sum x \right) \left(\sum y \right).$$

The right-hand side expressions are often found in older textbooks, because they require less computation. However, these expressions are numerically unstable, and should be avoided if possible.

The **error variance** (i.e., σ^2) is the variance of the points around the least-squares line, and can be estimated using

$$\widehat{\sigma}^2 = \frac{\text{SS}_e}{n - 2} \tag{3.3.7}$$

where

$$\text{SS}_e = \sum_{i=1}^{n} (y_i - \widehat{y}_i)^2 = \sum_{i=1}^{n} \widehat{\epsilon}_i^2 \tag{3.3.8}$$

which is referred to as the **error sum of squares**.

Example 3.3.1   Consider the set of bivariate data in Table 3.8. A scatter plot of these data (Figure 3.2) indicates model (3.3.1) is appropriate. For these data, $\bar{x} = 4$ and $\bar{y} = 6$, and the remaining calculations can be done as in Table 3.9.

Table 3.8. A Set of Bivariate Data

x	2	9	3	5	1
y	1	17	3	9	0

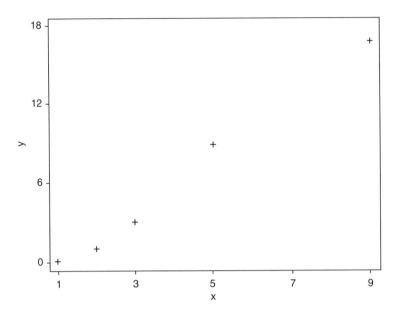

Figure 3.2. A Scatter Plot of the Bivariate Data in Example 3.3.1.

Table 3.9. Calculations of the Least-Squares Line in Example 3.3.1

x	y	$x - \overline{x}$	$y - \overline{y}$	$(x - \overline{x})(y - \overline{y})$	$(x - \overline{x})^2$
2	1	−2	−5	10	4
9	17	5	11	55	25
3	3	−1	−3	3	1
5	9	1	3	3	1
1	0	−3	−6	18	9
Sum	20	30	0	0	89	40

Therefore (equations (3.3.3)),

$$\widehat{\beta}_1 = \frac{89}{40} = 2.225$$

and

$$\widehat{\beta}_0 = 6 - 2.225(4) = 6 - 8.9 = -2.9.$$

The least-squares line is (equation (3.3.6)) $\widehat{y} = -2.9 + 2.225x$. Thus, if $x = 6$, we would predict that

$$\widehat{y} = -2.9 + 2.225(6) = 10.45.$$

Similarly, if $x = 8$, we would predict

$$\widehat{y} = -2.9 + 2.225(8) = 14.9.$$

Calculating SS_e (equation (3.3.8)) is easily done as in Table 3.10. Thus, $SS_e = 1.975$, and it follows from equation (3.3.7) that $\widehat{\sigma}^2 = 1.975/(5 - 2) = 0.6583$.

Table 3.10. Calculation of SS_e

x	Y	$\widehat{Y}$	$Y - \widehat{Y}$	$(Y - \widehat{Y})^2$
2	1	1.550	-0.550	0.3025
9	17	17.125	-0.125	0.015625
3	3	3.775	-0.775	0.600625
5	9	8.225	0.775	0.600625
1	0	-0.675	0.675	0.455625
Sum			0	1.975

Example 3.3.2 (*Repair Test Temperature Data, p. 12*)   The scatter plot of the data (p. 13) indicates model (3.3.1) is reasonable. Let $x =$ starting temperature and $Y =$ time to $-10°C$. For this data set we have $\overline{x} = -25.97$, $\overline{y} = 10.92$, $S_{xy} = -96.76$, and $S_{xx} = 51.91$. Therefore,

$$\widehat{\beta}_1 = \frac{-96.76}{51.91} = -1.864$$

$$\widehat{\beta}_0 = 10.92 - (-1.864)(-25.97) = -37.49$$

and the least-squares line is

$$\widehat{y} = -37.49 - 1.864x.$$

Thus, if a Repair panel had a starting temperature of $-26°C$, we would predict the time to $-10°C$ to be

$$\widehat{y} = -37.49 - 1.864(-26) = 10.97 \text{ seconds}.$$

Similarly, if a Repair panel had a starting temperature of $-23°C$, the predicted time to $-10°C$ is

$$\widehat{y} = -37.5 - 1.864(-23) = 5.38 \text{ seconds}.$$

Fitting Exponential Curves

So far, we have seen how to fit a straight line to a set of bivariate data. Of course, a straight line is only the simplest model for a set of bivariate data, and is not appropriate if the scatter plot of the data shows curvature. One class of models that accounts for curvature is the class of exponential curves. Some examples of exponential curves are

$$Y = \beta_0\, x^{\beta_1}$$

$$\text{(3.3.9)}$$

$$Y = \beta_0\, e^{\beta_1 x}$$

$$\text{(3.3.10)}$$

$$e^Y = \beta_0\, x^{\beta_1}.$$

$$\text{(3.3.11)}$$

Note: Obviously our observed values Y will still have some error associated with them, which we have not accounted for in the above models. This is because we will be transforming each of these models in order to fit it to data, and we are really only concerned about the form of the errors after carrying out the transformation.

These models have two convenient features. They are monotone (either increasing or decreasing), and they are easy to fit. Monotonicity is important because often a monotone relationship between two variables is the only one that makes sense. For example, if x is the level of traffic in a communications network and Y is the number of packets lost, it only makes sense for Y to increase as x increases. Models (3.3.9)–(3.3.11) are easy to fit using the technique for fitting lines. This is because, when transformed to a different scale, they are actually lines. Which, if any, of these models is appropriate for a particular set of data is best determined by drawing scatter plots of the transformed data to see which transformation makes the data look the most linear.

The Model $Y = \beta_0 x^{\beta_1}$

Taking logarithms on both sides of equation (3.3.9) (and then adding an error term), one obtains

$$\ln(Y) = \ln(\beta_0) + \beta_1 \ln(x) + \epsilon$$

or

$$Y^* = \beta_0^* + \beta_1 x^* + \epsilon$$

$$\text{(3.3.12)}$$

where

$$Y^* = \ln(Y)$$
$$x^* = \ln(x)$$
$$\beta_0^* = \ln(\beta_0).$$

Thus, model (3.3.12) is a line in the variables Y^* and x^*. In order to estimate the parameters, one can simply calculate $\ln(y)$ and $\ln(x)$ for all the data points, and then use least-squares to fit a straight line.

Example 3.3.3 Consider the set of data in Table 3.11. A graph of these data (Figure 3.3) shows that a line is not the appropriate model and suggests that model (3.3.9) might be reasonable. Taking logarithms of all the data, one obtains the values in Table 3.12. A scatter plot of the log-transformed data (Figure 3.4) looks quite linear.

Table 3.11. Data for Example 3.3.3

x	9	2	5	8	4	1	3	7	6
y	253.5	7.1	74.1	165.2	32.0	1.9	18.1	136.1	100.0

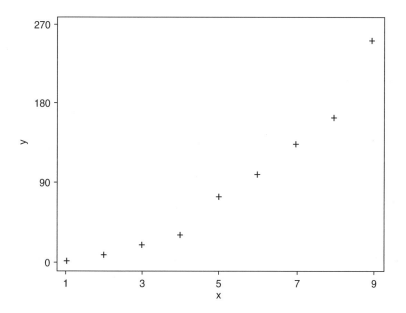

Figure 3.3. A Scatter Plot of the Data in Example 3.3.3.

Table 3.12. Logarithms of the Data in Table 3.11 (Example 3.3.3)

$x^* = \ln(x)$	2.197	0.693	1.609	2.079	1.386	0	1.099	1.946	1.792
$y^* = \ln(y)$	5.54	1.96	4.30	5.11	3.47	0.65	2.90	4.91	4.61

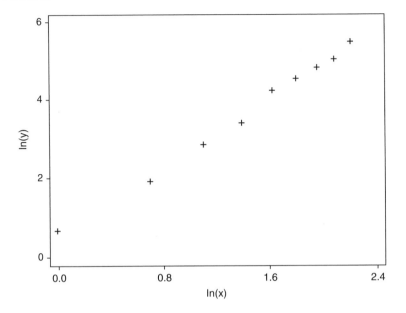

Figure 3.4. A Scatter Plot of the Log-Transformed Data in Example 3.3.3.

To obtain estimates for the parameters in the model we calculate

$$\overline{x^*} = 1.422$$

$$\overline{y^*} = 3.715$$

$$\sum (x^* - \overline{x^*})^2 = 4.138$$

$$\sum (x^* - \overline{x^*})(y^* - \overline{y^*}) = 9.315.$$

Finally,

$$\widehat{\beta_1} = 9.315/4.138 = 2.251$$

$$\widehat{\beta_0^*} = 3.715 - 2.251(1.422) = 0.51$$

and our fitted model is

$$\widehat{y^*} = 0.51 + 2.251 x^*$$

$$\widehat{\ln(y)} = 0.51 + 2.251 \ln(x)$$

or

$$\widehat{y} = 1.67 x^{2.251}.$$

Thus, when $x = 3.36$, we would predict

$$\widehat{y} = 1.67(3.36)^{2.251} = 25.6.$$

Residuals for this fitted line would be computed using equation (3.3.2). For example,

$$\widehat{\epsilon_1} = 5.54 - [0.51 + 2.251(2.197)] = 0.08.$$

The Model $Y = \beta_0\, e^{\beta_1 x}$

Taking logarithms on both sides of equation (3.3.10) (and adding in an error term), one obtains

$$\ln(Y) = \ln(\beta_0) + \beta_1 x + \epsilon$$

or

$$Y^* = \beta_0^* + \beta_1 x + \epsilon.$$

Thus, we can estimate the parameters of model (3.3.10) by calculating $\ln(y)$ for each data point and then using least squares with $\ln(y)$ as the dependent variable and x as the independent variable.

Example 3.3.4 Consider the set of data in Table 3.13. It is clear from the scatter plot of these data (Figure 3.5) that a line is not an appropriate model. However, if one transforms the y's by taking logs, we obtain the transformed values in Table 3.14, and a scatter plot of the transformed data (Figure 3.6) looks very linear.

Table 3.13. Data for Example 3.3.4

x	4.99	1.96	2.98	9.00	4.04
y	4.35×10^9	1.97×10^4	5.99×10^4	3.73×10^{17}	1.67×10^8

x	6.06	0.88	8.02	6.97
y	7.20×10^{10}	180	5.97×10^{14}	3.67×10^{12}

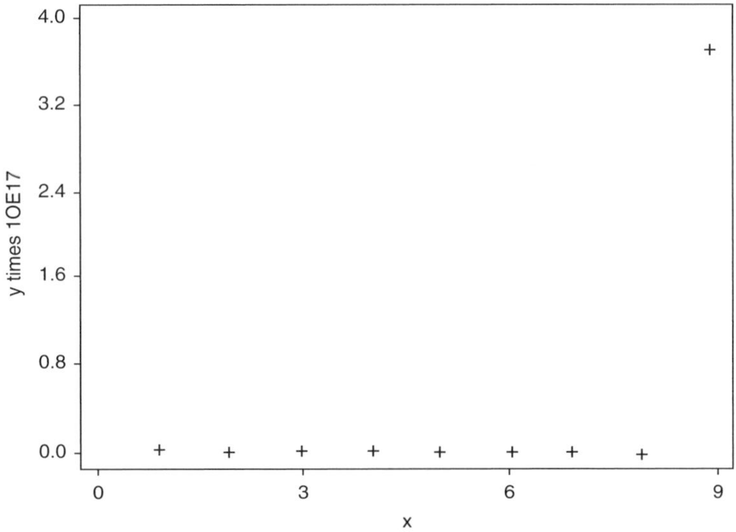

Figure 3.5. Scatter Plot for the Data in Example 3.3.4.

Table 3.14. Transformed Data Values for Example 3.3.4

x	4.99	1.96	2.98	9.00	4.04	6.06	0.88	8.02	6.97
$y^* = \ln(y)$	22.19	9.89	11.00	40.46	18.93	25.00	5.19	34.02	28.93

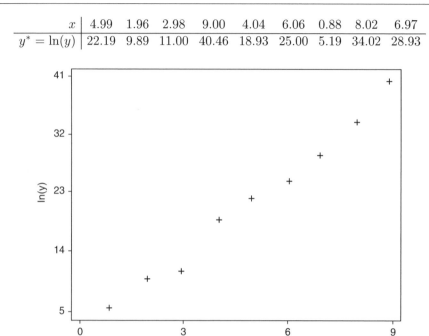

Figure 3.6. Scatter Plot for the Transformed Data in Example 3.3.4.

To estimate the parameters in the model, we compute

$$\bar{x} = 4.99$$

$$\overline{y^*} = 21.73$$

$$\sum (x - \bar{x})^2 = 61.34$$

$$\sum (x - \bar{x})(y^* - \overline{y^*}) = 258.2.$$

Then,

$$\widehat{\beta}_1 = 258.2/61.34 = 4.209$$

$$\widehat{\beta}_0^* = 21.73 - 4.209(4.99) = 0.73$$

so our fitted model is

$$\widehat{y}^* = 0.73 + 4.209x$$

or

$$\widehat{y} = e^{0.73 + 4.209x} = 2.075e^{4.209x}.$$

The Model $e^Y = \beta_0\, x^{\beta_1}$

Once again, taking logarithms, this time on both sides of equation (3.3.11) (and adding in an error term), results in

$$Y = \ln(\beta_0) + \beta_1 \ln(x) + \epsilon$$
$$= \beta_0^* + \beta_1 x^* + \epsilon.$$

Example 3.3.5　　Consider the set of data in Table 3.15.

Table 3.15. Data for Example 3.3.5

x	17,000	47	150	1000	14	4300	200	18
y	45	15	18	25	9	32	20	10

A scatter plot of these data (Figure 3.7) is not linear. However, if one transforms the data by taking the logs of the x's, one obtains Table 3.16.

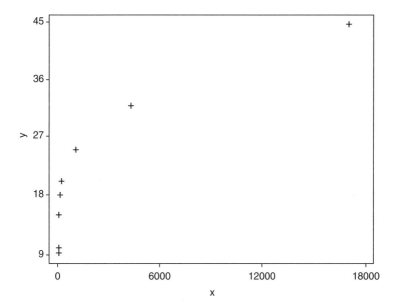

Figure 3.7. Scatter Plot for the Original Data in Example 3.3.5.

Table 3.16. Transformed Data for Example 3.3.5.

$x^* = \ln(x)$	9.74	3.85	5.01	6.91	2.64	8.37	5.30	2.89
y	45	15	18	25	9	32	20	10

A scatter plot of the transformed data (Figure 3.8) indicates that model (3.3.11) is appropriate.

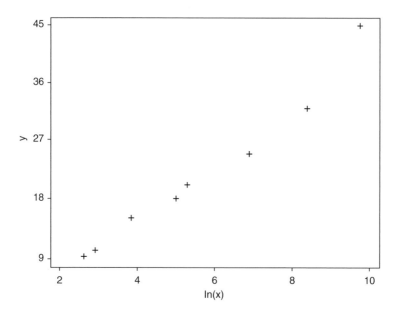

Figure 3.8. Scatter Plot of the Transformed Data in Example 3.3.5.

To estimate the parameters in the model, we compute

$$\overline{x^*} = 5.59$$

$$\overline{y} = 21.75$$

$$\sum (x^* - \overline{x^*})^2 = 46.12$$

$$\sum (x^* - \overline{x^*})(y - \overline{y}) = 213.0.$$

Then,

$$\widehat{\beta}_1 = 213.0/46.12 = 4.62$$

$$\widehat{\beta}_0^* = 21.75 - 4.62(5.59) = -4.1$$

and the fitted model is

$$\widehat{y} = -4.1 + 4.62x^*$$

$$= -4.1 + 4.62\ln(x).$$

Fitting Polynomial Curves

Polynomial models are models of the form

$$Y = \beta_0 + \beta_1 x + \beta_2 x^2 + \epsilon \qquad (3.3.13)$$

$$Y = \beta_0 + \beta_1 x + \beta_2 x^2 + \beta_3 x^3 + \epsilon \qquad (3.3.14)$$

and so forth. Model (3.3.13) is a parabola and model (3.3.14) is a cubic curve. Unlike exponential models, polynomial models are non-monotonic. A parabola has one inflection point where it changes from decreasing to increasing (or vice versa), a cubic polynomial has two inflection points, and in general an nth order polynomial has $n - 1$ inflection points.

Thus, a parabolic model is useful for cases where increasing the explanatory variable has one effect up to a point, and the opposite effect thereafter. For example, in a chemical experiment, increasing the reaction temperature might increase the yield up to some optimal temperature, after which a further increase in temperature would cause the yield to drop off. In an agricultural experiment, increasing fertilizer will probably increase plant growth up to a point, after which more fertilizer will cause a chemical imbalance in the soil and actually hinder plant growth.

It is easy to think of cases where parabolic models are appropriate, and there are also examples where cubic models can be useful. However, it is usually difficult to justify the use of a high-order polynomial because, generally, no physical explanation justifies the large number of inflection points. Thus, such models are seldom used.

The general procedure for obtaining estimates for the parameters of polynomial models is the same as for a line, although implementing it becomes more complicated. One finds estimates that minimize $\mathcal{E}$. For the parabolic model (3.3.13)

$$\mathcal{E} = \sum_{i=1}^{n} [y_i - (b_0 + b_1 x + b_2 x^2)]^2$$

and one must

- Find $\partial \mathcal{E}/\partial b_0$, $\partial \mathcal{E}/\partial b_1$, and $\partial \mathcal{E}/\partial b_2$.
- Set all three partial derivatives equal to zero.
- Replace b_0, b_1, and b_2 with $\widehat{\beta}_0$, $\widehat{\beta}_1$, and $\widehat{\beta}_2$; respectively.
- Solve simultaneously for $\widehat{\beta}_0$, $\widehat{\beta}_1$, and $\widehat{\beta}_2$.

In this case, we have a system of three equations and three unknowns to solve. A cubic model is fit similarly, resulting in a system of four equations and four unknowns. Doing this by hand is extremely tedious, so one generally uses a computer package.

Example 3.3.6 (*Particle Size Data*)   A process engineer is studying the relationship between a vacuum setting and particle size distribution for a granular product. A shaker test is one method for judging the distribution of particle sizes in such a product. A sample of material is passed through a series of screens (decreasing in mesh size) and the proportion of material remaining on each size screen is measured. In Table 3.17, x = vacuum setting and y = percent of material left on a 40-mesh screen (a measurement of large particles for this product).

Table 3.17. The Particle Size Data for Example 3.3.6

x	18	18	20	20	22	22	24	24	26	26
y	4.0	4.2	5.6	6.1	6.5	6.8	5.4	5.6	3.3	3.6

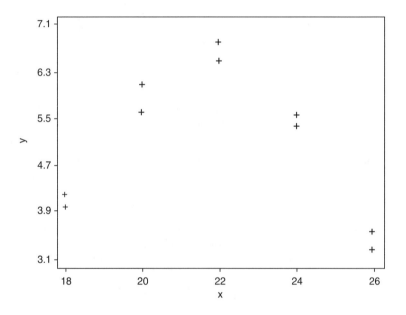

Figure 3.9. A Scatter Plot of the Particle Size Data.

A scatter plot of the data (Figure 3.9) shows definite evidence of polynomial curvature. A computer program was used to fit the parabolic model (3.3.13) to the data, resulting in the output in Table 3.18.

Table 3.18. Parabolic Model Fit to the Data in Example 3.3.6

```
UNWEIGHTED LEAST SQUARES LINEAR REGRESSION OF Y
```

PREDICTOR VARIABLES	COEFFICIENT	STD ERROR	STUDENT'S T	P
CONSTANT	-74.2500	5.24071	-14.17	0.0000
X	7.42107	0.48221	15.39	0.0000
X_SQUARED	-0.17054	0.01094	-15.58	0.0000

R-SQUARED	0.9731	RESID. MEAN SQUARE (MSE)	0.05365
ADJUSTED R-SQUARED	0.9654	STANDARD DEVIATION	0.23163

SOURCE	DF	SS	MS	F	P
REGRESSION	2	13.5734	6.78671	126.49	0.0000
RESIDUAL	7	0.37557	0.05365		
TOTAL	9	13.9490			

We see from Table 3.18 that the fitted model is

$$\hat{y} = -74.25 + 7.42x - 0.17x^2.$$

Residuals are computed using equation (3.3.2). For example,

$$\hat{\epsilon}_1 = 4.0 - [-74.25 + 7.42107(18) - 0.170536(18)^2]$$
$$= 4.0 - 4.075$$
$$= -0.075.$$

When $x = 23$, we can use the fitted model to predict the percentage of material left on a 40-mesh screen as

$$\hat{y} = -74.25 + 7.42107(23) - 0.170536(23)^2 = 6.22.$$

A parabolic curve such as this has a maximum, which we can find using calculus.

$$\frac{d\hat{y}}{dx} = 7.42 - 0.34x$$
$$\frac{d\hat{y}}{dx} = 0 \Rightarrow 7.42 - 0.34x = 0$$
$$\Rightarrow x = 21.82.$$

If the goal is to maximize the large particles in the product, then the best setting for the vacuum is at 21.8, with the predicted percentage of material left on a 40-mesh screen being $\hat{y} = 6.5$.

Problems

1. Use the data below to do the following.

x	7	12	1	10	5
y	11	20	5	16	8

(a) Make a scatter plot of the data.

(b) By hand, calculate $\widehat{\beta_0}$ and $\widehat{\beta_1}$ and report the resulting least-squares equation.

(c) Plot the resulting line on your scatter plot.

(d) Find $\widehat{y}$ when $x = 2$.

(e) Find $\widehat{y}$ when $x = 9$.

(f) Calculate the residuals.

(g) Calculate SS_e.

2. Using the data below, do the following.

x	1	2	3	4	5	6	7
y	3	2	5	8	7	10	15

(a) By hand, calculate $\widehat{\beta_0}$ and $\widehat{\beta_1}$ and report the resulting least-squares equation.

(b) Calculate the seven residuals.

(c) Calculate the error sum of squares.

(d) Use a computer to find the least-squares equation. On your output, identify $\widehat{\beta_0}$, $\widehat{\beta_1}$, and SS_e.

3. A lab received a new instrument to measure pH. To compare the new instrument to the old lab instrument 11 samples were measured with both pieces of equipment. Using the data in `phmeas.txt`, do the following.

(a) Use a computer to find the least-squares equation (x = old and Y = new).

(b) Find SS_e.

(c) Predict the value on the new instrument if the old instrument gave a pH of 6.

(d) Predict the value on the new instrument if the old instrument gave a pH of 6.3.

4. You have already plotted the data in `adhesion2.txt` (see Problem 2.1.3, p. 31). Using these data, do the following.

  (a) Use a computer to fit two regression lines, one for catalyst 1 and one for catalyst 2.

  (b) For each catalyst type predict the adhesion value for a pH level of 5.

  (c) Use the two regression equations to describe the difference that the catalyst type has on the effect of pH on adhesion levels.

  (d) What level of pH and catalyst would you want to use to maximize adhesion? Minimize adhesion?

5. Longwave (LW) and shortwave (SW) are two appearance measures used in the automotive industry to rate the quality of a paint job. These two measures are generally related. Values for 13 cars are given in `lwsw.txt`. Use these data to do the following.

  (a) Using SW as the independent variable, find the best-fitting line for this set of data.

  (b) Calculate SS_e.

6. For the data set `odor.txt` consider using the function (3.3.9) to model the relationship between odor and yellowing. Odor and yellowing are both performance measures of a monomer used in making optical lenses. Lower values of both are desired. Let $Y = $ odor.

  (a) If $\beta_0 > 0$, what will the sign of Y be? Does this make sense?

  (b) If $\beta_1 > 0$, what is the relationship (increasing, decreasing, etc.) of Y with respect to x? Does this make sense?

  (c) Make an appropriate scatter plot for assessing the suitability of this model. (You will have to transform the data.) Comment on your scatter plot. Are there any obvious outliers?

  (d) Remove any obvious outliers from your data set, and fit the model to the remaining data points. Report the fitted model.

  (e) Explain what β_1 represents in the context of the problem.

  (f) If a particular monomer has a yellowing value of 6, what is the predicted odor?

7. The data set `epoxy.txt` contains data from an experiment to determine the effect of the level of epoxy in an automotive paint formulation. The response considered is the appearance measure longwave (LW). Consider the function (3.3.11) to model this set of data, where $Y = $ LW and $x = $ epoxy level in the formulation.

  (a) Make two scatter plots of the data to assess the suitability of the model. Make one plot of the original data and a second after the data have been transformed.

(b) Use least squares to find the parameter estimates and report the fitted model.

(c) Predict the LW reading for a paint with an epoxy level of 1.25.

8. An experiment was performed to explore the effect of a marketing campaign on web site traffic. The data in `webtraff.txt` are the number of weeks into the marketing campaign and the web site traffic. The traffic measure is a daily average for two random 24-hour periods during the particular week. The day that the traffic was sampled was determined before the study began and only weekdays were considered. The values are in 1000's of hits per day.

(a) Make a scatter plot of the data and suggest a suitable polynomial model.

(b) Fit the model you chose in part (a).

(c) What does the data at time zero represent? Why is this important?

(d) Predict the web site traffic for the third week of the study.

(e) Plot your fitted model on your scatter plot (you may need to do this by hand), and comment on the fit. How precise do you think your predictions will be?

9. In 2002, Colorado experienced a drought and many towns on the front range issued either mandatory or voluntary water restrictions on outdoor watering. The data file `drought.txt` contains data from five front range towns on their June 2001 and June 2002 usage in millions of gallons.

(a) Using the data from the first four towns and only one independent variable, find the best model you can for predicting 2002 values based on 2001 values.

(b) Re-do part (a) using all five data points.

(c) Discuss how the models differ. Consider the small size of the data set in your discussion.

3.4 MODELS FOR MULTIVARIATE DATA

Just as we can fit a line or a curve to two-dimensional data, we can also fit a plane or curved surface to three-dimensional data, and a hyperplane or hypersurface to four- or higher-dimensional data.

Example 3.4.1 (*Chemical Process Yield Data*) An industrial engineer wants to develop a model for the yield of a chemical process based on two key variables. The two variables are measurements made on a slurry, which is then further processed into the final product. The two independent variables are x_1 = temperature of the slurry as it enters the next process step and x_2 = pH of the slurry. The dependent variable of interest is Y = process yield (in tons) from the batch of slurry.

A simple model for this situation is

$$Y = \beta_0 + \beta_1 x_1 + \beta_2 x_2 + \epsilon \qquad (3.4.1)$$

which is the equation for a plane in three dimensions. The process of fitting models for surfaces and hypersurfaces is the same as for lines. Write down an expression for $\mathcal{E}$, find the derivatives with respect to all the parameters, set the equations equal to zero, and solve simultaneously. This requires a large amount of algebra, and is usually done with a computer. In most computer packages you only need to specify the dependent variable and all the independent variables of interest.

Example 3.4.2 (*Continuation of Example 3.4.1*) The data for the chemical process yield study is given in Table 3.19 as well as in `yield.txt`.

Table 3.19. Chemical Process Yield Data (Example 3.4.2)

x_1 = Temperature	20.9	21.2	20.8	20.1	20.3	22.7	20.4	22.0	20.5	21.5
x_2 = pH	6.8	6.3	6.8	6.4	6.3	6.6	6.4	6.7	6.8	6.5
y = Yield	32.5	32.1	32.2	31.6	30.8	33.0	31.5	32.9	32.4	32.5

x_1 = Temperature	21.5	21.1	21.8	20.5	21.2	20.2	22.6	20.6	22.4	22.0
x_2 = pH	6.8	6.3	6.8	6.4	6.3	6.6	6.4	6.7	6.8	6.5
y = Yield	32.4	31.6	32.7	32.3	32.6	31.6	32.3	31.5	33.4	31.8

The output in Table 3.20 was obtained from a statistical package.

Table 3.20. Computer Output for Example 3.4.2

UNWEIGHTED LEAST SQUARES LINEAR REGRESSION OF YIELD

PREDICTOR VARIABLES	COEFFICIENT	STD ERROR	STUDENT'S T	P
CONSTANT	14.2751	3.57259	4.00	0.0009
TEMPERATU	0.47232	0.11618	4.07	0.0008
PH	1.20269	0.47293	2.54	0.0210

R-SQUARED	0.6223	RESID. MEAN SQUARE (MSE)	0.16540
ADJUSTED R-SQUARED	0.5779	STANDARD DEVIATION	0.40670

SOURCE	DF	SS	MS	F	P
REGRESSION	2	4.63367	2.31684	14.01	0.0003
RESIDUAL	17	2.81183	0.16540		
TOTAL	19	7.44550			

Thus, the fitted model is

$$\widehat{y} = 14.28 + 0.472x_1 + 1.203x_2$$

and (equation (3.3.2))

$$\widehat{\epsilon}_1 = 32.5 - [14.28 + 0.472(20.9) + 1.203(6.8)] = 32.5 - 32.33 = 0.17$$

$$\vdots$$

$$\widehat{\epsilon}_{20} = 31.8 - [14.28 + 0.472(22.0) + 1.203(6.5)] = 31.8 - 32.48 = -0.68.$$

If a batch has a temperature of 21.5 and a pH of 6.5 (i.e., $x_1 = 21.5$, $x_2 = 6.5$), we would expect the process yield for that batch to be

$$\widehat{y} = 14.28 + 0.472(21.5) + 1.203(6.5) = 32.25 \text{ tons.}$$

Problems

1. A chemist is working on a new coating for transparencies used in ink-jet printers. The current formulation causes the transparency to curl. There are two components the chemist has data on that are believed to affect curl. The data from this study are in `curl.txt`.

    (a) Make scatter plots of the response variable against each of the predictor variables and comment on the plots.

    (b) Fit model (3.4.1) and report the fitted model.

    (c) Discuss what β_1 and β_2 stand for in the context of this problem.

    (d) Compute the residuals.

2. A financial planner for a company wanted to model monthly sale on the previous month's capital expenses and power index (an industry specific index). The data for 48 months are given in `sales.txt`.

    (a) Fit model (3.4.1) and report the fitted model.

    (b) Explain what the estimated values of β_1 and β_2 represent in the context of this problem.

3. The application of powder coating (a type of paint used on appliances and, in limited settings, cars) is done by spraying the material through a "gun" that has an electric charge on it. There are three factors to consider in setting up an application booth. The gun distance from the target item, the charge, and the pressure (flow rate) of the material through the gun. Data from 18 test runs are given in `applicat.txt`. Fit the model $Y = \beta_0 + \beta_1 x_1 + \beta_2 x_2 + \beta_3 x_3 + \epsilon$ to the data. Report the fitted equation.

4. Fifty-five patients participated in the study of a new blood test for a particular disease. Their gender, age, duration that they have had the disease, level of a particular component in their blood (level A), and the value of the protein level obtained by the new test are in the data file `protein.txt`. The investigators of the new test want to know if it is influenced by the patients age, disease duration, or level A value.

    (a) Write a report to address this question.

    (b) Consider the gender of the patients and write a brief report.

5. The data set `moisture.txt` contains the results of an experiment to increase the moisture content of a silica product. Four factors were studied: temperature, speed, solids, and pH.

    (a) Fit a model of the form

$$Y = \beta_0 + \beta_1 x_1 + \beta_2 x_2 + \beta_3 x_3 + \beta_4 x_4 + \epsilon$$

    to the data.

(b) Exclude the pH term and fit a model of the form

$$Y = \beta_0 + \beta_1 x_1 + \beta_2 x_2 + \beta_3 x_3 + \epsilon$$

to the data.

3.5 ASSESSING THE FIT OF A MODEL

We will discuss three ways to assess how well a model fits a set of data. The first technique is applicable in general, the second is applicable to regression models (Sections 3.3 and 3.4), and the third is specifically for evaluating the fit of a straight line.

Coefficient of Determination

The **coefficient of determination** (denoted by R^2) measures the proportion of variation in the response variable that is accounted for by independent variables.

Notation: When we talked about single-factor experiments and two-factor factorial experiments, the responses were labeled as y_{ij}'s and y_{ijk}'s, respectively; and the independent variables were referred to as factors (or treatments). For bivariate and multivariate data the responses were y_i's. To simplify the notation we will refer to the responses here as just y_i's. That is, one must imagine that (for example) with N y_{ij}'s they have been relabeled as y_i's where $i = 1, \ldots, n(= N)$. Also for convenience, we will refer to the average of all the responses as $\overline{y}$.

The total variation of the data is

$$SS_{\text{total}} = \sum_{i=1}^{n} (y_i - \overline{y})^2 = (n-1)s^2 \tag{3.5.1}$$

which is called the **total sum of squares**. Since the variation **not** explained by the model is SS_e (equation (3.3.8)), it follows that

$$R^2 = \frac{SS_{\text{total}} - SS_e}{SS_{\text{total}}}. \tag{3.5.2}$$

Note that since R^2 is a proportion, it must satisfy $0 \le R^2 \le 1$. Values of R^2 close to 1 indicate a good fit, and values of R^2 close to 0 indicate a poor fit.

Example 3.5.1 In Example 3.3.1 (p. 51), we computed $SS_e = 1.975$ for the fitted line. The sample variance of the y_i's is $s^2 = 50$, and therefore (equation (3.5.1)), $SS_{\text{total}} = 4(50) = 200$ and (equation (3.5.2))

$$R^2 = \frac{200 - 1.975}{200} = 0.9901.$$

The line provides an extremely good fit to the data.

Example 3.5.2 (*Paint Formulation Data, p. 28*)   In Example 3.2.3 (p. 43), we computed the residuals for each of the 24 observed responses. From those we can compute

$$SS_e = (0.98)^2 + (-0.72)^2 + \cdots + (2.00)^2 = 25.1.$$

The sample variance of the Y's is 7.715, and therefore, $SS_{total} = 23(7.715) = 177.4$ and

$$R^2 = \frac{177.4 - 25.1}{177.4} = 0.859.$$

The model provides a good fit to the data.

Residual Plots

Each data point has associated with it a residual (equation (3.3.2)). A plot of the residuals against x can sometimes suggest ways to improve the model. In general, we expect the residuals to be randomly distributed, that is, without any pattern. Any pattern in the residual plot is an indication that a different model should be used. If there is more than one independent variable in the data set, one should make a residual plot against each independent variable.

Example 3.5.3 (*Test Temperature Data, p. 12*)   For the Repair panels a plot of the residuals versus the x values is given in Figure 3.10. The residual plot looks like random "noise", so there is no reason to think a more complicated model is needed.

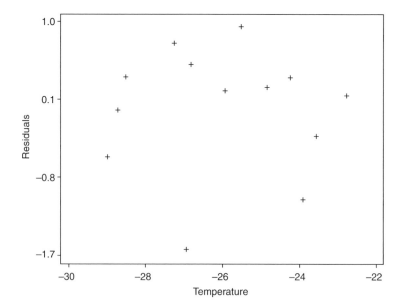

Figure 3.10. Scatter Plot of Residuals Versus Temperature for Example 3.5.3.

Example 3.5.4 (*Chemical Process Yield Data, p. 67*)   The residual plots against both independent variables are shown in Figures 3.11 and 3.12, and there is no evidence of any pattern. The model appears to be adequate.

Example 3.5.5   (*More Particle Size Data*)   In a continuation of the experiment

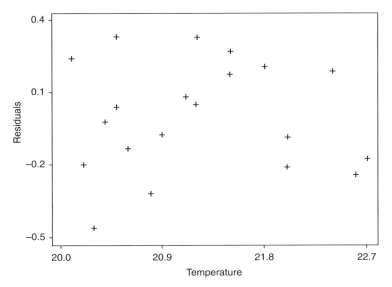

Figure 3.11. Scatter Plot of Residuals Versus Temperature for Example 3.5.4.

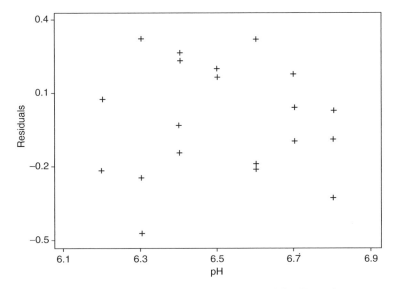

Figure 3.12. Scatter Plot of Residuals Versus pH for Example 3.5.4.

described in Example 3.3.6, an engineer studied the effects of both the flow rate and vacuum setting on the final product particle size. To study these two factors an experiment was run with nine observations. The data are given in Table 3.21, where Y = amount of material left on a 40-mesh screen from a shaker test (i.e., a measurement of particle size), x_1 = flow rate, and x_2 = vacuum setting.

Table 3.21. Particle Size Data for Example 3.5.5

i	y	x_1	x_2
1	6.3	85	20
2	6.1	90	20
3	5.8	95	20
4	5.9	85	22
5	5.6	90	22
6	5.3	95	22
7	6.1	85	24
8	5.8	90	24
9	5.5	95	24

Suppose we start with the model $Y = \beta_0 + \beta_2 x_2 + \epsilon$. Fitting this model, one obtains

$$\widehat{y} = 7.29 - 0.0667 x_2 \qquad \text{with } R^2 = 0.1308.$$

This R^2 value is quite low, and hopefully, we can do better. The residuals for this model are (see Problem 3.5.6)

$$\widehat{\epsilon}_1 = 6.3 - [7.29 - 0.0667(20)] = 0.344$$

$$\vdots$$

$$\widehat{\epsilon}_9 = 5.5 - [7.29 - 0.0667(24)] = -0.189$$

and a plot of the residuals versus Flow Rate is given in Figure 3.13.

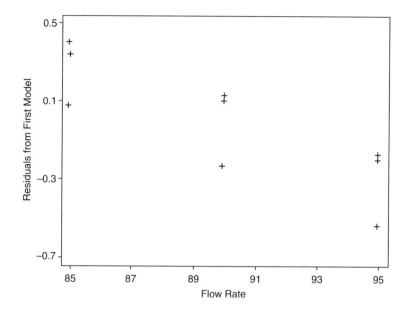

Figure 3.13. Scatter Plot of Residuals Versus Flow Rate for Example 3.5.5.

The clear linear trend in the plot indicates that x_1 should be added to the model. Fitting the model $Y = \beta_0 + \beta_1 x_1 + \beta_2 x_2 + \epsilon$ results in (see Problem 3.5.7)

$$\widehat{y} = 12.39 - 0.0567x_1 - 0.0667x_2 \qquad \text{with } R^2 = 0.7214.$$

Plots of the residuals for this model versus both x_1 and x_2 are given in Figures 3.14 and 3.15.

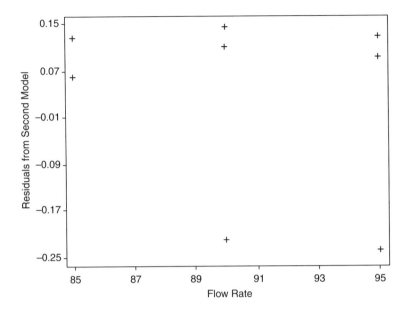

Figure 3.14. Scatter Plot of the Residuals Versus x_1 for Example 3.5.5.

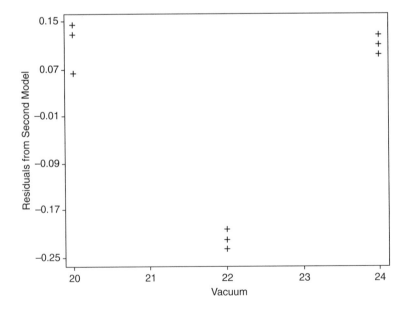

Figure 3.15. Scatter Plot of the Residuals Versus x_2 for Example 3.5.5.

There is no obvious pattern in the first plot, but the second plot shows possible curvature, indicating a need for an x_2^2 term in the model. Adding x_2^2 to the model, one obtains (see Problem 3.5.8)

$$\hat{y} = 52.5 - 0.0567x_1 - 3.733x_2 + 0.0833x_2^2 \qquad \text{with } R^2 = 0.9939.$$

Correlation Coefficient

The **correlation coefficient** measures how closely (x, y) pairs lie to a straight line. It is denoted by r and is related to R^2. It can be computed using either

$$\boxed{r = (\text{sign of } \hat{\beta}_1)\sqrt{R^2}} \tag{3.5.3}$$

or

$$\boxed{r = \frac{\sum(x - \bar{x})(y - \bar{y})}{\sqrt{\sum(x - \bar{x})^2 \sum(y - \bar{y})^2}} = \frac{S_{xy}}{\sqrt{S_{xx}SS_{\text{total}}}}.} \tag{3.5.4}$$

It is clear from the first of these two equations (and the fact that $0 \leq R^2 \leq 1$) that $-1 \leq r \leq 1$. The closer r is to ± 1, the more closely the (x, y) pairs lie to a straight line. Note that although R^2 is defined for any regression model, the correlation coefficient is only defined for bivariate data modeled with a straight line (or a model that can be transformed to a straight line, such as the exponential models).

Example 3.5.6 (*Repair Test Temperature Data, p. 12*)   In Example 3.3.2 (p. 53), we were given $S_{xy} = -96.76$ and $S_{xx} = 51.91$, from which one can compute $\hat{\beta}_1 = -1.86 < 0$. We could also compute $SS_{\text{total}} = 186.92$ and $SS_e = 6.52$, and we could then compute r using either equation (3.5.3)

$$r = -\sqrt{\frac{186.92 - 6.52}{186.92}} = -\sqrt{0.965} = -0.982$$

or alternatively, equation (3.5.4)

$$r = \frac{-96.76}{\sqrt{(51.91)(186.92)}} = -0.982.$$

This means that about $(0.982)^2 = 96\%$ of the variation in times is accounted for by differences in starting temperatures. The remaining 4% is either "random" or is caused by factors that we have not measured.

Problems

1. Using the data from Problem 3.3.1 (p. 64), calculate r and R^2. What do these indicate about the fitted model?

2. Using your output from Problem 3.3.2 (p. 64), find R^2. What does this indicate about the fitted model?

3. Calculate the residuals from the least-squares fitted model for the pH measurement data set `phmeas.txt` and plot them against the fitted values. What does this indicate about the fitted model?

4. For the LW and SW data set `lwsw.txt` plot the residuals from the least-squares model against the fitted values and comment on the plot.

5. For the adhesion data set `adhesion2.txt` plot the residuals for the least-squares model against the fitted values and comment on the plot.

6. In Example 3.5.5 (p. 73), compute all the residuals for the model $Y = \beta_0 + \beta_2 x_2 + \epsilon$.

7. Use a computer to verify that in Example 3.5.5 (p. 73), R^2 and the fitted values given for the parameters in the model $Y = \beta_0 + \beta_1 x_1 + \beta_2 x_2 + \epsilon$ are correct.

8. Use a computer to verify that in Example 3.5.5 (p. 73), R^2 and the fitted values given for the parameters in the model $Y = \beta_0 + \beta_1 x_1 + \beta_2 x_2 + \beta_3 x_2^2 + \epsilon$ are correct.

9. Refer to the model you fit in Problem 3.4.1 (p. 69).

    (a) Compute R^2 for the fitted model.

    (b) Construct residual plots to help assess the fit of the model.

10. Repeat Problem 9 for the model in Problem 3.4.2 (p. 69).

11. For the data in Problem 3.4.2 (p. 69) do the following.

    (a) Use R^2 to choose one independent variable and fit model (3.3.1).

    (b) Compare the results from part (a) with those for model (3.4.1), which you fit in Problem 3.4.2

12. Assess the fit of the model in Problem 3.4.3 (p. 69).

13. Consider the models fit in Problem 3.4.5 (p. 69) and assess the fit of each model. Which model would you recommend for use? Why?

3.6 CHAPTER PROBLEMS

1. An experiment was conducted to study the hardness Y of a particular kind of plastic as a function of curing time (x_1) and curing temperature (x_2). The resulting data are given below.

y	x_1	x_2
1	0	0
15	1	2
60	2	4
6	3	0
21	4	2
58	5	4

(a) If one fits a line to these data using least squares and just the independent variable x_2, one obtains an estimated line $\hat{y} = -0.92 + 13.9x_2$ with $R^2 = 0.920$. Find the residuals for this model.

(b) Plot the residuals versus the x_2 values.

(c) What term does your plot suggest should be added to the model? Why?

Several fitted models together with their R^2 values are given below.

Model 2: $\hat{y} = -2.4 + x_1 + 13.4x_2$ with $R^2 = 0.924$
Model 3: $\hat{y} = 3.5 + 0.625x_2 + 3.31x_2^2$ with $R^2 =?$
Model 4: $\hat{y} = 3.79 + 3.46x_2^2$ with $R^2 = 0.990$
Model 5: $\hat{y} = 2 + x_1 + 0.125x_2 + 3.31x_2^2$ with $R^2 = 0.994$

(d) Without doing any computation (and using the above information) find bounds on the R^2 value for Model 3.

(e) Which model would you choose for the above data? Why?

2. Tomato plants were grown in a greenhouse under treatment conditions consisting of combinations of three soil types (Factor B) and two fertilizers (Factor A). Two plants were grown under each of the six treatment combinations. The (coded) yields of tomatoes are recorded below.

	Factor B		
	1	6	10
Factor A	1	4	8
	8	4	1
	10	6	1

(a) Make an interaction plot for these data.

(b) Based on the plot in part (a), what model would you suggest?

(c) Find estimates for all the parameters in your model.

3. Three different alcohols can be used in a particular chemical process. The resulting yields (in %) from several batches using the different alcohols are given below.

Alcohol

1	2	3
93	95	76
95	97	77

(a) Using the model $Y_{ij} = \mu_i + \epsilon_{ij}$ where $\epsilon_{ij} \sim N(0, \sigma^2)$, find estimates of the four parameters.

(b) Compute the six residuals.

4. An experiment was performed to study the effect of gaugers and breakers on the strength of Portland cement (measured in psi). Three gaugers and three breakers were used, and $n = 4$ replicate observations were taken for each gauger/breaker combination. A table of the resulting cell means, row means, column means, and the grand mean is given below.

		Breaker			
		1	2	3	
	1	5340	4990	4815	5048.3
Gauger	2	5045	5345	4515	4968.3
	3	5675	5415	4917.5	5335.8
		5353.3	5250	4749.2	5117.5

(a) Compute the main effects for Breakers and Gaugers.

(b) Which factor appears to have a bigger effect on strength? Why?

5. The density of bricks is to be studied to see how it is effected by the grain size and the pressure used in the process. Two grain sizes and two pressures were used and $n = 3$ observations were taken at each grain size/pressure combination. For each trial the experimenter recorded the density of the brick. A table of the resulting cell means, row means, column means, and the grand mean is given below.

		Pressure		
		1	2	
Grain Size	1	2.6	2.8	2.7
	2	2.4	2.2	2.3
		2.5	2.5	2.5

(a) Compute the main effects for Grain Size and Pressure.

(b) Which factor appears to have a bigger effect on density? Why?

6. An experiment was done to determine the amount of heat loss for a particular type of thermal pane as a function of the outside temperature. The inside temperature was kept at a constant 68°F, and heat loss was recorded for three different outside temperatures.

$$\text{Outside temperature: } 80 \quad 50 \quad 20$$
$$\text{Heat loss: } \quad 5 \quad 10 \quad 12$$

   (a) Fit the simple linear model.

   (b) Compute an estimate of σ^2.

   (c) Compute R^2 and comment on what it tells you about the fit of this model.

7. The data given below are the number of hours (x) a student spent studying, their grade on a particular exam (y), and the corresponding residual $(\widehat{\epsilon})$.

$$x: \quad 0 \quad\quad 2.3 \quad 4 \quad\quad 6.7 \quad 7$$
$$y: \quad 6 \quad\quad 45 \quad 70 \quad 95 \quad 89$$
$$\widehat{\epsilon}: \; -6.92 \; \; 4.43 \; \; 9 \; \; 1.55 \; \; ?$$

Some useful summary statistics are: $\overline{x} = 4$, $\overline{y} = 61$, $S_{xx} = 35.18$.

   (a) Assuming the model $Y = \beta_0 + \beta_1 x + \epsilon$ is appropriate, compute the fitted line.

   (b) The residuals associated with the first four y values are given above. Compute the residual associated with $y = 89$, and compute SS_e.

   (c) Compute the coefficient of determination and comment on the fit of the line.

8. Consider the following multiple linear regression problem with two independent variables x_1 and x_2. At five (x_1, x_2) combinations the following measurements are obtained.

i	y	x_1	x_2
1	10	0	2
2	22	−1.4	1
3	34	−2	1.9
4	20	1.4	1
5	35	2	3.1

Starting with the simple linear model $Y = \beta_0 + \beta_1 x + \epsilon$, one obtains

$$\widehat{y} = 24.2 - 0.067x_1$$

with $R^2 = 0.0001$ (call this Model 1).

(a) Construct residual plots to assess the fit of Model 1.

(b) Based on the plots in part (a), what terms (if any) should be added to the model?

9. The iron content of slag can be estimated magnetically, or measured chemically. The magnetic method is quicker and easier, but the chemical method is more precise. Five slag samples are measured both ways.

(a) Use the data given below to estimate the linear relationship between $Y =$ chemical yield and $x =$ magnetic measurement.

$$\text{Magnetic:} \quad 22 \quad 19 \quad 25 \quad 21 \quad 28$$
$$\text{Chemical:} \quad 24 \quad 21 \quad 25 \quad 23 \quad 25$$

(b) Estimate σ^2 and explain what it measures in the context of this problem.

(c) Plot the residuals from the linear model against the magnetic readings and comment on the model.

10. One is interested in modeling the output of an electronic component as a function of the relative humidity. Observations were taken at five relative humidities, and the data are given below.

Relative Humidity	Output
10	3
20	7
30	13
40	21
50	35

(a) Fit the simple linear model.

(b) Compute the SS_e for this model.

(c) Compute the proportion of variation explained by the model.

(d) Plot the residuals and use the plot to suggest how would you improve the model?

11. A 2×3 factorial experiment was run to study the effect of a water bath on the final product's moisture content. The two factors studied were the temperature of the water bath and the rate at which the product moved through the water bath. For your convenience tables of sample means and variances are given below.

Means

			Rate		
		Fast	Medium	Slow	
Temperature	185	62.65	62.3	62.75	62.567
	210	63.2	62.95	63.5	63.217
		62.925	62.625	63.125	62.892

Variances

			Rate	
		Fast	Medium	Slow
Temperature	185	0.005	0.02	0.005
	210	0.02	0.005	0.02

(a) Construct an interaction plot, and use it to determine an appropriate model for the data.

(b) Find estimates for the parameters of the model you found in part (a).

(c) Which factor seems to have a bigger effect?

4

MODELS FOR THE RANDOM ERROR

In Chapter 3, we studied models for various types of experiments. In this chapter, we will study models for the random error that occurs because we must take samples, and in the following chapters we will combine the two types of models.

Models for the random error are called **random variables**. A random variable is a function that maps experimental outcomes to real numbers. For example, $Y =$ the number of heads obtained in tossing a coin once, is a random variable that maps the experimental outcomes of a head or a tail to the numbers 1 and 0, respectively. If the experimental outcome is already a real number, then the outcome itself is a random variable. For example, $Y =$ the yield for the next batch of a chemical process is a random variable.

4.1 RANDOM VARIABLES

Every random variable has associated with it a **distribution function** that assigns probabilities to the possible values of the random variable. Distribution functions are specified by either a **probability density function** or a **probability mass function**, which emulate a histogram (that has been scaled so its area is equal to one) for a random sample of y values. If the response variable Y is continuous (i.e., Y can take on any value in some range), then its probabilities are specified by a probability density function $f(y)$, which is continuous. If the response variable Y is discrete (i.e., Y can take on only a finite or countably infinite number of values k), then its probabilities are specified by a probability mass function $P(Y = k)$, which discrete.

*Note: We use uppercase letters to denote random variables: the **theoretical outcomes** of random events. We use lowercase letters to denote **actual outcomes** that are specific instances of those random variables. This allows us to write things like "Let Y be the number of defective items in a lot of size 1000, and find $P(Y=k)$, the probability of k defectives in the lot."*

Density Functions and Probability Functions

We will refer to probability density functions simply as density functions (since they tell us how concentrated the probabilities are in different regions of experimental outcomes), and we will refer to probability mass functions simply as probability functions (since they tell us directly the probabilities associated with experimental outcomes).

Any function $P(Y = k) = p_k$ such that

(i) $p_k \geq 0$ for all k

(ii) $\sum_k p_k = 1$

is a legitimate probability function. For example, if one performs the simple experiment of tossing a coin and defines Y to be the number of heads that appear, then one possible probability function to describe the behavior of Y is

$$P(Y = 0) = 0.5 \quad \text{and} \quad P(Y = 1) = 0.5. \tag{4.1.1}$$

Note that $p_0 = 0.5 \geq 0, p_1 = 0.5 \geq 0$, and $p_0 + p_1 = 1$.

Similarly, any continuous function $f(y)$ such that

(i) $f(y) \geq 0$ for all y

(ii) $\int_{-\infty}^{\infty} f(y)\, dy = 1$

is a legitimate density function. For example, if one performs the experiment of choosing a college student at random and defines Y to be his/her current GPA, then one possible density function for Y is

$$f(y) = \begin{cases} 0.25 & \text{for } 0 \leq y \leq 4 \\ 0 & \text{otherwise.} \end{cases} \tag{4.1.2}$$

Note that $f(y) \geq 0$ for all y and

$$\int_{-\infty}^{\infty} f(y)\, dy = \int_0^4 0.25\, dy = \left.\frac{y}{4}\right|_0^4 = 1.$$

The density function (4.1.2) is an example of what is called a uniform density, which says that GPA's are equally likely to be in any interval of equal length between 0 and 4. The uniform distribution is discussed in more detail in Section 4.3.

The Mean and Variance of a Random Variable

Random variables can be thought of as models for populations that one is trying to characterize based on a random sample. Thus, random variables have means and variances, which correspond to the mean and variance of the underlying population.

Discrete Random Variables

The mean of a discrete random variable is defined as

$$E(Y) = \sum_k k\,P(Y = k). \tag{4.1.3}$$

The mean is a weighted average of the different values the random variable can take on, weighted according to their probabilities of occurrence. The mean is also called the **expected value**, which is why the symbol $E(Y)$ is used. We often use the Greek letter μ to denote the mean of a population. Generally, μ and $E(Y)$ are used interchangeably.

The variance of a discrete random variable is defined as

$$\mathrm{Var}(Y) = \sum_k [k - E(Y)]^2\,P(Y = k). \tag{4.1.4}$$

We also use the symbol σ^2 to denote the population variance.

Example 4.1.1   The mean and variance of the random variable (see equations (4.1.1)) used in the coin tossing example are (using equations (4.1.3) and (4.1.4))

$$E(Y) = (0)P(Y = 0) + (1)P(Y = 1)$$
$$= p_0 = 0.5$$

and

$$\mathrm{Var}(Y) = (0 - 0.5)^2 P(Y = 0) + (1 - 0.5)^2 P(Y = 1)$$
$$= (0.25)(0.5) + (0.25)(0.5) = 0.25.$$

Continuous Random Variables

The mean and variance of a continuous random variable are defined as

$$E(Y) = \int_{-\infty}^{\infty} y f(y)\,dy \tag{4.1.5}$$

and

$$\mathrm{Var}(Y) = \int_{-\infty}^{\infty} [y - E(Y)]^2 f(y)\,dy. \tag{4.1.6}$$

Example 4.1.2 The mean and variance of the random variable (see equation (4.1.2)) used to describe GPA's are (using equations (4.1.5) and (4.1.6))

$$E(Y) = \int_{-\infty}^{\infty} y f(y)\, dy$$

$$= \int_{0}^{4} y(0.25)\, dy = \left.\frac{y^2}{8}\right|_{0}^{4} = \frac{16}{8} = 2$$

and

$$\text{Var}(Y) = \int_{-\infty}^{\infty} (y - 2)^2 f(y)\, dy$$

$$= \int_{0}^{4} (y - 2)^2 (0.25)\, dy$$

$$= \int_{0}^{4} (y^2 - 4y + 4)(0.25)\, dy$$

$$= \left.\frac{y^3}{12} - \frac{y^2}{2} + y\right|_{0}^{4}$$

$$= \frac{64}{12} - \frac{16}{2} + 4 = \frac{4}{3}.$$

Properties of Expected Values

Two properties of expected values that will be useful to us are:

> **Property 1**: The expected value of a constant is that constant. For example, $E(\mu) = \mu$.
>
> **Property 2**: The expected value of the sum of two random variables is the sum of their expected values. That is, $E(X + Y) = E(X) + E(Y)$.

These two properties tell us, for example, that with model (3.3.1) the expected value of the response is

$$E(Y) = E[\beta_0 + \beta_1 x + \epsilon]$$
$$= E[\beta_0 + \beta_1 x] + E(\epsilon)$$

and then if on average the random error ϵ is zero,

$$E(Y) = \beta_0 + \beta_1 x.$$

Properties of Variances

There are also two properties of variances that will be useful to us. One of these depends on the concept of independence. Two random variables (or two random samples) are said to be **independent** if the outcome of one random variable (or random sample) does not affect the outcome of the other random variable (or random sample).

Property 1: The variance of a constant times a random
variable is the square of the constant times
the variance of the random variable.
That is, $\mathrm{Var}(cY) = c^2\,\mathrm{Var}(Y)$.

Property 2: The variance of the sum of two independent
random variables is the sum of their variances.
That is, $\mathrm{Var}(X + Y) = \mathrm{Var}(X) + \mathrm{Var}(Y)$
if X and Y are independent.

Property 1 tells us, for example, that $\mathrm{Var}(-Y) = \mathrm{Var}(Y)$, which agrees with our intuition that multiplying values by -1 should not affect how far a set of values is spread out.

Definitions (4.1.4) and (4.1.6) for the variance of a random variable can be rewritten in a form that is sometimes more convenient for computation. This is fairly easily done using the two properties of expected values and the fact that a real-valued function of a random variable is also a random variable (e.g., if Y is a random variable, then so is Y^2). Note that (using μ instead of $E(Y)$ to simplify the notation) both equations (4.1.4) and (4.1.6) could be written as

$$\mathrm{Var}(Y) = E[(Y - \mu)^2].$$

Using the two properties of expected values and a little algebra, one then obtains

$$\begin{aligned}
\mathrm{Var}(Y) &= E[(Y - \mu)^2] \\
&= E[Y^2 - 2\mu Y + \mu^2] \\
&= E(Y^2) - E(2\mu Y) + E(\mu^2) \\
&= E(Y^2) - 2\mu E(Y) + \mu^2 \\
&= E(Y^2) - \mu^2.
\end{aligned} \qquad (4.1.7)$$

Example 4.1.3   The variance for the random GPA value computed in Example 4.1.2 (p. 87) could be more easily computed using equation (4.1.7). We had already obtained $E(Y) = 2$. Now

$$E(Y^2) = \int_{-\infty}^{\infty} y^2 f(y)\, dy$$

$$= \int_0^4 y^2 (0.25) \, dy$$

$$= \frac{y^3}{12} \Big|_0^4 = \frac{64}{12}$$

and therefore (equation (4.1.7)),

$$\text{Var}(Y) = \frac{64}{12} - 2^2 = \frac{4}{3} \, .$$

Determining Probabilities

Probabilities can be thought of as the long-term relative frequencies of the outcomes for an experiment if it is performed over and over again. Density functions and probability functions determine the probabilities associated with the possible outcomes of the particular random variable (experiment) they describe. For simplicity, we will refer to an experimental outcome (or any collection of possible outcomes) as an **event**.

Note: Since probabilities are long-term relative frequencies, the probability of an event A (denoted P(A)) must satisfy

$$0 \leq P(A) \leq 1. \qquad (4.1.8)$$

For a discrete random variable the probability of a particular event is simply the sum of all the probabilities associated with the corresponding outcomes.

Note: We are thinking of outcomes as being **mutually exclusive** *events. That is, no two outcomes can occur at the same time.*

Example 4.1.4   Suppose that one performs the experiment of tossing a fair die, and let Y denote the number that appears. Since the die is fair, the probability function for Y is

$$P(Y = i) = \frac{1}{6} \quad \text{for } i = 1, \ldots, 6.$$

If A is the event that an even number appears, then

$$P(A) = P(Y = 2) + P(Y = 4) + P(Y = 6)$$
$$= \frac{1}{6} + \frac{1}{6} + \frac{1}{6} = \frac{1}{2} \, .$$

For a continuous random variable probabilities are defined over intervals rather than for specific points. The probability that a random variable Y takes on values in the interval $[a, b]$ is the area under its density function between a and b. That is,

$$P(a \leq Y \leq b) = \int_a^b f(y)\,dy.$$

(4.1.9)

It should now be clear that the two conditions on a density function are needed in order to satisfy the restriction (4.1.8) on probabilities. Note that with definition (4.1.9) the probability of any specific value is zero. Therefore, $P(Y = a) = P(Y = b) = 0$, and it follows that

$$P(a \leq Y \leq b) = P(a < Y < b)$$
$$P(a \leq Y \leq b) = P(a \leq Y < b)$$
$$P(a \leq Y \leq b) = P(a < Y \leq b).$$

Example 4.1.5   For the randomly chosen GPA value discussed in Example 4.1.2 (p. 87), the probability that the GPA is between 2.5 and 3.5 is

$$P(2.5 \leq Y \leq 3.5) = \int_{2.5}^{3.5} (0.25)\,dy$$
$$= \frac{y}{4}\Big|_{2.5}^{3.5}$$
$$= \frac{3.5}{4} - \frac{2.5}{4} = \frac{1}{4}.$$

Distribution Functions

Rather than always integrating density functions, it is often more convenient to compute probabilities using the **distribution function** of the random variable. The distribution function for a random variable Y is

$$F(y) = P(Y \leq y).$$

(4.1.10)

Note: Distribution functions are defined for both discrete and continous random variables. They are particularly useful for computing probabilities for continuous random variables.

For a continuous random variable Y the distribution function is

$$F(y) = \int_{-\infty}^y f(u)\,du$$

(4.1.11)

and for any a and b where $a < b$

$$P(a < Y \leq b) = F(b) - F(a). \qquad (4.1.12)$$

Example 4.1.6   The distribution function for the randomly chosen GPA discussed in Example 4.1.2 (p. 87) is

$$F(y) = \int_0^y (0.25) \, dy = \frac{y}{4} \quad \text{if } 0 \leq y \leq 4.$$

Therefore, the probability of obtaining a GPA between 2.5 and 3.5 is

$$P(2.5 \leq Y \leq 3.5) = F(3.5) - F(2.5)$$
$$= \frac{3.5}{4} - \frac{2.5}{4} = \frac{1}{4}.$$

Problems

1. A discrete uniform random variable Y defined on the integers 1 to 5 has probability function $P(X = i) = 0.2$ for $i = 1, 2, \ldots, 5$.

    (a) Find $E(Y)$.

    (b) Find $\text{Var}(Y)$.

2. Let $f(y) = 3y^2$ for $y \in (0, 1)$, and $f(y) = 0$ elsewhere.

    (a) Show that $f(y)$ is a legitimate density function.

    (b) If Y is a random variable with density $f(y)$ (from part (a)), find $E(y)$ and $\text{Var}(y)$.

3. Let $f(x) = \left[\frac{x}{k+1}\right]^k$ for $k = $ some positive constant and $x \in (0, k+1)$.

    (a) Show that $f(x)$ is a legitimate density function.

    (b) If X is a random variable with density $f(x)$ (from part (a)), find $E(X)$ and $\text{Var}(X)$.

4. A function $g(x) = 1 - x$ is defined for values of x between 0 and 1 (i.e., $g(x) = 0$ if $x \notin (0, 1)$).

    (a) Find the constant c such that $f(x) = cg(x)$ is a density function.

    (b) If X is a random variable with density $f(x)$ (from part (a)), find $E(X)$ and $\text{Var}(X)$.

4.2 IMPORTANT DISCRETE DISTRIBUTIONS

The Bernoulli Distribution

The Bernoulli distribution is a model for an experiment that has only two possible outcomes. These outcomes are generally referred to as "success" and "failure," but they could actually be, for example, pass and fail if one is considering a grade as the random variable, or male and female if sex is the random variable. A Bernoulli random variable counts the number of successes in one trial of the experiment. In general, experimental trials that have only two possible outcomes are referred to as **Bernoulli trials**.

Denoting the probability of a success as p, the probability function for a Bernoulli random variable Y is

$$\boxed{\begin{aligned} P(Y = 0) &= 1 - p \\ P(Y = 1) &= p. \end{aligned}} \tag{4.2.1}$$

For the Bernoulli distribution (using equations (4.1.3) and (4.1.7))

$$E(Y) = 0(1 - p) + 1(p) = p \tag{4.2.2}$$

$$E(Y^2) = 0^2(1 - p) + 1^2 p = p$$

and

$$\mathrm{Var}(Y) = E(Y^2) - [E(Y)]^2 = p - p^2 = p(1 - p). \tag{4.2.3}$$

The Binomial Distribution

A simple, but very useful, statistic associated with many processes is the number of defective items found in a random sample. When the process is stable, that is, when the proportion of defective items being produced does not change over time, the binomial distribution is the appropriate model for the number of defectives in a random sample.

The binomial distribution can also be described as counting the number of successes (defectives in the process example) in a sequence of independent Bernoulli trials that all have the same probability of a success. Bernoulli random variables (and Bernoulli trials) are named after James Bernoulli (1654–1705), who derived the binomial distribution for the case $p = r/(r + s)$ where r and s are positive integers. This derivation, however, was not published until eight years after his death (Bernoulli, 1713).

The probability function for a binomial random variable depends on two parameters:

n = the number of trials (e.g., the number of items sampled),
p = the probability of success (e.g., the proportion of defective items in the population).

Notation: If Y is a binomial random variable with parameters n and p, we denote this by $Y \sim \mathrm{BIN}(n, p)$.

In order to determine the probability function for a binomial random variable we will need to use the following two laws of probability.

> **Multiplication Rule**: The probability of several independent events all occurring is the product of their individual probabilities.
>
> **Addition Rule**: The probability of several mutually exclusive events all occurring is the sum of their individual probabilities.

One way to obtain exactly k successes in n trials is to have the first k trials all result in successes and the remaining $n - k$ trials all result in failures. This particular sequence of events would occur with probability (using the Multiplication Rule)

$$p^k (1 - p)^{n-k}. \tag{4.2.4}$$

All the other ways to obtain exactly k successes would simply be rearrangements of the order of the k successes and $n - k$ failures. Rearrangement would not affect the probability, so each arrangement would have probability (4.2.4).

Now all that remains is to count the number of possible arrangements. More specifically, we want to count the number of ways k positions for the successes can be chosen from among the n trials. This is done using the **binomial coefficient** $\binom{n}{k}$, which is computed with the formula

$$\binom{n}{k} = \frac{n!}{k!(n-k)!} \tag{4.2.5}$$

where $i!$ (read "i factorial") is defined as

$$i! = i(i-1)(i-2)\cdots(2)1$$

and by convention, $0! \equiv 1$.

Example 4.2.1   There are

$$\binom{4}{2} = \frac{4!}{2!(4-2)!} = \frac{24}{2(2)} = 6$$

ways to assign two passing and two failing grades to four students (see Problem 4.2.1).

Therefore, since there are $\binom{n}{k}$ arrangements with exactly k successes, and each one has probability (4.2.4), the probability function for $Y \sim \text{BIN}(n, p)$ is (using the Addition Rule)

$$P(Y = k) = \binom{n}{k} p^k (1 - p)^{n-k} \qquad \text{for } k = 0, 1, 2, \ldots, n. \qquad (4.2.6)$$

Example 4.2.2   A manufacuter of computer mice obtains the tracking balls from an outside vendor. As part of its quality control program the manufacturer routinely samples 50 balls from each shipment received. The probability that a ball is defective (too small, too big, or blemished) is 0.03. What is the probability of finding less than two defective balls in a sample?

Let Y = the number of defective balls in a sample of 50. Then, $Y \sim$ BIN$(50, 0.03)$, and

$$\begin{aligned}
P(Y < 2) &= P(Y = 0) + P(Y = 1) \\
&= \binom{50}{0}(0.03)^0(0.97)^{50} + \binom{50}{1}(0.03)^1(0.97)^{49} \\
&= (0.97)^{50} + 50(0.03)(0.97)^{49} \\
&= 0.218 + 0.337 \\
&= 0.555.
\end{aligned}$$

The expected value and variance for a binomial random variable can be easily found using the properties of expected values and variances and the fact that a binomial random variable can be written in terms of Bernoulli random variables. Consider n independent Bernoulli random variables $X_1, X_2, \ldots, X_n$ where

$$X_i = \begin{cases} 1 & \text{if the } i^{\text{th}} \text{ trial is a success} \\ 0 & \text{if the } i^{\text{th}} \text{ trial is a failure} \end{cases}$$

and the probability of success is p at each trial. Then,

$$Y = X_1 + \cdots + X_n$$

is a BIN(n, p) random variable. Using Property 2 for expected values (p. 87) and the expected value for a Bernoulli random variable (equation (4.2.2)), one obtains

$$\begin{aligned}
E(Y) &= E(X_1 + \cdots + X_n) \\
&= E(X_1) + \cdots + E(X_n) \\
&= p + \cdots + p \\
&= np \qquad\qquad (4.2.7)
\end{aligned}$$

and using Property 2 for variances (p. 88) and the variance for a Bernoulli random variable (equation (4.2.3)) one obtains

$$
\begin{aligned}
\operatorname{Var}(Y) &= \operatorname{Var}(X_1 + \cdots + X_n) \\
&= \operatorname{Var}(X_1) + \cdots + \operatorname{Var}(X_n) \\
&= p(1-p) + \cdots + p(1-p) \\
&= np(1-p).
\end{aligned}
\tag{4.2.8}
$$

Example 4.2.3   For the tracking balls discussed in Example 4.2.2, the average number of defectives in the sample is $50(0.03) = 1.5$, and the variance is $50(0.03)(0.97) = 1.455$ (see equations (4.2.7) and (4.2.8)).

Parameter Estimation

It follows from equation (4.2.7) that we could estimate p based on a sample using

$$
\boxed{\widehat{p} = y/n.}
\tag{4.2.9}
$$

Example 4.2.4   Suppose that the manufacturer is working with a new vendor and does not know the defective rate of the tracking balls. To estimate the rate, a random sample of 200 balls is taken (from multiple shipments), and it is found that $y = 49$ of them are defective. The estimated defective rate is $\widehat{p} = 49/200 = 0.245$. The manufacturer quickly cancels the contract with this new vendor.

The Poisson Distribution

The Poisson distribution is another distribution for modeling count data. It is used for a response variable that counts the number of events that occur in a given space or time, such as the number of cars that drive through an intersection, or the number of sub-atomic particles emitted in a physics experiment, or the number of defects on a windshield. This distribution is named for the French mathematician Siméon Poisson (1781–1840), who derived it in 1830 (Poisson, 1830). Interestingly, it was actually first derived more than 100 years earlier by De Moivre (1711).

The Poisson distribution is appropriate when:

1. The interval of interest (e.g., space or time) can be divided into small equal units of opportunity h in such a way that at most one event can occur in each opportunity unit.

2. For any opportunity unit h the probability of one event occurring is λh.

3. The number of events occurring in non-overlapping units of opportunity are independent.

(4.2.10)

Example 4.2.5   A classical example of the use of the Poisson distribution is due to Bortkiewicz (1898), who used it to model the number of Prussian officers killed annually between 1875 and 1894 by horse kicks.

Example 4.2.6   It might be reasonable to assume that:

1. No more than one car per second can pass through an intersection ($h = 1$ sec.).

2. The probability of a car passing through the intersection during any 1 second time interval is 0.1 ($\lambda = 0.1$).

3. The number of cars going through the intersection during any time interval does not affect the number going through the intersection in any non-overlapping time intervals.

With the assumptions (4.2.10) the random variable Y = the number of events in an opportunity unit of size t is said to have a Poisson distribution with parameter λt, and its probability function is

$$P(Y = k) = \frac{e^{-\lambda t}(\lambda t)^k}{k!} \qquad \text{for } k = 0, 1, 2, \ldots. \qquad (4.2.11)$$

The parameter λ is referred to as the intensity parameter (e.g., $\lambda = 0.1$ cars/sec.), and t is size of the opportunity unit of interest (e.g., 1 hour).

Notation: We denote a random variable Y having probability function (4.2.11) by $Y \sim$ Poisson(λt).

Example 4.2.7   Under the conditions described in Example 4.2.6, what is the probability that no more than two cars will pass through the intersection in 30 seconds?

$$\lambda = 0.1 \text{ cars/sec.} \qquad t = 30 \text{ sec.} \qquad \lambda t = (0.1)(30) = 3$$

$$P(Y = 0) = e^{-3}3^0/0! = e^{-3}$$

$$P(Y = 1) = e^{-3}3^1/1! = 3e^{-3}$$
$$P(Y = 2) = e^{-3}3^2/2! = \frac{9}{2}e^{-3}$$

and the desired probability is

$$P(\text{no more than two cars in 30 seconds}) = \left[1 + 3 + \frac{9}{2}\right]e^{-3} = 0.423.$$

Starting with the definition of expected value (equation (4.1.3)), the expected value of $Y \sim \text{Poisson}(\lambda t)$ is

$$E(Y) = \sum_{k=0}^{\infty} k\frac{e^{-\lambda t}(\lambda t)^k}{k!}$$
$$= \lambda t \sum_{k=1}^{\infty} \frac{e^{-\lambda t}(\lambda t)^{k-1}}{(k-1)!}$$
$$= \lambda t \sum_{j=0}^{\infty} \frac{e^{-\lambda t}(\lambda t)^j}{j!}$$

and since

$$\sum_{j=0}^{\infty} \frac{e^{-\lambda t}(\lambda t)^j}{j!} = 1$$

(because we are summing a probability function over all its possible values)

$$E(Y) = \lambda t. \tag{4.2.12}$$

A similar, but lengthier derivation results in

$$\text{Var}(Y) = \lambda t. \tag{4.2.13}$$

Example 4.2.8   For the intersection discussed in Example 4.2.7, the expected number of cars going through the intersection in 1 minute ($t = 60$ sec.) would be $\lambda t = 0.1(60) = 6$.

Parameter Estimation

From equation (4.2.12), it follows that an estimate of λ is

$$\widehat{\lambda} = y/t. \tag{4.2.14}$$

Note, however, that this would not be a particularly good estimate unless the quantity t of opportunity units is fairly large. Alternatively, if one had observations

$$y_1, y_2, \ldots, y_n$$

from a number of smaller (non-overlapping) opportunity amounts

$$t_1, t_2, \ldots, t_n$$

then λ could be estimated as

$$\widehat{\lambda} = \frac{y_1 + \cdots + y_n}{t_1 + \cdots + t_n}. \tag{4.2.15}$$

This follows from the fact that if $Y_1 \sim \text{Poisson}(\lambda t_1)$ and $Y_2 \sim \text{Poisson}(\lambda t_2)$, where t_1 and t_2 are non-overlapping opportunity units, then (see Problem 4.2.15)

$$Y_1 + Y_2 \sim \text{Poisson}(\lambda(t_1 + t_2)). \tag{4.2.16}$$

Example 4.2.9   Suppose λ is the frequency at which α-particles are emitted when *polonium* (a radioactive element) decays. In a famous experiment, Rutherford and Geiger (1910) counted 10,097 α-particles in 52.16 hours (or 187,776 sec.). They estimated that the decay frequency was $\widehat{\lambda} = 10{,}097/187{,}776 = 0.0538$ particles/second.

Relation of the Poisson to the Binomial

The Poisson distribution is actually a limiting form of the binomial distribution and can be used to approximate binomial probabilities, and this is how Poisson (1830) derived it. If the number of trials for the binomial is large and the probability of success is small, then

$$\binom{n}{k} p^k (1-p)^{n-k} \doteq \frac{e^{-np}(np)^k}{k!}. \tag{4.2.17}$$

Example 4.2.10   In Example 4.2.2 (p. 95), we found the probability of $Y < 2$ defective balls in a sample of 50 to be 0.555 using the fact that $Y \sim \text{BIN}(50, 0.03)$. Using $np = 50(0.03) = 1.5$, we could have approximated this probability as

$$\begin{aligned}
P(Y < 2) &= P(Y = 0) + P(Y = 1) \\
&\doteq \frac{e^{-1.5}(1.5)^0}{0!} + \frac{e^{-1.5}(1.5)^1}{1!} \\
&= 0.2231 + 0.3347 \\
&= 0.558.
\end{aligned}$$

The Geometric Distribution

The geometric distribution models the position of the first success in a sequence of independent Bernoulli trials each with probability p of success. It has one parameter, the probability of success p. The geometric distribution was first discussed by Pascal (1679).

Notation: We will use $Y \sim \text{GEOM}(p)$ to denote that Y is a geometric random variable with parameter p.

The probability function for a geometric random variable is obtained as follows. In order for the first success to occur at the ith trial, the first $i - 1$ trials must result in failures. Applying the Multiplication Rule, this occurs with probability $(1 - p)^{i-1}$. The probability that the ith trial is a success is p, and combining these two probabilities (again using the Multiplication Rule) results in the probability function

$$P(Y = i) = p(1 - p)^{i-1} \qquad \text{for } i = 1, 2, \dots . \tag{4.2.18}$$

Example 4.2.11   The manufacturer of computer mice discussed in Example 4.2.2 was interested in trying to assure the quality of tracking balls without doing as much sampling. Instead of routinely sampling 50 balls from each shipment, the quality control department suggested routinely sampling only until the first defective ball was found. If the probability of a ball being defective is 0.03, what is the probability of finding the first defective within the first three balls sampled?

Let $Y = $ the position of the first defective ball

Then, $Y \sim \text{GEOM}(0.03)$, and

$$\begin{aligned} P(Y \leq 3) &= P(Y = 1) + P(Y = 2) + P(Y = 3) \\ &= 0.03 + 0.03(0.97) + 0.03(0.97)^2 \\ &= 0.087327. \end{aligned}$$

This probability is small, so if a defective ball was found within the first three sampled, one would question the quality of the shipment.

Using equation (4.1.3), the expected value of a geometric random variable Y with parameter p is

$$\begin{aligned} E(Y) &= \sum_{i=1}^{\infty} ip(1 - p)^{i-1} \\ &= p \sum_{i=1}^{\infty} iq^{i-1} \end{aligned} \tag{4.2.19}$$

where (for ease of notation) $q = 1 - p$. This can be simplified by noting that if we define

$$S = \sum_{i=1}^{\infty} iq^{i-1} = 1 + 2q + 3q^2 + \cdots$$

then

$$qS = \sum_{i=1}^{\infty} iq^i = q + 2q^2 + \cdots$$

and (subtracting the two equations)

$$S(1 - q) = \sum_{i=0}^{\infty} q^i. \qquad (4.2.20)$$

Further, since

$$p \sum_{i=0}^{\infty} q^i = \sum_{i=1}^{\infty} p(1 - p)^{i-1} = 1$$

(because we are summing a probability function over all its possible values)

$$\sum_{i=0}^{\infty} q^i = \frac{1}{p}. \qquad (4.2.21)$$

Combining equations (4.2.19)–(4.2.21) results in

$$S = \frac{1}{p^2}$$

and

$$\boxed{E(Y) = \frac{1}{p}.} \qquad (4.2.22)$$

A similar, but lengthier, derivation would show that

$$\boxed{\mathrm{Var}(Y) = \frac{q}{p^2}.} \qquad (4.2.23)$$

Example 4.2.12   For the tracking-ball quality-control scheme discussed in Example 4.2.11, the expected number of balls sampled until one finds the first defective ball is $1/0.03 = 33.3$ and the variance is $0.97/(0.03)^2 = 1077.8$ (equations (4.2.22) and (4.2.23)).

Parameter Estimation

Based on equation (4.2.22), if we knew the positions of the first successes from several samples, we could estimate p using

$$\boxed{\hat{p} = 1/\bar{y}.} \qquad (4.2.24)$$

Example 4.2.13   As an alternative estimate of the defective rate of tracking balls for the new vendor discussed in Example 4.2.4, it was noted that in five shipments the first defective balls were found in positions $6, 4, 2, 6,$ and 7. For these samples $\overline{y} = 5$, and the defective rate was estimated to be $\widehat{p} = 1/5 = 0.2$.

Problems

1. List all the ways four students (Tom, Ed, Carol, and Nancy) can be assigned grades if two pass and two fail.

2. Suppose that the random variable $Y \sim \text{BIN}(5, 0.1)$. Calculate $P(Y = k)$ for $k = 0, 1, \ldots, 5$; and make a graph of the resulting probability function.

3. Repeat Problem 2 for $p = 0.2, 0.5$, and 0.7.

4. If $Y \sim \text{BIN}(5, 0.1)$, find
    (a) $P(Y > 1)$
    (b) $P(Y < 3)$
    (c) $P(2 \leq Y < 4)$.

5. If $Y \sim \text{BIN}(20, 0.3)$ find
    (a) $P(Y = 0)$
    (b) $P(Y > 1)$
    (c) $P(Y = 5)$
    (d) $P(Y \leq 2)$
    (e) $E(Y)$
    (f) $\text{Var}(Y)$.

6. A quality engineer is monitoring a lens coating process, which on the average produces 1% defective lenses. Each hour the engineer samples 20 lenses and counts $Y =$ the number of defectives.
    (a) What is the distribution of Y?
    (b) Use a spreadsheet or other computer package (or a programmable calculator) to make a table of the possible values of Y and the probability associated with each of them.
    (c) How many lenses (out of 20) would the engineer expect to be defective?
    (d) Using the table from part (b) as a guide, when (i.e., how many defects out of the 20) should the engineer become concerned that the process may not be running as desired?

7. A new employee is supposed to be sorting coated lenses into one of three bins: ship, rework, or scrap. However, this employee was not well trained and sorts 50 lenses by randomly putting each lens in one of the three bins.
    (a) Let Y be the number of lenses the employee sorts correctly. What is the distribution of Y?
    (b) Find $P(Y < 10)$.
    (c) Find $E(Y)$.
    (d) Find $\text{Var}(Y)$.

8. The new employee from the previous problem was retrained and put back to work sorting lenses into three bins: ship, rework, and scrap. A well-trained worker will sort with 98% accuracy. To check to see if the new employee was now well trained a supervisor re-checked 50 lenses sorted by the new employee. If the employee is now well trained, how many lenses would the supervisor expect to find sorted incorrectly?

9. A consumer products company surveys 30 shoppers at a mall and asks each one if he/she likes the packaging of a new soap product. The results of the survey are:

Yes: 10; No: 19; No opinion: 1.

   (a) Use these data to estimate the proportion of shoppers who would like the package for the soap.

   (b) Suppose the company gathered these data by talking to customers in one of their stores. What impact might this have on the results. Suggest a better way to gather the data.

10. Suppose Y is Poisson with $\lambda = 2$ and the opportunity unit of interest is $t = 1$.
    (a) Find $P(Y = k)$ for $k = 0, 1, \ldots, 10$ and make a graph of this probability function.

    (b) Find $E(Y)$ and $\mathrm{Var}(Y)$.

    (c) Find $P(Y < 3)$.

    (d) Find $P(Y > 4)$.

11. The number of cars passing through an intersection has a Poisson distribution with $\lambda = 5$ cars/minute.
    (a) What is the probability that more than four cars will pass through the intersection in a given minute?

    (b) What is the probability that the intersection will be completely free of cars for 2 minutes?

12. The number of minor defects on a coated lens follows a Poisson distribution with $\lambda = 0.2/\text{lens}$.
    (a) What is the probability that a lens will be defect free?

    (b) A lens can be shipped as long as it has two or fewer minor defects. What is the probability that a lens can be shipped?

13. Production facilities must keep track of injuries on the job by their employees, known as recordables (injuries severe enough that they must be reported to the appropriate government agency). A large chemical plant maintains a sign at the entrance to the plant with the current number of injury-free work days. Assume that on any one day a recordable injury occurs with probability 0.02, and assume that this probability is the same for every day of the week.

    (a) What is the probability that the sign at the plant entrance reads 5 days.

    (b) What is the probability that the number on the sign is less than 5?

    (c) What is the probability that the number on the sign is greater than 5?

    (d) What is the average number of days that the plant goes without an injury?

    (e) In an attempt to decrease the injury rate at the plant, a safety initiative is undertaken. The next five strings of days without injuries at the plant were 92, 85, 103, 48, and 72. Was the initiative successful in lowering the injury rate? Justify your answer.

14. The engineer in Problem 6 was having a busy day, so instead of testing 20 lenses per hour he simply tested lenses until a defective lens was found.

    (a) What is the probability that the engineer tested fewer than four lenses?

    (b) What is the expected number of lenses that the engineer would test using this procedure?

    (c) Write up a proposal for a quality inspection plan based on testing until a defective is found. Include both a warning and a critical (i.e., stop production) limit for the inspection team. Hint: Base your limits on the minimal number of defect-free lenses that an inspector must find at each inspection. You may want to use a spreadsheet (or other such tool) to calculate a series of probabilities to help you in defining limits.

15. Justify formula (4.2.16).

4.3 IMPORTANT CONTINUOUS DISTRIBUTIONS

The Uniform Distribution

A **uniform distribution** (also referred to as a **rectangular distribution**) is a model where the probability of an outcome falling in any interval depends only on the length of the interval. The random numbers generated by a calculator are based on this distribution, and we used the uniform distribution as a model for GPA's in Example 4.1.2. The uniform distribution has two parameters, which are the end points of the (finite) interval of possible values. The density function for a uniform random variable on the interval $[a, b]$ is

$$f(y) = \frac{1}{b - a} \qquad \text{for } a \leq y \leq b. \tag{4.3.1}$$

A graph of this density for $[a, b] = [1, 3]$ is shown in Figure 4.1.

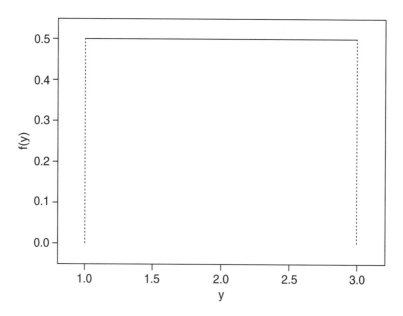

Figure 4.1. The Density Function of a UNIF$[1, 3]$ Random Variable.

Notation: A random variable U having a uniform distribution on the interval $[a, b]$ is denoted by $U \sim \text{UNIF}[a, b]$.

The distribution function for a UNIF$[a, b]$ random variable is (see Problem 4.3.1)

$$F(x) = \begin{cases} 0 & \text{if } x < a \\ \frac{x-a}{b-a} & \text{if } a \le x \le b \\ 1 & \text{if } x > b. \end{cases} \qquad (4.3.2)$$

The uniform distribution is the simplest continuous distribution, and integrating under its density function to obtain probabilities amounts only to computing areas of rectangles. Early use of the uniform distribution is found in the works of Bayes (1763) and Laplace (1812), but it was probably in use prior to this.

Example 4.3.1   The uniform distribution is often used to model the distribution of roundoff errors when values are rounded to d decimal places. In that case

$$a = -0.5 \times 10^{-d} \qquad \text{and} \qquad b = 0.5 \times 10^{-d}.$$

If E is the roundoff error, then

$$\begin{aligned} P[E < 0] &= F(0) \\ &= \frac{0 - (-0.5 \times 10^{-d})}{0.5 \times 10^{-d} - (-0.5 \times 10^{-d})} \\ &= 0.5 \end{aligned}$$

and this model implies that errors are just as likely to be positive as to be negative.

Example 4.3.2   If $U_1, \ldots, U_n$ is a random sample of size n from a population with a UNIF$[0, 1]$ distribution, and $U_{(1)}$ is the smallest observation (this is referred to as the **first-order statistic**), then

$$\begin{aligned} P[U_{(1)} \le x] &= 1 - P[\text{all the } U_i\text{'s are} > x] \\ &= 1 - P[U_i > x \text{ for } i = 1, \ldots, n] \\ &= 1 - \{P[U_i > x]\}^n \quad \text{(the Multiplication Rule)} \\ &= 1 - \{1 - x\}^n. \end{aligned}$$

Therefore, the probability that the smallest of a set of 5 UNIF$[0, 1]$ random numbers is less than 0.5 is

$$P[U_{(1)} < 0.5] = 1 - (1 - 0.5)^5 = 1 - 0.03125 = 0.96875.$$

Probability Integral Transformation

The uniform distribution is also important because any continuous random variable with a strictly increasing distribution function can be transformed to a uniform random variable. Any random variable X can be transformed using a real-valued function g to a new random variable $Y = g(X)$. For example, if $g(x) = x^2$, then $Y = X^2$ is a new random variable. When the function g is a strictly increasing

distribution function of a continuous random variable, we have the following very useful result.

> **Probability Integral Transformation**: If the continuous random variable X has a strictly increasing distribution function $F(x)$, then the random variable $U = F(X)$ has a UNIF$[0, 1]$ distribution.

It is easy to see that this is true by obtaining the distribution function of U. Note that a distribution function takes on only values between 0 and 1, so that possible values of U are in the interval $[0, 1]$. It follows that for $0 \leq x \leq 1$

$$
\begin{aligned}
P[U \leq x] &= P[F(X) \leq x] \\
&= P[X \leq F^{-1}(x)] \quad (F^{-1} \text{ is well defined since } F \text{ is strictly increasing}) \\
&= F(F^{-1}(x)) \\
&= x
\end{aligned}
$$

which is the distribution function of a UNIF$[0, 1]$ random variable (see equation (4.3.2)).

Example 4.3.3　　If $X_{(1)}$ is the smallest value of a sample of n observations from a population with continuous strictly increasing distribution function $F(x)$ (i.e., it is the first order statistic), then $F(X_{(1)})$ is distributed as $U_{(1)}$, the first order statistic of a sample of size n from a UNIF$[0, 1]$ distribution. The distribution function of $X_{(1)}$ is (see Problem 4.3.2)

$$
P[X_{(1)} \leq x] = 1 - \{1 - F(x)\}^n
$$

and

$$
\begin{aligned}
P[F(X_{(1)}) \leq x] &= P[X_{(1)} \leq F^{-1}(x)] \\
&= 1 - \{1 - F(F^{-1}(x))\}^n \\
&= 1 - (1 - x)^n
\end{aligned}
$$

which is the distribution function of $U_{(1)}$ (see Example 4.3.2). In a similar fashion, it is possible to show that for any order statistic $X_{(i)}$, $F(X_{(i)})$ has the same distribution as $U_{(i)}$.

The Exponential Distribution

When data are strongly right skewed with zero as the smallest value, such as the Rail Car Data in Example 2.1.1 (p. 6) or lifetime data, the **exponential distribution** sometimes provides a reasonable model. The density function for the exponential distribution is

P_5 $$f(y) = \frac{1}{\mu}e^{-y/\mu} \qquad \text{for } y \geq 0 \tag{4.3.3}$$

and has one parameter, $\mu > 0$. A graph of this density function with $\mu = 1000$ is given in Figure 4.2.

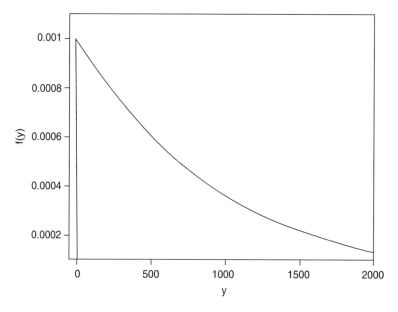

Figure 4.2. The Density Function of an Exponential Random Variable with Mean of 1000.

Note that $f(0) = 1/\mu$. Thus, for small values of μ, the density will be high at 0, and drop off rapidly. For large values of μ, the distribution will be much lower at 0 and drop off more slowly. With this flexibility the exponential distribution can be used to model many different kinds of data with exponential-type decay simply by changing the value of μ.

The distribution function for an exponential random variable is (see Problem 4.3.11)

$$F(y) = 1 - e^{-y/\mu} \qquad \text{for } y \geq 0. \tag{4.3.4}$$

Example 4.3.4 The lifetimes of compact flourecent light bulbs are exponentially distributed with a mean lifetime of 7 years. The probability that a light bulb will last less than 3.5 years (using equations (4.1.11) and (4.3.4)) is

$$P(Y < 3.5) = F(3.5) = 1 - e^{-3.5/7}$$
$$= 1 - 0.6065 = 0.39.$$

There is a 39% chance that a light bulb will fail before 3.5 years. Equivalently, we could say that 39% of compact fluorescent bulbs fail in the first 3.5 years of use.

The probability that a bulb will last no more than 16 years is

$$P(Y \le 16) = 1 - e^{-16/7}$$
$$= 1 - 0.1017 = 0.8982.$$

Example 4.3.5 (*Continuation of Example 4.3.4*)   The probability that a light bulb will last more than 6 years (using equation (4.3.4)) is

$$P(Y > 6) = 1 - P(Y \le 6)$$
$$= 1 - F(6)$$
$$= 1 - (1 - e^{-6/7})$$
$$= e^{-6/7} = 0.42.$$

That is, 42% of light bulbs last more than 6 years. ·

The probability that a light bulb will last between 3.5 and 9 years is

$$P(3.5 < Y < 9) = F(9) - F(3.5)$$
$$= (1 - e^{-9/7}) - (1 - e^{-3.5/7})$$
$$= e^{-3.5/7} - e^{-9/7} = 0.6065 - 0.2765 = 0.33.$$

About 33% of this brand of compact fluorescent light bulbs last between 3.5 and 9 years.

You can use equations (4.1.5) and (4.1.6) to check for yourself (see Problem 4.3.10.) that for any exponential random variable Y

$$\boxed{E(Y) = \mu} \qquad (4.3.5)$$

and

$$\boxed{\text{Var}(Y) = \mu^2.} \qquad (4.3.6)$$

Thus, μ represents the mean of the exponential population, and an exponential population always has a variance that is the square of the mean.

Percentiles and Quantiles

A **percentile** is a value that a specified portion of the population falls below. For example, the 95th percentile of light bulb lifetimes is the lifetime at which 95% of the bulbs will have failed. A **quantile** is the same number expressed as a proportion rather than a percentage. That is, the 95th percentile is the same as the 0.95 quantile. We use Q(0.95) to denote this. In general, the pth quantile $Q(p)$ is defined by

$$F(Q(P)) = p. \qquad (4.3.7)$$

Percentiles and quantiles are useful for describing populations in the same way that means and standard deviations are. In fact, in some cases percentiles (or quantiles) may be more useful. It may be more useful to know upper and lower limits on the lifetime of your light bulbs rather than knowing the average and standard deviation. Later in this chapter we will use quantiles to help assess the fit of a distribution to a set of data.

We have already found the 90th percentile of light bulb lifetimes in Example 4.3.4 (it is 16 years). However, we did this more or less by accident. How could we set out to find, say, the 80th percentile of lifetimes (i.e., $Q(0.8)$)? The quantile Q(0.80) for an exponential distribution with mean of 7 is shown in Figure 4.3. The $Q(0.8)$ quantile is defined by (equation (4.3.7))

$$0.80 = F(Q(0.80))$$
$$= 1 - e^{-Q(0.80)/7}.$$

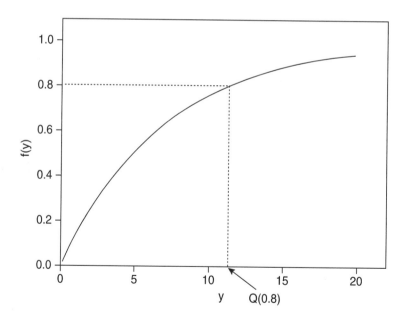

Figure 4.3. The 0.8 Quantile for an Exponential Distribution with $\mu = 7$.

Solving this equation for Q(0.80), we have

$$1 - e^{-Q(0.80)/7} = 0.80$$

$$e^{-Q(0.80)/7} = 0.20$$
$$-Q(0.80)/7 = \ln(0.20)$$
$$Q(0.80) = -7\ln(0.20) = 11.27.$$

This means that 80% of the light bulbs will fail in less than 11.27 years.

Generalizing the above, one finds that for an exponential population with mean μ, the quantile corresponding to a proportion p is

$$Q(p) = -\mu\ln(1 - p). \tag{4.3.8}$$

Example 4.3.6   The 0.95 quantile of an exponential distribution with a mean of 75 is

$$Q(0.95) = -75\ln(1 - 0.95) = 224.7.$$

Parameter Estimation

The above probability and quantile calculations were performed using a *known* mean. That is, we assumed the value of μ was known or given. In many situations we will only have a set of data that looks like it might come from an exponential distribution, and we need to use the data to estimate the population parameter μ. Since the population mean is $E(Y) = \mu$, we can estimate μ using

$$\widehat{\mu} = \overline{y}. \tag{4.3.9}$$

Notice that equation (4.3.9) does not say that the population mean and the sample mean are the same. It says we can use the sample mean as an *estimate* of the population mean. When we write "$\mu = 5$", we are claiming that 5 *is* the population mean. When we write "$\widehat{\mu} = 5$", we are only claiming that 5 is our best guess of the population mean. It is good scientific practice to always distinguish between things we know and things we are estimating.

Example 4.3.7 (*More Rail Car Data*)   Data on rail-car hold times for a second product from the company discussed in Example 2.1.1 (p. 6) are given in Table 4.1 (and in `railcar2.txt`). Note the zero hold times. These occur when the rail car is unloaded as soon as it arrives at the customer's site. A histogram for this set of data is given in Figure 4.4. Judging from this histogram it seems plausible that hold times for this product might follow an exponential distribution. In order to determine exactly which exponential distribution we would need to estimate the parameter μ of the distribution. Using equation (4.3.9), we would estimate μ as $\widehat{\mu} = \overline{y} = 5.16$. Thus, the estimated density function for rail-car hold times for this particular product is

$$\widehat{f}(y) = \frac{1}{5.16}e^{-y/5.16} \qquad \text{for } y \geq 0.$$

Table 4.1. Data on Rail Car Hold Times for Example 4.3.7

37	0	0	0	2	15	0	4	0	2	7	7	8	8	2
2	2	1	1	3	7	3	13	15	4	4	1	5	13	4
10	6	6	5	4	8	1	0	2	1	6	3	0		

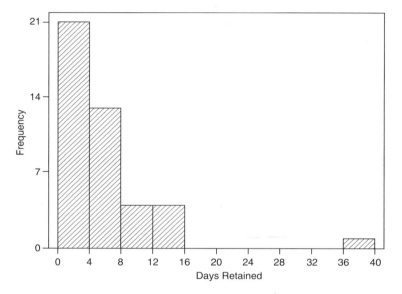

Figure 4.4. A Histogram of the Rail Car Hold Times Data in Example 4.3.7.

Any probability calculations that use an estimated parameter value are also estimates. Thus, from our data, we would estimate the distribution function for hold times to be

$$\widehat{F}(y) = 1 - e^{-y/5.16}$$

and the probability that a rail car is held less than 2 days as

$$P(Y < 2) \doteq \widehat{F}(2)$$
$$= 1 - e^{-2/5.16} = 1 - 0.6787 = 0.3213.$$

About 32% of the rail cars are returned to the system in less than 2 days. Similarly, we could estimate the 90th percentile of rail car hold times as

$$Q(0.90) = -\mu \ln(0.10) \doteq -\overline{y} \ln(0.10) = -5.16 \ln(0.10) = 11.88.$$

So we estimate that 90% of the rail cars will be held for less than 2 weeks.

Relation of the Exponential to the Poisson

If the number of cars arriving at an intersection has a Poisson distribution, one would say that they arrive according to a Poisson process. When events (like the arrival of cars) follow a Poisson process with intensity λ, then the time between successive events has an exponential distribution with mean $1/\lambda$. Let T be the time between successive events, and let Y represent the number of events occurring in a time period of length t. Then

$$
\begin{aligned}
P(T > t) &= P[\text{no arrivals in the time interval } (0, t)] \\
&= P(Y = 0) \\
&= \frac{e^{-\lambda t}(\lambda t)^0}{0!} \\
&= e^{-\lambda t}
\end{aligned}
$$

and

$$
\begin{aligned}
P(T \leq t) &= 1 - P(T > t) \\
&= 1 - e^{-\lambda t} \\
&= 1 - e^{-t/(1/\lambda)}.
\end{aligned}
$$

Example 4.3.8   E-mails arrive at a server according to a Poisson process with intensity of three e-mails per minute.

(a) What is the average time between e-mails?

(b) What is the probability that there will be less than one minute between successive e-mails?

Letting T represent the time between successive e-mails, one obtains

(a) $E(T) = 1/\lambda = 1/3$ min. $= 20$ sec.

(b) $P(T < 1) = 1 - e^{-1(3)} = 1 - 0.0498 = 0.95$.

The Normal Distribution

Many real-life sets of data (particularly physical measurements on a linear scale, such as heights, weights, diameters, etc.) follow a normal distribution. The density function for the normal distribution depends on two parameters, $\infty < \mu < \infty$ and $\sigma^2 > 0$, and is

$$
f(y) = \frac{1}{\sqrt{2\pi}\sigma} e^{-\frac{1}{2}\left(\frac{y-\mu}{\sigma}\right)^2} \qquad \text{for } -\infty < y < \infty. \tag{4.3.10}
$$

The parameters μ and σ^2 turn out to be the mean and variance of the distribution.

Notation: We use the notation $Y \sim N(\mu, \sigma^2)$ to denote that Y has a normal distribution with mean μ and variance σ^2.

An important special case, referred to as the **standard normal distribution**, is when $\mu = 0$ and $\sigma^2 = 1$. We will reserve Z to represent a random variable having a $N(0, 1)$ distribution. The standard normal distribution is important enough that its density function has its own special notation. Rather than the generic $f(y)$, the standard normal density function is denoted by

$$\phi(z) = \frac{1}{\sqrt{2\pi}} \, e^{-z^2/2} \qquad \text{for } -\infty < z < \infty \qquad (4.3.11)$$

and a graph of it is given in Figure 4.5.

Note that $\phi(z)$ approaches 0 as $z \to \pm\infty$. It is also clear from the graph (as with the sample mean, the expected value of a random variable is the balance point for its density function) that $E(Z) = 0$.

The standard normal distribution function also has its own notation. It is denoted by

$$\Phi(z) = \frac{1}{\sqrt{2\pi}} \int_{-\infty}^{z} e^{-u^2/2} \, du. \qquad (4.3.12)$$

Unfortunately, there is no antiderivative for the integral in equation (4.3.12), and therefore, numerical methods are needed to obtain standard normal probabilities. Fortunately, standard normal probabilities $\Phi(z)$ have been tabulated, and values are provided in Table B1. Standard normal probabilities are also widely available on many scientific calculators, spreadsheets, and other statistical software.

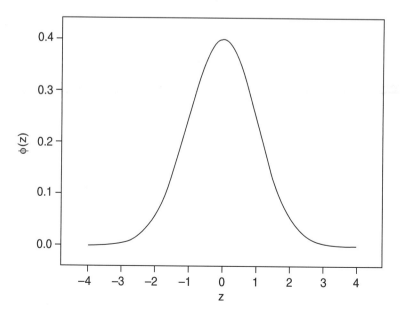

Figure 4.5. The Standard Normal Density Function $\phi(z)$.

Historically, the first derivation of the normal distribution was due to De Moivre (1733), who obtained it as an approximation to the binomial distribution. Laplace (1774) was the first person to evaluate the integral

$$\int_0^\infty e^{-z^2/2} \, dz = \sqrt{\frac{\pi}{2}}$$

from which it follows immediately that $\phi(z)$ is a legitimate density. Gauss (1809, 1816) introduced statistical techniques based on the normal distribution, which are still today standard methods.

Example 4.3.9   The probability that a standard normal random variable is less than $z = 0.22$ (i.e., $\Phi(0.22)$) is found in the row labeled 0.2 (the value of z through the first decimal place) and the column labeled 0.02 (the second decimal place of z). See the small portion of Table B1 reproduced in Table 4.2. Therefore, $\Phi(0.22) = 0.5871$.

Table 4.2. Part of Table B1

z	0.00	0.01	0.02	0.03
0.0	0.5000	0.5040	0.5080	0.5120
0.1	0.5398	0.5438	0.5478	0.5517
0.2	0.5793	0.5832	0.5871	0.5910
0.3	0.6179	0.6217	0.6255	0.6293
0.4	0.6554	0.6591	0.6628	0.6664

Since linear transformations of a random variable (i.e., $X \to aX + b$) do not affect the general form of the random variable's distribution, $Y = \sigma Z + \mu$ is also a normal random variable. Using the properties of expected values and variances, and the intuitive fact that shifting a random variable's location does not affect its variance, one finds

$$\begin{aligned} E(Y) &= E(\sigma Z + \mu) \\ &= \sigma E(Z) + \mu \\ &= \mu \end{aligned}$$

and

$$\begin{aligned} \mathrm{Var}(Y) &= \mathrm{Var}(\sigma Z + \mu) \\ &= \mathrm{Var}(\sigma Z) \\ &= \sigma^2 \mathrm{Var}(z) \\ &= \sigma^2. \end{aligned}$$

Therefore,

$$\boxed{Y = \sigma Z + \mu \sim N(\mu, \sigma^2)} \tag{4.3.13}$$

and probabilities for any random variable $Y \sim N(\mu, \sigma^2)$ can be found by noting

$$P(Y \leq y) = P(\sigma Z + \mu \leq y)$$
$$= P\left[Z \leq \frac{y - \mu}{\sigma}\right]$$
$$= \Phi\left(\frac{y - \mu}{\sigma}\right).$$

For a normal distribution most of the probability is concentrated within the first three standard deviations around the mean.

If Y is any normally distributed random variable, then

Y is within 1 standard deviation of the mean with 68% probability
Y is within 2 standard deviations of the mean with 95% probability
Y is within 3 standard deviations of the mean with 99.7% probability.

Example 4.3.10 (*pH Values*)   The pH of a chemical process is measured hourly. These values are normally distributed with a mean of 6.5 and a standard deviation of 0.78. (From above, this means that almost all of the pH values fall in the range 4.16 to 8.84.) What is the probability that a pH value is less than 7?

$$P(Y < 7) = \Phi\left(\frac{7 - 6.5}{0.78}\right) = \Phi(0.64) = 0.7389.$$

Thus, about 74% of the pH values are less than 7.

Example 4.3.11 (*Daily Yields*)   The daily yield from a production line is a $N(450, 900)$ random variable. Find the proportion of days with yield (a) below 500, (b) above 550, and (c) between 400 and 500.

Since $\sigma = \sqrt{900} = 30$,

(a) $P(Y < 500) = \Phi\left(\dfrac{500 - 450}{30}\right) = \Phi(1.67) = 0.9525$

(b) $P(Y > 550) = 1 - \Phi\left(\dfrac{550 - 450}{30}\right) = 1 - \Phi(3.33) = 1 - 0.9996 = 0.0004$

(c) $P(400 < Y < 500) = \Phi\left(\dfrac{500 - 450}{30}\right) - \Phi\left(\dfrac{400 - 450}{30}\right)$
$$= \Phi(1.67) - \Phi(-1.67)$$
$$= 0.9525 - (1 - 0.9525)$$
$$= 0.9050.$$

Percentiles and Quantiles

Percentiles and quantiles are defined the same way for all distributions. If the daily yield on your day of supervising the process was at the 0.89 quantile (or 89th percentile), this means that 89% of the daily yields were lower than your yield. For the normal distribution we obtain quantiles using the normal table (Table B1).

Example 4.3.12   What is $Q(0.9)$ for a $N(0,1)$ random variable? The quantity $Q(0.90)$ is defined by the equation

$$0.90 = \Phi(Q(0.90)).$$

Alternatively, $Q(0.9)$ can be represented in terms of the density function $\phi(z)$ as shown in Figure 4.6.

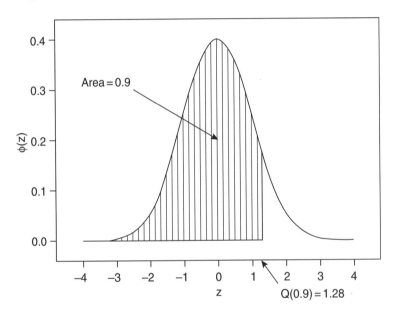

Figure 4.6. The $Q(0.9)$ Quantile for a Standard Normal Distribution.

Table 4.3. A Portion of Table B1

z	0.00	0.01	0.02	0.03	0.04	0.05	0.06	0.07	0.08	0.09
1.0	0.8413	0.8438	0.8461	0.8485	0.8508	0.8531	0.8554	0.8577	0.8599	0.8621
1.1	0.8643	0.8665	0.8686	0.8708	0.8729	0.8749	0.8770	0.8790	0.8810	0.8830
1.2	0.8849	0.8869	0.8888	0.8907	0.8925	0.8944	0.8962	0.8980	0.8997	0.9015
1.3	0.9032	0.9049	0.9066	0.9082	0.9099	0.9115	0.9131	0.9147	0.9162	0.9177
1.4	0.9192	0.9207	0.9222	0.9236	0.9251	0.9265	0.9279	0.9292	0.9306	0.9319

Thus, we want to find the value closest to 0.9 in the body of Table B1, and see what z value corresponds to it. The closest value to 0.9 is 0.8997, which corresponds to $z = 1.28$ (see the portion of Table B1 reproduced in Table 4.3), and therefore $Q(0.9) = 1.28$.

Example 4.3.13 What is $Q(0.1)$ for a $N(0, 1)$ random variable? Since all the probabilities in Table B1 are greater than or equal to 0.5, we cannot look for a value close to 0.1 in the body of that table. Instead, we must make use of the symmetry of the normal density. The $Q(0.1)$ quantile is shown in Figure 4.7. Note that the marked region in Figure 4.7 has exactly the same area as the unmarked region in Figure 4.6. Therefore, because of the symmetry of the normal density function, $Q(0.1) = -Q(0.9) = -1.28$.

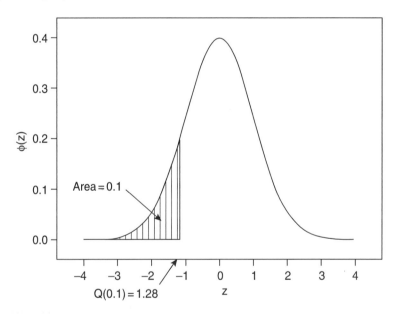

Figure 4.7. The $Q(0.1)$ Quantile of a Standard Normal Distribution.

The technique described in Example 4.3.13 works in general for finding standard normal quantiles $Q(p)$ when $p < 0.5$. Algebraically,

$$Q_Z(p) = -Q_Z(1 - p) \qquad (4.3.14)$$

where the subscript Z has been added to emphasize that this equality is for standard normal quantiles. With a little algebra (see Problem 4.3.16) one can show that in general the pth quantile of $Y \sim N(\mu, \sigma^2)$ is

$$Q_Y(p) = \sigma Q_Z(p) + \mu. \qquad (4.3.15)$$

Example 4.3.14 (*Daily Yields, p. 117*) What is the 0.9 quantile for the daily yields? From Example 4.3.11, we know $\mu = 450$ and $\sigma = 30$. Using equation (4.3.15) and Table B1,

$$Q_Y(0.9) = 30Q_Z(0.9) + 450 = 30(1.28) + 450 = 488.4.$$

Daily yields of less than 488.4 occur 90% of the time.

Example 4.3.15 (*Daily Yields, p. 117*) What is the 0.25 quantile for daily yields?

$$
\begin{aligned}
Q_Y(0.25) &= 30Q_Z(0.25) + 450 \\
&= 30[-Q_Z(1 - 0.25)] + 450 \\
&= 30[-Q_Z(0.75)] + 450 \\
&= 30(-0.67) + 450 \\
&= 429.9.
\end{aligned}
$$

Parameter Estimation

Because μ is the mean of the normal distribution and σ^2 is its variance, we can estimate these parameters using

$$\boxed{\hat{\mu} = \overline{y} \quad \text{and} \quad \hat{\sigma}^2 = s^2.} \tag{4.3.16}$$

Notation: In order to study the properties of different estimates, one needs to look at how they behave when used over and over again, rather than just the specific value obtained with a particular data set. When studying this long term behavior, one refers to the formula for computing the estimate as an **estimator**. For example, the random variable $\overline{Y}$ is an estimator of the population mean; and $\overline{y}$, the specific value of $\overline{Y}$ for a particular data set, is an estimate of the population mean.

Note: The estimators $\overline{Y}$ and S^2, corresponding to the estimates given in equation (4.3.16) are not the only possible estimators for the two parameters of a normal distribution. For example, we could also use the sample median $\widetilde{Y}$ as an estimator for μ. However, $\overline{Y}$ and S^2 are the "best" estimators for the parameters of the normal distribution in the sense that among all the estimators that are correct on average (e.g., $E(\overline{Y}) = \mu$ and $E(S^2) = \sigma^2$), they are the ones having the smallest variance.

Example 4.3.16 (*Clear-Coat Thickness Data*) When paint is applied to a car it is applied at a particular film build (i.e., thickness). The thickness of the coating has an effect on the properties of the paint so it is important to monitor the coating process to maintain the correct film build. The final layer of paint on a car is called the clear coat, the film build of this layer was to be 65 microns. The film build data on the clear-coat layer of 40 cars are in `ccthickn.txt` and are given in Table 4.4. From the histogram of data shown in Figure 4.8 it appears that the normal distribution may be a good model.

Table 4.4. Clear-Coat Thickness Data for Example 4.3.16

64.7	64.5	71.3	61.4	60.6	66.4	64.2	62.6	61.4	64.8
61.6	65.4	62.2	67.9	64.2	67.0	66.4	61.2	63.9	62.3
62.6	64.0	58.2	64.4	66.4	66.2	64.4	64.8	67.2	70.2
66.5	61.1	64.9	62.7	60.7	65.5	64.4	60.5	64.1	67.6

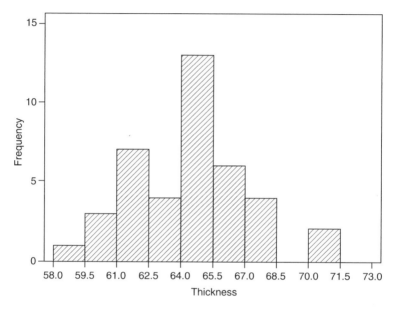

Figure 4.8. A Histogram for the Clear-Coat Thickness Data in Example 4.3.16.

The sample mean and variance of the clear-coat thickness data are $\bar{y} = 64.3$ and $s^2 = 7.29$ (hence, $s = 2.7$). So our estimated mathematical model for the clear-coat data (using equations (4.3.16)) is

$$\widehat{f}(y) = \frac{1}{\sqrt{2\pi}(2.7)} e^{-\frac{1}{2}\left(\frac{y-64.3}{2.7}\right)^2}.$$

As we did with the exponential distribution, we can use our estimated distribution function to estimate probabilities. If Y is the film build on a randomly chosen car, what is the probability that $Y < 60$?

$$P(Y < 60) \doteq \Phi\left(\frac{60 - 64.3}{2.7}\right) = \Phi(-1.59) = 0.0559.$$

An Important Property

The normal distribution has many interesting and important properties. We will make extensive use of one of these. Specifically, we will use the fact that the sum of two independent normal random variables is again a normal random variable. This fact together with the properties of expected values and variances tells us that if $Y_1 \sim N(\mu_1, \sigma_1^2)$ and $Y_2 \sim N(\mu_2, \sigma_2^2)$ are independent, then

$$E(Y_1 + Y_2) = E(Y_1) + E(Y_2) = \mu_1 + \mu_2$$
$$\mathrm{Var}(Y_1 + Y_2) = \mathrm{Var}(Y_1) + \mathrm{Var}(Y_2) = \sigma_1^2 + \sigma_2^2$$

and

$$Y_1 + Y_2 \sim N(\mu_1 + \mu_2, \sigma_1^2 + \sigma_2^2). \tag{4.3.17}$$

Example 4.3.17   The daily yield from Production Line 1 is an $N(450, 900)$ random variable, and the daily yield from Production Line 2 is an $N(400, 1225)$ random variable. What is the probability that the daily yield for both production lines is less than 1000?

$$P(Y_1 + Y_2 < 1000) = \Phi\left(\frac{1000 - (450 + 400)}{\sqrt{900 + 1225}}\right)$$
$$= \Phi(3.25) = 0.9994.$$

The Lognormal Distribution

The lognormal distribution is an appropriate model when the logarithms of observed values have a normal distribution. Some examples of situations where this occurs are the distribution of particle size in naturally occurring aggregates, the distribution of the amount of drug necessary to cause a reaction, and the distribution of dust concentration in industrial atmospheres. The lognormal distribution can also be used to model lifetime distributions for manufactured products when (unlike those lifetime distributions modeled by the exponential distribution) initially the product's lifetime tends to increase with age. It can also be used as an alternative to the normal distribution for modeling characteristics such as height, weight, and density; which cannot take on negative values.

The density function of a lognormal random variable is

$$f(y) = \frac{1}{\sqrt{2\pi}\,(y\sigma)}\, e^{-\frac{1}{2}\left(\frac{\ln(y) - \mu}{\sigma}\right)^2} \qquad \text{for } y > 0 \tag{4.3.18}$$

where the two parameters μ and σ^2 are the mean and variance of the corresponding normal distribution.

Notation: A lognormal random variable Y with parameters μ and σ^2 will be denoted by $Y \sim \mathrm{LN}(\mu, \sigma^2)$.

A graph of three different lognormal densities is given in Figure 4.9. The lognormal distribution was first discussed by Galton (1879) and McAlister (1879).

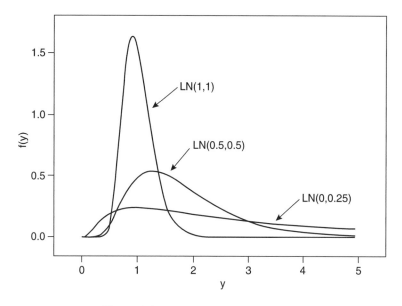

Figure 4.9. Three Lognormal Densities.

Because of its relationship to the normal distribution, the distribution function of a lognormal random variable $Y \sim \text{LN}(\mu, \sigma^2)$ is easily obtained in terms of the standard normal distribution function Φ as

$$
\begin{aligned}
P(Y \leq y) &= P[\ln(Y) \leq \ln(y)] \\
&= P[X \leq \ln(y)] \quad \text{where } X \sim N(\mu, \sigma^2) \\
&= \Phi\left(\frac{\ln(y) - \mu}{\sigma}\right).
\end{aligned}
\tag{4.3.19}
$$

Formulas for the expected value and variance of $Y \sim \text{LN}(\mu, \sigma^2)$ are

$$
E(Y) = e^{\mu + \sigma^2/2}
\tag{4.3.20}
$$

and

$$
\text{Var}(Y) = e^{2\mu + \sigma^2}(e^{\sigma^2} - 1).
\tag{4.3.21}
$$

Formulas for the parameters μ and σ^2 in terms of the mean and variance of Y are

$$
\mu = \ln\left\{\frac{[E(Y)]^2}{\sqrt{\text{Var}(Y) + [E(Y)]^2}}\right\}
\tag{4.3.22}
$$

and

$$\sigma^2 = \ln\left\{\frac{\text{Var}(Y) + [E(Y)]^2}{[E(Y)]^2}\right\}. \tag{4.3.23}$$

Example 4.3.18   The density of bricks from a particular production process is believed to have a lognormal distribution with mean of 0.165 gm/cc and a variance of 0.00255 (gm/cc)^2. What proportion of bricks from this process will have a density greater than 0.19 gm/cc? Using equations (4.3.22) and (4.3.23), one can compute the parameters of the lognormal distribution as

$$\mu = \ln\left\{\frac{(0.165)^2}{\sqrt{0.00255 + (0.165)^2}}\right\} = -1.85$$

$$\sigma^2 = \ln\left\{\frac{0.00255 + (0.165)^2}{(0.165)^2}\right\} = 0.09.$$

Then, using equation (4.3.19),

$$P(Y > 0.19) = 1 - \Phi\left(\frac{\ln(0.19) - (-1.85)}{0.3}\right) = 1 - \Phi(0.63) = 1 - 0.7357 = 0.2643.$$

A little more than 26% of the bricks will have a density greater than 0.19 gm/cc.

Percentiles and Quantiles

Again, because of its relationship to the normal distribution, quantiles of a lognormal random variable $Y \sim \text{LN}(\mu, \sigma^2)$ can easily be obtained in terms of the standard normal quantiles. Using the definition of quantile (equation (4.3.7)) and equation (4.3.19), one has

$$p = P(Y \le Q_Y(p))$$
$$= \Phi\left(\frac{\ln[Q_Y(p)] - \mu}{\sigma}\right)$$

and therefore,

$$\frac{\ln[Q_Y(p)] - \mu}{\sigma} = Q_Z(p)$$

and

$$Q_Y(p) = e^{\mu + \sigma Q_Z(p)}. \tag{4.3.24}$$

Example 4.3.19   For the process in Example 4.3.18, 90% of the bricks have a density higher than what value? Using equation (4.3.24),

$$Q_Y(0.1) = e^{-1.85 + 0.00255 Q_Z(0.1)} = e^{-1.85 + 0.00255(-1.28)} = 0.157.$$

Parameter Estimation

Estimates of the parameters μ and σ^2 for a $LN(\mu, \sigma^2)$ distribution are

$$\widehat{\mu} = \frac{1}{n} \sum_{i=1}^{n} \ln(y_i) \tag{4.3.25}$$

and

$$\widehat{\sigma}^2 = \frac{1}{n} \sum_{i=1}^{n} [\ln(y_i) - \widehat{\mu}]^2 . \tag{4.3.26}$$

Example 4.3.20   Over a 2-hour period twenty five 200-gm samples were drawn at random from a process that recycles plastic, and the amount of aluminum impurities in ppm was determined for each sample. The values are given in Table 4.5 as well as in `alum.txt`. If the amount of aluminum impurities follows a lognormal distribution, what is the probability of obtaining a sample with less than 100 ppm aluminum impurities, and only 10% of the sample will have more than what amount of impurities? The logarithms of the data values are given in the Table 4.6. Using equations (4.3.25), (4.3.26), and (4.3.19); one obtains

$$\widehat{\mu} = \frac{4.98 + 5.88 + \cdots + 5.68}{25} = 5.1$$

$$\widehat{\sigma}^2 = 0.4842$$

and

$$P(Y < 100) \doteq \Phi\left(\frac{\ln(100) - 5.1}{\sqrt{0.4842}}\right) = \Phi(-0.771) = 1 - 0.7794 = 0.2206.$$

Table 4.5. Aluminum Impurities in ppm from 25 Samples from a Plastic Recycling Process

145	358	103	32	54	97	234	99	106	72
139	334	98	239	329	379	167	144	395	106
122	579	22	237	293					

Table 4.6. Logarithms of the Values in Table 4.5

4.98	5.88	4.63	3.47	3.99	4.57	5.46	4.60	4.66	4.28
4.93	5.81	4.58	5.48	5.80	5.94	5.12	4.97	5.98	4.66
4.80	6.36	5.40	5.47	5.68					

There is only a 22% chance of obtaining a sample with less than 100 ppm aluminum impurities. From equation (4.3.24), one finds

$$Q_Y(0.9) \doteq e^{5.1+0.6958Q_Z(0.9)} = e^{5.1+0.6958(1.28)} = e^{5.99} = 399.7$$

and only 10% of the sample will have more than 399.7 ppm aluminum impurities.

The Weibull Distribution

The Weibull distribution, which includes the exponential distribution as a special case, is named after the Swedish physicist Waloddi Weibull because of his extensive work with it. Weibull (1939a, b) first used this distribution to model the breaking strength of materials, and later Weibull (1951) used it in a variety of other applications, including the fiber strength of Indian cotton and the stature of adult males born in the British Isles. Prior to that, however, Rosen and Rammler (1933) had used this distribution to model the fineness of powdered coal, and Frechét (1927) had shown that it was the limiting distribution for the maximum value in a sample.

A random variable W has a Weibull distribution if the random variable

$$Y = \left(\frac{W}{\beta}\right)^{\delta} \tag{4.3.27}$$

has an exponential distribution with mean of 1. The parameter δ is referred to as the **shape parameter**, and the parameter β is referred to as the **scale parameter**.

Notation: A Weibull random variable W with shape parameter δ and scale parameter β is denoted by $W \sim \text{WEIB}(\delta, \beta)$.

Definition (4.3.27) says that for $W \sim \text{WEIB}(\delta, \beta)$

$$P\left[\left(\frac{W}{\beta}\right)^{\delta} \leq y\right] = 1 - e^{-y}$$

and, therefore,

$$P\left[W \leq y^{1/\delta}\beta\right] = 1 - e^{-y}.$$

Letting $w = y^{1/\delta}\beta$ (or $y = (w/\beta)^{\delta}$), one obtains the distribution function of a $\text{WEIB}(\delta, \beta)$ random variable as

$$F(w) = 1 - e^{-(w/\beta)^{\delta}} \qquad \text{for } w \geq 0. \tag{4.3.28}$$

Differentiating, one obtains the corresponding density function (see Problem 4.3.17)

$$f(w) = \frac{\delta}{\beta}\left(\frac{w}{\beta}\right)^{\delta-1} e^{-(w/\beta)^{\delta}} \qquad \text{for } w \geq 0. \tag{4.3.29}$$

Graphs of the Weibull density for three different shape parameters (all with scale parameter of 1) are shown in Figure 4.10.

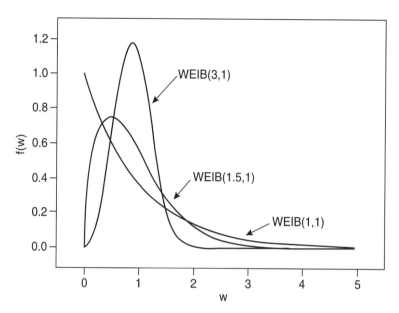

Figure 4.10. A Graph of Weibull Densities for Three Different Shape Parameters.

Note: A WEIB(1, β) random variable is an exponential random variable with mean β.

Example 4.3.21   The failure time of a particular electronic component (measured in years) is known to be WEIB(1.5, 3). The probability that a randomly selected component of this particular type will fail before 2 years is (using equation (4.3.28))

$$P(W \leq 2) = 1 - e^{-(2/3)^{1.5}}$$
$$= 1 - 0.5802 = 0.4198.$$

Almost 42% of these components will fail before 2 years.

Equations (4.1.3) and (4.1.4) can be used with the Weibull density (4.3.29) to obtain expressions for the mean and variance of a $W \sim \text{WEIB}(\delta, \beta)$ random variable. However, the integration is more complicated than for the exponential distribution, and the resulting formulas are not simple functions of the parameters, so they would not be particularly useful to us. What will be useful are the mean and variance of $\ln(W)$, which are simple functions of the parameters. These are

$$E[\ln(W)] = \ln(\beta) - \frac{\gamma}{\delta} \qquad (4.3.30)$$

where $\gamma = 0.57722$ is Euler's constant, and

$$\text{Var}[\ln(W)] = \frac{\pi^2}{6\delta^2} . \tag{4.3.31}$$

Percentiles and Quantiles

The quantile $Q(p)$ of a WEIB(δ, β) distribution is obtained using definition (4.3.7) and equation (4.3.28). Namely,

$$
\begin{aligned}
1 - e^{-(Q(p)/\beta)^\delta} &= p \\
e^{-(Q(p)/\beta)^\delta} &= 1 - p \\
-(Q(p)/\beta)^\delta &= \ln(1 - p) \\
(Q(p)/\beta)^\delta &= -\ln(1 - p) \\
Q(p)/\beta &= [-\ln(1 - p)]^{1/\delta}
\end{aligned}
\tag{4.3.32}
$$

and, finally,

$$Q(p) = \beta[-\ln(1 - p)]^{1/\delta}. \tag{4.3.33}$$

Example 4.3.22   The strength of a new synthetic fabric has a WEIB$(3, 5)$ distribution. Ninety percent of the fabric will have strength less than (using equation (4.3.33))

$$Q(0.9) = 5[-\ln(1 - 0.9)]^{1/3} = 6.6.$$

Parameter Estimation

Many different estimators have been proposed for the parameters of a WEIB(δ, β) distribution. Most of them, however, cannot be expressed in closed form. Simple estimates of the parameters δ and β that can be expressed in closed form are obtained by replacing $E[\ln(W)]$ and $\text{Var}[\ln(W)]$ in equations (4.3.30) and (4.3.31) with their sample counterparts. This results in

$$\widehat{\delta} = \frac{\pi}{s\sqrt{6}} \tag{4.3.34}$$

and

$$\widehat{\beta} = \exp\left(\overline{x} + \frac{\gamma}{\widehat{\delta}}\right) \tag{4.3.35}$$

where $\overline{x}$ and s are the sample mean and standard deviation of the logs of the original data values.

Example 4.3.23   Fourteen samples of steel were stress tested, and their breaking strengths are recorded in Table 4.7 and in `break.txt`. The breaking strength is believed to have a Weibull distribution. Taking logs of all the observations one obtains the values in Table 4.8. The sample mean and variance of these values are $\overline{x} = 5.07$ and $s^2 = 0.493$. Using equations (4.3.34) and (4.3.35), one would estimate the parameters as

$$\widehat{\delta} = \frac{\pi}{\sqrt{0.493}\sqrt{6}} = 1.83$$

and

$$\widehat{\beta} = \exp\left(\overline{x} + \frac{\gamma}{\widehat{\delta}}\right) = \exp\left(5.07 + \frac{0.57722}{1.83}\right) = 218.2.$$

Table 4.7. Breaking Strengths of Steel Samples

34	53	84	107	137	156	176
207	227	254	276	295	330	342

Table 4.8. The Logarithms of the Data values in Table 4.7

3.526	3.970	4.431	4.673	4.920	5.050	5.170
5.333	5.425	5.537	5.620	5.687	5.799	5.835

Problems

1. Show that the distribution function of a UNIF$[a, b]$ random variable is given by equation (4.3.2).

2. Using Example 4.3.2 as a model, show that the distribution function of $X_{(1)}$ is $P[X_{(1)} \leq x] = 1 - \{1 - F(x)\}^n$.

3. Using equations (4.1.5) and (4.1.6), find the expected value and variance for a UNIF$[0, \theta]$ random variable.

4. Using the results from Problem 3, find the expected value and variance for a UNIF$[a, b]$ random variable.

5. Find an expression for the pth quantile of a UNIF$[a, b]$ random variable.

6. Let X represent the lifetime of a component in a piece of lab equipment, and suppose that X is exponentially distributed with a mean of $\mu = 1000$ hours.

   (a) Sketch the density function of X.

   (b) Find $P(X < 1000)$.

   (c) Find $P(X < 5000)$.

   (d) Find $P(X > 100)$.

   (e) Find $P(50 < X < 250)$.

   (f) Find the 90th percentile of X.

   (g) Why might this information be important to the lab personnel?

7. The stability of a paint was tested by subjecting it to increasing times at a high temperature. The viscosity of the paint was used to determine the point of failure (when the viscosity is too high the paint is said to have gelled and is no longer usable). The time until the material gelled is given for 17 samples in `viscosity.txt`.

   (a) Construct a histogram for these data.

   (b) The test was designed assuming the time to failure would be exponentially distributed. Estimate the parameter of the exponential distribution, and write down the resulting estimated distribution function.

   (c) Use your answer from part (b) to estimate the probability that the paint would gel in less than 4.5 days.

   (d) Find the 10th percentile of the distribution, and explain what it stands for.

   (e) Another engineer in looking over the results says "These numbers must be wrong. With normal use, we would expect the paint to last for years, while most of these samples failed in a matter of days." Explain the seeming paradox.

8. The operational lifetimes of 18 components (in hours until failure) were measured as

$$
\begin{array}{ccccccccc}
3139 & 1557 & 2618 & 579 & 6314 & 3863 & 4302 & 7133 & 3063 \\
1606 & 4502 & 1762 & 6444 & 4730 & 6716 & 8905 & 1406 & 5913.
\end{array}
$$

   (a) Construct a histogram for these data.

   (b) The component lifetimes are supposed to be exponentially distributed. Does this seem like a reasonable assumption? Explain.

   (c) Estimate the parameter of the exponential distribution and explain what it means.

9. For the components in Problem 8 assume that all the components started their life test at the same time, and do the following.

   (a) Compute the times between successive failures and the average time between successive failures.

   (b) Using the results of part (a), estimate the probability of two failures in 1 month (30 days).

10. Using equations (4.1.5) and (4.1.6), show that for an exponential random variable Y with parameter μ

   (a) $E(Y) = \mu$

   (b) $\mathrm{Var}(Y) = \mu^2$.

11. Verify equation (4.3.4).

12. Suppose Y has a normal distribution with $\mu = 70$ and $\sigma = 11$.

   (a) Sketch the density function of Y, labeling the mean and the standard deviation.

   (b) Find $P(Y \leq 80)$.

   (c) Find $P(Y > 72)$.

   (d) Find $P(Y < 55)$.

   (e) Find $P(40 \leq Y \leq 75)$.

   (f) Find $P(Y > 46)$.

   (g) Find $P(75 < Y < 85)$.

   (h) Find $Q(0.0735)$.

   (i) Find $Q(0.7967)$.

   (j) Find $Q(0.10)$.

   (k) Find $Q(0.80)$.

13. If $Y \sim N(14.2, (3.75)^2)$, then find

   (a) $P(Y < 16)$

(b) $P(Y \leq 13)$

(c) $P(12.2 < Y < 15.84)$

(d) $Q(0.20)$

(e) $Q(0.75)$

(f) $Q(0.90)$.

14. Let Y = diameter of a tracking ball used in a particular brand of computer mouse, and $Y \sim N(2, 0.01)$.

   (a) Sketch the density function of Y, labeling the mean and the standard deviation.

   (b) Find a range in which almost all tracking ball diameters should fall.

   (c) Find $P(Y \leq 3)$.

   (d) Find $P(Y > 2)$.

   (e) Find $P(2.1 < Y < 2.7)$.

   (f) Find the 90th and 99th percentiles of the tracking ball diameters.

15. Let W = a drum tare weight in pounds (i.e., the weight of the empty drum). Suppose $W \sim N(58, 0.13)$.

   (a) Find $P(W > 58.5)$.

   (b) Find $P(57.8 < W < 58.2)$.

   (c) Find the 99th percentile of W.

   (d) Find the 10th percentile of W, and explain what it stands for.

16. Use the fact that an $N(\mu, \sigma^2)$ random variable Y can be written as $Y = \sigma Z + \mu$ to show that $Q_Y(p) = \sigma Q_Z(p) + \mu$.

17. By differentiating the distribution function of a WEIB(δ, β) random variable (equation (4.3.28)), show that its density is given by equation (4.3.29).

4.4 ASSESSING THE FIT OF A DISTRIBUTION

We have seen that the random error in a data set can be modeled with many different distributions. Once a distribution has been chosen, we need some way to decide if it is a reasonable fit, or if we need to search for a different distribution. We will consider two graphical methods for assessing the fit of a distribution. One obvious way is to plot the density function (or probability function) on top of a histogram scaled so that the area of the histogram is 1. If the density function is a good description of the data, we would expect the two plots to coincide fairly closely.

A somewhat more sophisticated plot is a **probability plot**. A probability plot is based on the idea that each data point in the set estimates a certain quantile of the population. In particular, using definition (2.2.4) that

$$q(p) = y_{(np+\frac{1}{2})}$$

observation $y_{(i)}$ would be the

$$p_i = \frac{i - \frac{1}{2}}{n} \qquad (4.4.1)$$

sample quantile. We can obtain the same quantiles p_i for our chosen distribution. If the distribution is a good choice, the distribution quantiles and the corresponding sample quantiles will all be close to one another. Thus, if the distribution is a good fit, a plot of the sample quantiles versus the corresponding distribution quantiles should look like a straight line (with a slope of 1 if both axes use the same scale).

Exponential Probability Plots

Example 4.4.1 (*Cell-Phone Lifetime Data*)   The lifetimes (in years) of 10 cell phones are given in Table 4.9, their frequency distribution is given in Table 4.10, and the corresponding histogram is given in Figure 4.11.

Table 4.9. Lifetimes of 10 Cell Phones

4.23	1.89	10.52	6.46	8.32	8.60	0.41	0.91	2.66	35.71

Table 4.10. A Frequency Distribution for the Cell Phone Lifetimes in Table 4.9

Class	Frequency	Rel. Freq.
0–5	5	0.5
5–10	3	0.3
10–15	1	0.1
15–20	0	0.0
20–25	0	0.0
25–30	0	0.0
30–35	0	0.0
35–40	1	0.1

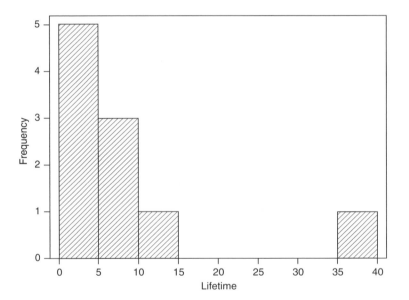

Figure 4.11. A Histogram Corresponding to the Frequency Distribution in Table 4.10.

For this data set $\bar{y} = 7.97$, so our exponential model is

$$\widehat{f}(y) = \frac{1}{7.97}e^{-y/7.97} = 0.125e^{-0.125y} \qquad \text{for } y \geq 0.$$

It is difficult to judge from such a small sample, but from the plot of the estimated density and histogram given in Figure 4.12, it appears that the exponential distribution might be a good fit to this set of data.

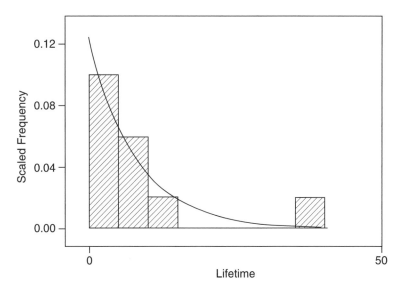

Figure 4.12. The Histogram of the Cell Phone Data Together with a Plot of the Estimated Exponential Density.

To obtain more information, we can construct an exponential probability plot of the data. Equation (4.4.1) is used to determine which quantile each data point corresponds to, and those quantiles for the exponential distribution are obtained using equation (4.3.8) with μ replaced by its estimate $\overline{y}$. For example,

$$Q(0.05) = -7.97 \ln(1 - 0.05) = 0.41.$$

This results in Table 4.11. This information is more easily viewed as a scatter plot of $(Q(p_i), y_{(i)})$ pairs (or $(y_{(i)}, Q(p_i))$ pairs) shown in Figure 4.13.

The plotted points do not all seem to fall on a straight line, so we would conclude that the exponential distribution is not a very good choice to describe this set of data.

Table 4.11. The Quantiles Associated with the Data Points in Example 4.4.1

i	$y_{(i)}$	$p_i = \frac{i-0.5}{10}$	$Q(p_i)$
1	0.41	0.05	0.41
2	0.91	0.15	1.30
3	1.89	0.25	2.29
4	2.66	0.35	3.43
5	4.23	0.45	4.76
6	6.46	0.55	6.36
7	8.32	0.65	8.37
8	8.60	0.75	11.05
9	10.52	0.85	15.12
10	35.71	0.95	23.88

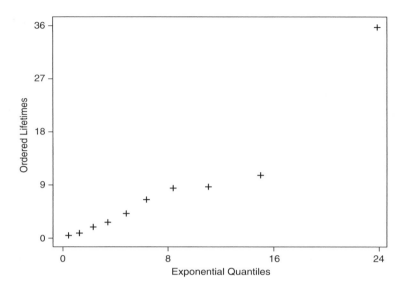

Figure 4.13. An Exponential Probability Plot for the Data in Example 4.4.1.

It is not actually necessary to compute $\bar{y}$ in order to construct an exponential probability plot. One can simply plot the ordered data values versus the quantiles for an exponential distribution with a mean of 1. That is, one can use the quantiles

$$Q(p) = -\ln(1-p). \qquad (4.4.2)$$

This results in a plot that will still appear linear when an exponential distribution is a good fit, but will not in general have a slope of 1. There are two advantages to using the quantiles (4.4.2). First, it is easier since less computation is required; and second, the exponential probability plot can then be used to obtain another estimate of μ.

Example 4.4.2   One is interested in determining if the 10 observation in Table 4.12 could have come from an exponential distribution.

Table 4.12. Ten Observations that Might Have Come from an Exponential Distribution

1.76 5.71 1.17 0.49 1.09 5.56 6.82 9.48 1.54 1.88

Using the quantiles (4.4.2), one obtains Table 4.13, and the exponential probability plot in Figure 4.14.

Table 4.13. The Quantiles Associated With the Data Points in Example 4.4.2

i	$y_{(i)}$	$p_i = \frac{i-0.5}{10}$	$Q(p_i)$
1	0.49	0.05	0.05
2	1.09	0.15	0.16
3	1.17	0.25	0.29
4	1.54	0.35	0.43
5	1.76	0.45	0.60
6	1.88	0.55	0.80
7	5.56	0.65	1.05
8	5.71	0.75	1.39
9	6.82	0.85	1.90
10	9.48	0.95	3.00

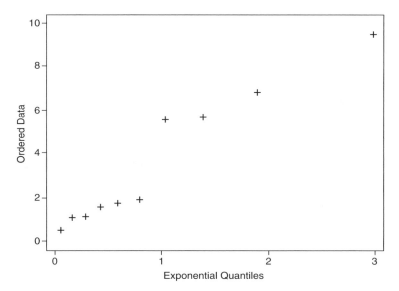

Figure 4.14. An Exponential Probability for the Data in Example 4.4.2.

Since the plot appears linear, it is reasonable to assume the data came from an exponential distribution. One would have obtained exactly the same picture if the quantiles (4.3.8) had been used with μ replaced by $\bar{y} = 3.55$ (see Problem 4.4.10). Further, fitting a line to these points one obtains Table 4.14, and the slope of this line $(\widehat{\beta}_1 = 3.24)$ is an estimate of μ. Note that this estimate is close to, but not exactly the same as, $\bar{y} = 3.55$.

Table 4.14. The Result of Fitting a Line to the Data in Example 4.4.2

UNWEIGHTED LEAST SQUARES LINEAR REGRESSION OF Y

PREDICTOR VARIABLES	COEFFICIENT	STD ERROR	STUDENT'S T	P
CONSTANT	0.41557	0.40648	1.02	0.3365
Q	3.24139	0.31229	10.38	0.0000

R-SQUARED	0.9309	RESID. MEAN SQUARE (MSE)	0.74030
ADJUSTED R-SQUARED	0.9222	STANDARD DEVIATION	0.86040

SOURCE	DF	SS	MS	F	P
REGRESSION	1	79.7538	79.7538	107.73	0.0000
RESIDUAL	8	5.92237	0.74030		
TOTAL	9	85.6762			

Normal Probability Plots

Example 4.4.3 (*Rubber-Strip Length Data*)   A lab technician routinely cuts strips of rubber to be used in a strength test. The samples are supposed to be 4 in. long. To check on the consistency of the lengths another technician selects 12 samples at random. The lengths of the 12 samples are given in Table 4.15. A histogram of the data (Figure 4.15) is not very clear with regard to the underlying distribution. The data might come from a normal distribution, or they might come from a skewed distribution. The plot of the histogram and density together is shown in Figure 4.16 and is not particularly informative (note that this time the density was scaled to match the histogram).

Table 4.15. Lengths (in inches) of 12 Strips of Rubber

4.03 4.04 4.16 4.02 4.18 4.14 4.11 4.13 4.19 3.94 4.21 4.25

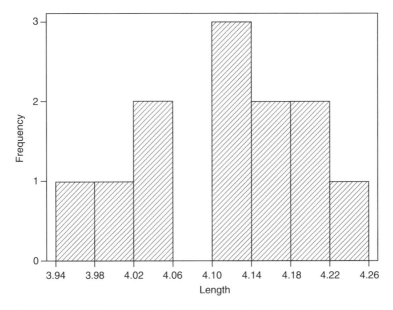

Figure 4.15. A Histogram of the Rubber-Strip Length Data in Table 4.15.

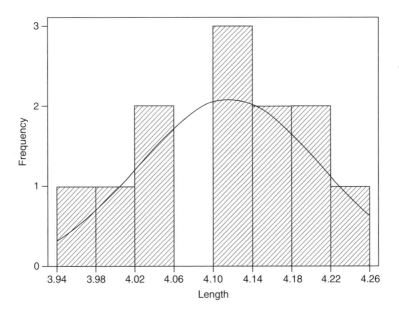

Figure 4.16. The Histogram Together with the Estimated Normal Density for the Data in Table 4.15.

A probability plot would provide better information. Because we are using the normal distribution as our model, the quantiles we want must be obtained using Table B1. There is no need to transform these quantiles to the quantiles for a normal distribution with the same mean and variance as the sample. Such a transformation would not affect the linearity (or lack of it) for the resulting plot (see Problem 4.4.8), but would only change the slope, should the plot be linear.

Again using equation (4.4.1) to determine which quantile each data point corresponds to, and obtaining standard normal quantiles as described in Example 4.3.12, one obtains Table 4.16. Note that if one starts at the bottom with the largest quantiles, they can be obtained directly from the normal table, and then the smaller quantiles are easily obtained by inserting minus signs.

Table 4.16. The Standard Normal Quantiles Corresponding to the Data in Table 4.15.

i	$y_{(i)}$	$\frac{i-0.5}{12}$	$Q_Z\left(\frac{i-0.5}{12}\right)$
1	3.94	0.0417	-1.73
2	4.02	0.125	-1.15
3	4.03	0.2083	-0.81
4	4.04	0.2917	-0.55
5	4.11	0.375	-0.32
6	4.13	0.4583	-0.10
7	4.14	0.5417	0.10
8	4.16	0.625	0.32
9	4.18	0.7083	0.55
10	4.19	0.7916	0.81
11	4.21	0.875	1.15
12	4.25	0.9583	1.73

The normal probability plot given in Figure 4.17 shows a reasonably straight line, so we would conclude that a normal distribution provides a reasonable fit for this set of data. The sample mean and variance for these data are $\bar{y} = 4.12$ and $s^2 = 0.00841$, so a $N(4.12, 0.00841)$ distribution would provide a reasonable fit to the data.

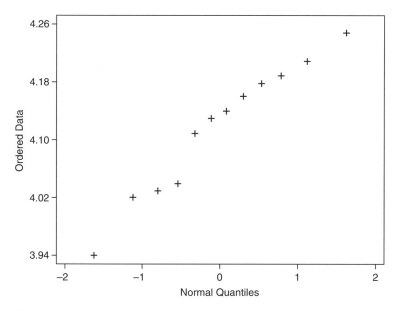

Figure 4.17. A Normal Probability Plot of the Rubber Strip-Length Data.

If a normal probability plot appears to be linear, then the slope of a line fitted to the plotted points is an estimate of σ.

Note: If one were to plot $(y_{(i)}, Q(p_i))$ pairs, then slope of the fitted line would be an estimate of $1/\sigma$.

Example 4.4.4   Fitting a line to the points in the normal probability plot of the strip-length data in Example 4.4.3, one obtains Table 4.17 and $\widehat{\beta}_1 = 0.0907$ is an estimate of σ, which is quite close to $s = 0.0917$.

Table 4.17. Result of Fitting a Line to the Points in Figure 4.17

UNWEIGHTED LEAST SQUARES LINEAR REGRESSION OF LENGTH

PREDICTOR VARIABLES	COEFFICIENT	STD ERROR	STUDENT'S T	P
CONSTANT	4.11667	0.00549	749.47	0.0000
Q	0.09074	0.00579	15.67	0.0000

R-SQUARED	0.9608	RESID. MEAN SQUARE (MSE)	3.620E-04
ADJUSTED R-SQUARED	0.9569	STANDARD DEVIATION	0.01903

SOURCE	DF	SS	MS	F	P
REGRESSION	1	0.08885	0.08885	245.40	0.0000
RESIDUAL	10	0.00362	3.620E-04		
TOTAL	11	0.09247			

The Shapiro–Wilk Test for Normality

Using a normal probability plot to obtain an estimate of σ is the basic idea behind a test of normality proposed by Shapiro and Wilk (1965). If the estimate of σ from the normal probability plot is close enough to s (an estimate of σ that does not depend on the assumption of normality), then the normality assumption is reasonable. In order to account for the fact that the data values must be ordered for the plot, and they are then no longer independent, the **Shapiro–Wilk test** actually employs a more complicated method of fitting the line than the least squares procedure we have discussed. Thus, obtaining the actual test statistic requires a good deal of computation. If, however, the computation is done by a computer package, the interpretation is straightforward.

Example 4.4.5   The normal probability plot given in Figure 4.18 for the strip-length data in Example 4.4.3 shows the Shapiro–Wilk statistic W, together with its *p*-**value**.

Note: A p-value is a probability, and a small p-value (something around 0.05 or smaller) indicates the data do not appear to be normal. We will discuss p-values in greater detail in Section 5.4.

From Figure 4.18, one finds that the *p*-value is 0.7508, and there is no indication that the data are not normal.

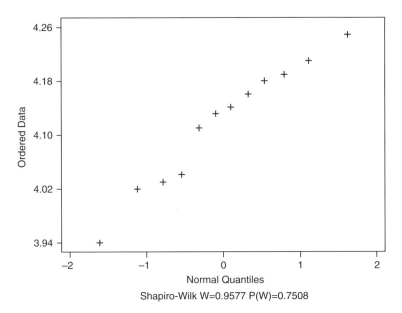

Figure 4.18. A Normal Probability Plot of the Strip-Length Data from Example 4.4.3 Showing the Shapiro–Wilk W Statistic.

Lognormal Probability Plots

Rather than using lognormal data directly and comparing it with quantiles obtained using equation (4.3.24), it is easier to consider the logs of the original data values and construct a normal probability plot.

Example 4.4.6   The cell-phone data in Example 4.4.1 (p. 133) did not appear to come from an exponential population. To check if it can be modeled with a lognormal distribution one could construct a lognormal probability plot. The ordered logs of the data values and the corresponding normal quantiles are given in Table 4.18. The normal probability plot (Figure 4.19) shows no indication of non-normality (the Shapiro–Wilk p-value is 0.9405), and therefore, the assumption that the original data came from a lognormal distribution appears reasonable.

Table 4.18. Normal Quantiles for Use in Constructing a Normal Probability of the Cell-Phone Data in Example 4.4.1

i	$\ln(y_{(i)})$	$p_i = \frac{i-0.5}{10}$	$Q_Z(p_i)$
1	−0.89	0.05	−1.64
2	−0.09	0.15	−1.04
3	0.64	0.25	−0.67
4	0.98	0.35	−0.39
5	1.44	0.45	−0.13
6	1.87	0.55	0.13
7	2.12	0.65	0.39
8	2.15	0.75	0.67
9	2.35	0.85	1.04
10	3.58	0.95	1.64

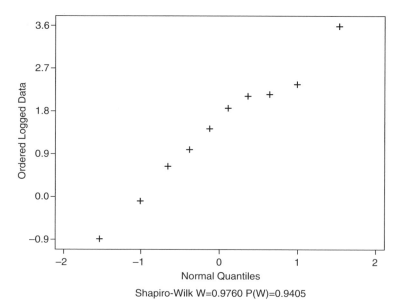

Shapiro-Wilk W=0.9760 P(W)=0.9405

Figure 4.19. A Normal Probability Plot to Check if the Data in Example 4.4.1 is Lognormal.

Weibull Probability Plots

Rather than plotting the ordered values versus the corresponding quantiles obtained using equation (4.3.33) with δ and β replaced by their estimates, a more informative plot that does not require initial estimates of the parameters can be constructed. Taking logs on both sides of equation (4.3.32), one obtains

$$\delta \ln[Q(p)] - \delta \ln(\beta) = \ln[-\ln(1-p)] \tag{4.4.3}$$

which indicates there is a linear relationship between

$$x = \ln[Q(p)] \qquad \text{and} \qquad y = \ln[-\ln(1-p)].$$

It follows that if the data actually come from a Weibull distribution, the same linear relationship should hold for

$$x = \ln[w_{(i)}] \qquad \text{and} \qquad y = \ln[-\ln(1-p_i)].$$

Thus, the usual Weibull probability plot consists plotting the ordered pairs

$$\boxed{\big(\ln[w_{(i)}], \ln[-\ln(1-p_i)]\big)} \tag{4.4.4}$$

and the plot will appear linear if the data come from a Weibull distribution.

Example 4.4.7   For the data in Example 4.3.23 (p. 129), the ordered values $\ln[w_{(i)}]$ and the corresponding values $\ln[-\ln(1-p_i)]$) are given in Table 4.19. From the Weibull probability plot given in Figure 4.20 there is no reason to doubt that the data came from a Weibull population.

Table 4.19. Values for Constructing a Weibull Probability of the Data in Example 4.3.23

i	$\ln[w_{(i)}]$	$p_i = \frac{i-0.5}{14}$	$\ln[-\ln(1-p_i)]$
1	3.53	0.036	-3.31
2	3.97	0.107	-2.18
3	4.43	0.179	-1.63
4	4.67	0.250	-1.25
5	4.92	0.321	-0.95
6	5.05	0.393	-0.70
7	5.17	0.464	-0.47
8	5.33	0.536	-0.26
9	5.42	0.607	-0.07
10	5.54	0.679	0.13
11	5.62	0.750	0.33
12	5.69	0.821	0.54
13	5.80	0.893	0.80
14	5.83	0.964	1.20

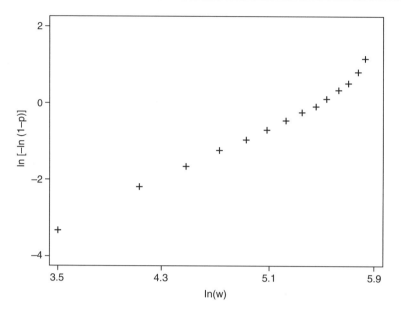

Figure 4.20. A Weibull Probability for the Data in Example 4.3.23.

It is also possible to estimate the Weibull parameters from the Weibull probability plot. From equation (4.3.32) one sees that the slope of the line in the probability plot (if one obtains a line) is an estimate of δ, and the intercept is an estimate of $-\delta \ln(\beta)$. Thus, by fitting the least-squares line to the probability plot one obtains the estimates

$$\boxed{\widehat{\delta} = \widehat{\beta}_1} \tag{4.4.5}$$

and

$$\boxed{\widehat{\beta} = \exp\left[\frac{-\widehat{\beta}_0}{\widehat{\delta}}\right].} \tag{4.4.6}$$

Example 4.4.8   For the data in Example 4.3.23 (p. 129) one obtains Table 4.20. From the Weibull probability plot given in Figure 4.20 there is no reason to doubt that the data came from a Weibull population.

Table 4.20. The Least-Squares Line for the Probability Plot in Example 4.3.23

UNWEIGHTED LEAST SQUARES LINEAR REGRESSION OF Y

PREDICTOR VARIABLES	COEFFICIENT	STD ERROR	STUDENT'S T	P
CONSTANT	-9.37428	0.36347	-25.79	0.0000
LNW	1.73907	0.07106	24.47	0.0000

R-SQUARED	0.9804	RESID. MEAN SQUARE (MSE)	0.03234
ADJUSTED R-SQUARED	0.9787	STANDARD DEVIATION	0.17985

SOURCE	DF	SS	MS	F	P
REGRESSION	1	19.3712	19.3712	598.89	0.0000
RESIDUAL	12	0.38814	0.03234		
TOTAL	13	19.7593			

Problems

1. Consider again the data in Problem 4.3.7 (p. 130), which are the failure times, in days, of paint subjected to a high temperature.

   (a) By hand, make a normal probability plot for this set of data.

   (b) Write a paragraph commenting on the plot, and the likely shape of the set of data. Be specific! It is not sufficient to say "The data appear to be normally distributed." or "The data do not appear to be normally distributed." Especially if the data appear non-normal, you should give some details about how and where the data differ from the normal distribution curve.

2. Using the hardness data from the odor data set `odor.txt`, do the following.

   (a) Make a normal probability plot of the data.

   (b) Do the data appear to be normally distributed? Explain.

3. Use the web site data `webvisit.txt` to do the following.

   (a) Make a normal probability plot for the entire data set.

   (b) Make normal probability plots for the two apparent groups in the data.

   (c) Calculate the mean and standard deviation for each group.

   (d) Assuming the group with the larger values is normally distributed, use your calculated mean and standard deviation to estimate the proportion of days that have more than 15,300 visits.

   (e) Write a paragraph describing this set of data and the conclusions you have drawn from it.

4. Make a normal probability plot of the delta haze data `dhaze.txt` discussed in Example 2.3.1 (p. 26) and comment on it.

5. Make a normal probability plot of the `surfarea.txt` data and comment on it.

6. Make an exponential probability plot of the rail car lifetime data `railcar2.txt` and comment on it.

7. Using the computer mice tracking-ball-diameter data `diameter.txt`, make a normal probability plot for each of the two production lines and comment on the plots.

8. For the data in Example 4.4.3 (p. 138), construct a normal probability plot using the quantiles from the fitted normal model, and compare this plot with the one given in Example 4.4.3.

9. Construct an exponential probability plot for the times between successive failures in Problem 4.3.9 to see if they appear to have an exponential distribution.

10. Construct an exponential probability plot for the data in Example 4.4.2 (p. 136) using the quantiles (4.3.8) with μ replaced by $\bar{y} = 3.55$, and compare it with the plot in the example.

11. In Example 4.3.20 (p. 125), the `alum.txt` data were assumed to have come from a lognormal distribution.
    (a) Check this by constructing a normal probability plot of the logs of the data.
    (b) Plot the ordered original data values from Example 4.3.20 versus the corresponding quantiles obtained from equation (4.3.24) with $\mu = 0$ and $\sigma = 1$, and compare the plot with the one in part (a).

12. Construct a Weibull probability plot to ascertain if the 10 data values

$$91.82 \quad 35.04 \quad 147.25 \quad 306.48 \quad 131.87$$
$$142.74 \quad 164.72 \quad 208.80 \quad 119.19 \quad 311.17$$

can be modeled with a Weibull distribution.

13. (a) If the plot in Problem 12 appears linear, use it to estimate the Weibull parameters.
    (b) Compare the estimates in part (a) with those obtained using equations (4.3.34) and (4.3.35).

14. In Example 4.4.6, we found the cell-phone data could be modeled with a lognormal distribution.
    (a) Check to see if a Weibull distribution could also be used.
    (b) Estimate the parameters for the Weibull distribution using both the Weibull probability plot and equations (4.3.34) and (4.3.35).

4.5 CHAPTER PROBLEMS

1. Machines used to fill food cartons are usually set to slightly overfill the boxes so that the contents will not weigh less than the advertised net weight. This takes into account the fact that the actual weight of each box varies somewhat from the machine set value. Suppose a machine is set to put 12.2 ounces of food in boxes whose advertised weight is 12 ounces, and the filling process has a distribution that is normal with a standard deviation of 0.1 ounces.

    (a) What percentage of boxes would you expect to be under weight?

    (b) Ninety percent of the boxes should have at least how many ounces of food?

2. The length of a continuous nylon filament that can be drawn without a break occurring is an exponential random variable with mean 5000 feet.

    (a) What is the probability that the length of a filament chosen at random lies between 4750 and 5550 feet?

    (b) Find the 10th percentile of the distribution of filament length.

3. A particular electronic system has three electronic components. Let A_i represent the event that component i is functioning. Let $P(A_1) = 0.4$, $P(A_2) = 0.6$, and $P(A_3) = 0.5$. The events A_i are independent, and at least two components must be functioning in order for the system to function.

    (a) In how many different ways can the system be functioning?

    (b) What is the probability that the system is functioning?

    (c) In how many ways could the system be functioning if there were n components and at least k of them had to function in order for the system to function.

4. A national survey of engineering school graduates indicates that only 10% of them have had a course in statistics.

    (a) Find the probability that in a random sample of five engineering graduates exactly two have had a course in statistics.

    (b) Find the probability that in a random sample of five engineering graduates at least two have had a course in statistics.

    (c) Suppose one took a random sample of size 100, where 100 is a small sample relative to the total number of engineering graduates. Write down the formula for the probability of exactly 10 of those sampled having had a course in statistics.

    (d) Write down the formula for the Poisson approximation to the probability in part (c).

    (e) Obtain a numerical value for the expression in part (d).

5. Consider the function $f(x) = cx^3$ for $x = 1, 2, 3, 4$; where c is a constant. Let X be a discrete random variable with probability function $f(x)$.

(a) Determine the value of c that makes $f(x)$ a probability function.

(b) Find $P(X < 2)$.

(c) Find $P(2 < X \leq 3)$.

(d) Find $E[X]$.

(e) Find $\text{Var}[X]$.

6. Defects on computer disks are known to occur at a rate of two defects per $(\text{mm})^2$ and they occur according to a Poisson model.

   (a) Find the probability that at least one defect will be found if 1 $(\text{mm})^2$ is examined.

   (b) Find the probability of less than two defects if 1 $(\text{mm})^2$ is examined.

7. Defective computer disks are known to occur with a probability of 0.1. A random sample of five disks is taken.

   (a) Find the probability that at least one disk is defective.

   (b) Find the probability that exactly two disks are defective.

8. The lifetime of a particular brand of computer is exponentially distributed with an average lifetime of 3.2 years.

   (a) Find the probability that a randomly selected computer of this brand will have failed within 3 years.

   (b) At what lifetime would 90% of all of this brand of computer have failed?

9. A person taking a multiple choice exam has a probability of 0.8 of correctly answering any particular question and there are ten questions on the exam.

   (a) What is the probability of scoring 90% or more?

   (b) On the average what would a person score?

10. Electrical insulation was tested by subjecting it to continuously increasing voltage stress. The resulting failure times (in minutes) were:

    219.3 79.4 86.0 150.2 21.7 18.5 121.9 40.5 147.1 351.0 42.3 48.7

    (a) Assuming the failure times are exponentially distributed, write down the estimated density function.

    (b) The assumption in part (a) can be checked using an exponential probability plot. Find the coordinates on such a plot of the point corresponding to the data value 150.2.

11. Suppose that in Problem 10 rather than using an exponential model one decided to use the normal distribution as a model.

    (a) What normal distribution would you use?

    (b) Using the model in part (a), estimate the probability that insulation would fail before 2.5 hours.

12. The number of defects on the surface of a computer disk occurs at an average rate of 3 defects per square inch and at most one defect can occur in a square mm.

    (a) What is the probability of more than two defects in 1.5 square inches?

    (b) On the average how many defects would you expect in 1.5 square inches?

13. The number of people entering a particular coffee shop is never more than one in any small time interval, and on the average the rate is two people per minute.

    (a) What is the probability that at least two people arrive in a 5 minute period?

    (b) If 60% of the people order coffee, what is the probability that out of ten people selected at random exactly two will have ordered coffee?

14. At the present we are able to fit continuous data with several different models. One way to check the adequacy of the fit is with a probability plot. Consider that following small data set consisting of five observations.

$$1.1 \quad 2.2 \quad 1.6 \quad 1.3 \quad 3$$

    (a) Construct a probability plot for each possible model.

    (b) Based on the plots in part (a), which model would you prefer? Explain.

    (c) Fit the model you preferred (e.g., if you picked the exponential, exactly which exponential distribution would you use to describe the data?)

    (d) Using the model in part (c), estimate the probability of obtaining a value from this population that is less than 1.

15. Suppose that the probability of making an A on a particular exam is 0.1.

    (a) If five students who took the exam are chosen at random, what is the probability that no more than one made an A?

    (b) Suppose that one was unsure about the 0.1 probability given above and want to estimate the probability of making an A. If a random sample of 30 students resulted in six who made A's, what would you estimate the probability of making an A to be?

16. Let $Y \sim \text{UNIF}(0, \theta_y)$ and let X have an exponential distribution with parameter θ_x, where the parameters θ are chosen such that the means of the distributions are 50.

    (a) Find $P(30 < X < 70)$.

    (b) Find $P(30 < Y < 70)$.

    (c) Which distribution would you prefer as a model for exam grades? Why?

17. A vendor has historically had a 15% defective rate on a particular part. They claim to have improved their process and provide their customer 10 units to be tested. The test is a pass/fail destructive test. All units pass the test.

   (a) What is the probability of observing no defectives in this sample under the historical 15% defective rate?

   (b) Suppose the defective rate had been reduced to 10%, now what is the probability of no defectives in this sample?

   (c) Suppose the defective rate had been reduced to 5%, now what is the probability of no defectives in this sample?

   (d) Is a sample of size 10 large enough to provide any practical information regarding the vendors supposed process improvement?

18. Purified water is used in one step in the production of a medical device. The water is tested daily for bacteria. The results of 50 days of testing are contained in `bacteria.txt`. The values are the count of a particular strain of bacteria in a 100 ml sample of water. The process engineers would like to set up a warning limit and an action limit for future testing based on this data. If a sample were to test above the warning limit, then the engineers would be aware of potential system problems, such as a filter that might need changing. An action limit would indicate that the process should be stopped. Although there are regulatory limits on the bacteria level (which all of the samples fell below), the engineers wanted data-based warning and action limits on the bacteria level. They felt that a warning limit which 80% of the data fell below and an action limit which 95% of the data fell below would be reasonable.

   (a) The data here are counts, so they are discrete, however, one might try to approximate the distribution with a continuous distribution. What distribution(s) might approximate the distribution of the data? What distribution(s) would not approximate the distribution of this data?

   (b) Using quantiles, set up warning and action limits for the engineers.

   (c) Examine these limits on a runs chart of the data (it is in the order in which it was collected).

   (d) Do you think that the engineers could reduce their sampling from every day to every other day, or once a week? What information should you take into account in making such a decision?

19. Suppose that the number of major defects on a windshield from a particular production line follows a Poisson distribution with $\lambda = 0.01$/windshield.

   (a) What is the probability that a windshield will be defect free?

   (b) What is the expected number of defects per windshield?

   (c) The production line is stopped when a windshield has two or more defects. What is the probability of the line stopping?

20. A manufacturer of plastic bags tested a random sample of 43 bags from a production line and found that on average the bags could hold 18.5 lbs before failing (i.e., breaking). The raw data are in `weight.txt`.

    (a) Determine what distribution these data follow, justify your conclusion.

    (b) One grocery store chain demands that its bags be able to hold at least 17.5 lbs. What is the probability that any one bag can hold 17.5 lbs? Do you believe that the manufacturer should agree to a contract with this grocery chain given its current process capabilities?

    (c) A second grocery store chain has less stringent requirements. They only need bags that can hold up to 15 lbs. Evaluate the manufacturer's ability to supply this chain.

5

INFERENCE FOR A SINGLE POPULATION

So far we have discussed choosing mathematical models for both experimental outcomes and for the random error, and we have found out how to estimate the unknown parameters in our models. In this and subsequent chapters we combine the two types of models in order to determine the precision of our estimates. That way we will be able to infer exactly how far away the estimates might be from the true values. In order to study the precision of an estimate, we need to study the distribution of the estimator used to obtain the estimate. Since many of our parameter estimators are actually sample means, we will make use of an important statistical result called the **Central Limit Theorem**. In this chapter, we will discuss inference for parameter estimates from a single population.

5.1 CENTRAL LIMIT THEOREM

Central Limit Theorem: *If $Y_1, Y_2, \ldots, Y_n$ is a random sample from a distribution with $\sigma^2 < \infty$, then for large samples $\overline{Y}$ is approximately normally distributed, regardless of the original underlying distribution.*

Note: Since a linear transformation does not effect the general form of a random variable's distribution (see Section 4.3), if $\overline{Y}$ is (approximately) normal, then so is $n\overline{Y} = \sum_{i=1}^{n} Y_i$, and the Central Limit Theorem applies to sums as well as averages.

It is important to note that the Central Limit Theorem (CLT) is a statement about the (asymptotic) behavior of the sample mean, *not* of the sample itself. The CLT is *not* saying "large samples are normally distributed". As the sample size gets larger, a histogram of the sample will start to look like the density function of the underlying distribution (i.e., the population from which the sample was taken) – which is not necessarily a normal density function!

155

The CLT tells us that for large enough samples the distribution of the sample mean will be approximately normal. Using the properties of expected values and variances together with the expected value and variance of the underlying distribution, we can determine the expected value and variance for the approximate normal distribution. If our random sample comes from a distribution with mean μ and variance σ^2, then

$$
\begin{aligned}
E(\overline{Y}) &= E\left[\frac{1}{n}\sum_{i=1}^{n}Y_i\right] \\
&= \frac{1}{n}E\left[\sum_{i=1}^{n}Y_i\right] \\
&= \frac{1}{n}\sum_{i=1}^{n}E(Y_i) \\
&= \frac{1}{n}\sum_{i=1}^{n}\mu \\
&= \mu
\end{aligned}
$$

and

$$
\begin{aligned}
\mathrm{Var}(\overline{Y}) &= \mathrm{Var}\left[\frac{1}{n}\sum_{i=1}^{n}Y_i\right] \\
&= \frac{1}{n^2}\mathrm{Var}\left[\sum_{i=1}^{n}Y_i\right] \\
&= \frac{1}{n^2}\sum_{i=1}^{n}\mathrm{Var}(Y_i) \\
&= \frac{1}{n^2}\sum_{i=1}^{n}\sigma^2 \\
&= \frac{\sigma^2}{n}.
\end{aligned}
$$

This information together with the CLT tells us that for large enough n the distribution of the sample mean from a population with mean μ and variance σ^2 satisfies (at least approximately)

$$
\boxed{\overline{Y} \sim N\left(\mu, \sigma^2/n\right).}
\tag{5.1.1}
$$

Exactly how large is "large enough"? The answer depends somewhat on how "non-normal" the underlying population is. If the underlying population has a normal or "nearly normal" distribution, then $\overline{Y}$ will be at least approximately normal even for small samples. For continuous distributions that are not already approximately normal, values of $n > 30$ are generally sufficient.

We can get an idea of how "non-normal" our population is by looking at a normal probability plot of the data. If the plot is severely curved, we may need a larger sample (i.e., more than 30); if the plot is reasonably linear, a smaller sample will suffice.

Example 5.1.1   Suppose that compact fluorescent light bulb lifetimes are exponentially distributed, as in Example 4.3.4 (p. 109), with mean $\mu = 7$. (Recall that this implies the variance is 49.) For a random sample of 50 light bulbs the distribution of $\overline{Y}$ is approximately normal with mean 7 and variance $49/50 = 0.98$ (formula (5.1.1)).

Example 5.1.2   (*continuation of Example 5.1.1*)   What is the probability that the average of 50 light bulb lifetimes is less than 8 years?

$$P\left(\overline{Y} < 8\right) \doteq \Phi\left(\frac{8-7}{\sqrt{0.98}}\right) = \Phi(1.01) = 0.8438.$$

What is the probability that $\overline{Y}$ will be within 1 year of the actual population mean?

$$\begin{aligned}
P\left(|\overline{Y} - \mu| < 1\right) &= P\left(|\overline{Y} - 7| < 1\right) \\
&= P\left(6 < \overline{Y} < 8\right) \\
&\doteq \Phi\left(\frac{8-7}{\sqrt{0.98}}\right) - \Phi\left(\frac{6-7}{\sqrt{0.98}}\right) \\
&= \Phi(1.01) - \Phi(-1.01) \\
&= 2\Phi(1.01) - 1 \\
&= 2(0.8438) - 1 = 0.6876.
\end{aligned}$$

Example 5.1.3   Suppose the distribution of compact fluorescent light bulb lifetimes is unknown, but we know the variance of lifetimes is 8 years. How many bulbs would one have to sample in order to be 95% certain that $\overline{Y}$ will be within 1 year of the unknown population lifetime mean?

$$\begin{aligned}
0.95 &= P\left(|\overline{Y} - \mu| < 1\right) \\
&= P\left(\mu - 1 < \overline{Y} < \mu + 1\right) \\
&\doteq \Phi\left(\frac{\mu + 1 - \mu}{\sqrt{8/n}}\right) - \Phi\left(\frac{\mu - 1 - \mu}{\sqrt{8/n}}\right) \\
&= \Phi\left(\frac{1}{\sqrt{8/n}}\right) - \Phi\left(\frac{-1}{\sqrt{8/n}}\right) \\
&= 2\Phi\left(\frac{1}{\sqrt{8/n}}\right) - 1.
\end{aligned}$$

Therefore,

$$\Phi\left(\sqrt{\frac{n}{8}}\right) = \frac{1.95}{2} = 0.975$$

$$\sqrt{\frac{n}{8}} = Q_Z(0.975) = 1.96$$

and

$$n = 8(1.96)^2 = 30.7 \rightarrow 31.$$

One would need to sample 31 bulbs to be 95% certain that $\overline{Y}$ was within 1 year of the population mean lifetime.

Note: By using the normal approximation we are assuming that we have a large enough value of n (i.e., $n > 30$). Therefore, had we obtained a sample size smaller than 30, the result would not have been valid.

Normal Approximation to the Binomial

Recall from Section 4.2 that a binomial random variable $Y \sim \text{BIN}(n, p)$ can be written as a sum of n independent Bernoulli random variables, each having probability of success p. That is,

$$Y = X_1 + \cdots + X_n$$

where

$$P(X_i = 1) = p \quad \text{and} \quad P(X_i = 0) = 1 - p = q.$$

Therefore, the CLT implies that for large enough n the binomial distribution can be approximated with the normal distribution. Specifically, for large enough n,

$$\boxed{Y \sim \text{BIN}(n, p) \mathrel{\dot\sim} N(np, npq).} \qquad (5.1.2)$$

Notation: We use $Y \mathrel{\dot\sim} N(\mu, \sigma^2)$ to denote that Y is approximately normal with mean μ and variance σ^2.

For discrete distributions, "large enough" tends to require larger samples than continuous distributions before the distributions of the sample means start to look normal. For binomial distributions a general rule of thumb is that n is large enough if

$$\boxed{np > 5 \quad \text{and} \quad n(1 - p) > 5.} \qquad (5.1.3)$$

Example 5.1.4   Suppose we are sampling items from a manufacturing process and classifying them as defective or non-defective, and the proportion of defectives in the population is $p = 0.07$. If we take a sample of $n = 100$ items, what is the probability of obtaining more than 9 defectives? Since

$$np = 100(0.07) = 7 > 5 \quad \text{and} \quad n(1 - p) = 100(0.93) = 93 > 5$$

we can apply the normal approximation, and

$$P(Y > 9) = 1 - P(Y \leq 9)$$
$$\doteq 1 - \Phi\left(\frac{9 - 7}{\sqrt{6.51}}\right)$$
$$= 1 - \Phi(0.78)$$
$$= 1 - 0.7823 = 0.2177.$$

Continuity Corrections

When using the normal distribution to approximate a discrete distribution, the approximation can be improved by applying a correction referred to as a **continuity correction**. The problem with approximating a discrete distribution with a continuous distribution is illustrated in Figure 5.1.

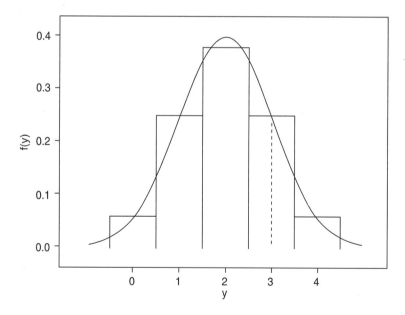

Figure 5.1. An Example of Why Continuity Corrections are Needed.

Figure 5.1 is a graph of a $Y \sim \text{BIN}(4, 0.5)$ probability function with a $W \sim N(2, 1)$ density curve superimposed. (While $np = 4(0.5) = 2 \not> 5$, so a $N(2, 1)$ distribution would not provide a particularly good approximation, it is easier to see the problem when there are fewer binomial probabilities to plot.) If, for example, one wanted to approximate the probability that Y was less than or equal to 3 with W, then using

$$P(Y \leq 3) \doteq P(W \leq 3)$$

neglects almost half of the probability (area) for $Y = 3$. In order to include all of the probability associated with $Y = 3$, one would use the approximation

$$P(Y \leq 3) \doteq P(W \leq 3 + 0.5).$$

The 0.5 is the continuity correction.

Example 5.1.5   In Example 5.1.4, we used the normal approximation to obtain $P(Y > 9) \doteq 0.2177$. If we had included a continuity correction, we would have obtained

$$
\begin{aligned}
P(Y > 9) &= 1 - P(Y \leq 9) \\
&\doteq 1 - \Phi\left(\frac{9.5 - 7}{\sqrt{6.51}}\right) \\
&= 1 - \Phi(0.98) \\
&= 1 - 0.8365 = 0.1638
\end{aligned}
$$

which is closer to the actual probability of 0.1620.

Normal Approximation to the Poisson

It follows from formula (4.2.16) that we can think of $Y \sim \text{Poisson}(\lambda t)$ as

$$Y = Y_1 + \cdots + Y_n$$

where

$$Y_i \sim \text{Poisson}(\lambda t_i)$$
$$t_i = t/n \text{ for } i = 1, \ldots, n$$

the t_i's are non-overlapping.

The CLT then implies that if λt is large enough,

$$\boxed{Y \sim \text{Poisson}(\lambda t) \overset{\cdot}{\sim} N(\lambda t, \lambda t).}$$ (5.1.4)

A general rule of thumb for large enough λt is

$$\boxed{\lambda t > 5.}$$ (5.1.5)

Example 5.1.6 For the intersection discussed in Example 4.2.7, where $\lambda = 0.1$, what is the probability of no more than 10 cars going through the intersection in $t = 60$ seconds? Since $\lambda t = 0.1(60) = 6 > 5$,

$$P(Y \leq 10) \doteq \Phi\left(\frac{10 - 6}{\sqrt{6}}\right)$$
$$= \Phi(1.63) = 0.9484.$$

Example 5.1.7 If in Example 5.1.6 we had employed a continuity correction, we would have obtained

$$P(Y \leq 10) \doteq \Phi\left(\frac{10.5 - 6}{\sqrt{6}}\right)$$
$$= \Phi(1.84) = 0.9671$$

which in this case is not closer to the actual probability of 0.9574.

Problems

1. A random sample of 35 items is drawn from a population with mean 75 and standard deviation 15.

   (a) What is the distribution of $\overline{Y}$?

   (b) Sketch the density function of $\overline{Y}$, labeling the mean and standard deviation.

   (c) Give a range of values that is almost certain to contain $\overline{Y}$.

   (d) What is the probability that $\overline{Y}$ will be greater than 80?

   (e) What is the probability that $\overline{Y}$ will be less than 72?

   (f) What is the probability that $\overline{Y}$ will be greater than 68?

2. A random sample of 47 items is drawn from a population with mean 40 and standard deviation 10.

   (a) Sketch the density function of $\overline{Y}$, labeling the mean and standard deviation.

   (b) Give a range of values that is almost certain to contain $\overline{Y}$.

   (c) What is the probability that $\overline{Y}$ will be greater than 50?

   (d) What is the probability that $\overline{Y}$ will be less than 38?

   (e) What is the probability that $\overline{Y}$ will be greater than 45?

3. A random sample of 60 items is drawn from a population with mean 500 and standard deviation 100.

   (a) What is the distribution of $\overline{Y}$? Sketch the density function and label the horizontal axis.

   (b) Give a range of values that is 95% likely to contain $\overline{Y}$.

   (c) Find the probability that $\overline{Y}$ will be less than 530.

   (d) For what proportion of such samples will $\overline{Y}$ be less than 480?

   (e) Find $P(\overline{Y} > 510)$.

   (f) Find the 99th percentile of $\overline{Y}$.

   (g) Find the 10th percentile of $\overline{Y}$.

4. A random sample of 250 items is drawn from a Bernoulli population with $p = 0.23$.

   (a) What is the distribution of $\widehat{p}$? Sketch the density function.

   (b) Find $P(\widehat{p} > 0.25)$.

   (c) Find $P(\widehat{p} < 0.30)$.

   (d) Find $P(0.20 < \widehat{p} < 0.25)$.

   (e) Find $Q(0.90)$.

(f) Find $Q(0.15)$.

5. A random sample of 50 items is taken from an exponential distribution with mean 72.

   (a) What is the distribution of $\overline{Y}$?

   (b) Find a range of values that will contain $\overline{Y}$ with probability 0.95.

   (c) Find the probability that $\overline{Y}$ exceeds 75.

   (d) Find the 90th, 95th, and 99th percentiles of $\overline{Y}$.

6. A random sample of three items is taken from a normal distribution with mean 12 and standard deviation 1.

   (a) What is the distribution of $\overline{Y}$?

   (b) What proportion of the time will $\overline{Y}$ be greater than 13?

   (c) What proportion of the time will $\overline{Y}$ be more than 1 unit away from the population mean?

   (d) What proportion of the time will $\overline{Y}$ be more than 2 units away from the population mean?

7. Times between e-mails are exponentially distributed with $\mu = 15$ minutes.

   (a) If 35 times are sampled, what is the distribution of $\overline{Y}$?

   (b) If 35 times are sampled, what is the probability that $\overline{Y}$ will be greater than 17?

   (c) If 35 times are sampled, what is the 10th percentile of $\overline{Y}$?

   (d) If 35 times are sampled, what is the probability that $\overline{Y}$ will be within 1 minute of the true mean?

   (e) If 50 times are sampled, what is the distribution of $\overline{Y}$?

   (f) If 50 times are sampled, what is the probability that $\overline{Y}$ will be greater than 17?

   (g) If 50 times are sampled, what is the 10th percentile of $\overline{Y}$?

   (h) If 50 times are sampled, what is the probability that $\overline{Y}$ will be within 1 minute of the true mean?

   (i) How many items must be sampled to ensure that the $\overline{Y}$ is within 1 minute of the true mean with at least 99% probability?

   (j) How many items must be sampled to ensure that the $\overline{Y}$ is within 30 seconds of the true mean with at least 95% probability?

8. Suppose we are planning to take a random sample from a population with unknown mean and $\sigma = 5$.

   (a) If 35 items are sampled, how far can $\overline{Y}$ be from the unknown mean?

   (b) If 50 times are sampled, what is the maximum error that can occur when using $\overline{Y}$ to estimate the unknown population mean?

(c) How many items must be sampled in order to be virtually certain that $\overline{Y}$ will be within 2 units of the unknown population mean?

(d) How many items must be sampled in order to be 98% certain that $\overline{Y}$ will be within 1.5 units of the unknown population mean?

9. A manufacturer of a gel product packages the product in 14 oz. tubes. The filling machine is actually set to fill the tubes to 14.3 oz. Historical data show that the weight of the tubes follow a normal distribution with a mean of 14.25 oz. and a standard deviation of 0.1 oz.

(a) What is the probability of a tube weighing below the claimed weight of 14 oz.? Comment on this probability in terms of what it means to the manufacturer. Would you change the filling target?

(b) What is the probability of a tube weighing above 14.5 oz.?

(c) Between what two weights would we expect 95% of the tube weights to fall?

5.2 A CONFIDENCE INTERVAL FOR μ

An estimate such as $\overline{y}$ that provides a specific value for an unknown parameter is called a **point estimate**. By adding and subtracting an amount d from $\overline{y}$, we would obtain an **interval estimate** or a **confidence interval** for μ. The larger the value of d, the greater the chances that the interval will actually contain μ. We would like to be able to specify the chances the interval will contain μ and use this to determine the appropriate value of d.

Notation: For convenience we define $z(\alpha) = Q_Z(1 - \alpha)$, which is called the upper α quantile for a standard normal distribution. It is the point under the standard normal density having area α to the right, as is depicted in Figure 5.2.

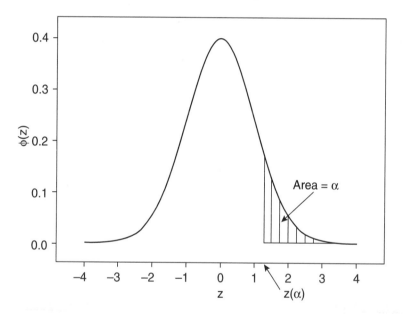

Figure 5.2. The Upper α Quantile of a Standard Normal Distribution.

For $Z \sim N(0, 1)$ (see Problem 5.2.1)

$$P\left[-z(\alpha/2) \leq Z \leq z(\alpha/2)\right] = 1 - \alpha.$$

Therefore, if

$$\overline{Y} \sim N\left(\mu, \sigma^2/n\right)$$

(for large n the CLT would imply this) it follows that

$$P\left[-z(\alpha/2) \leq \frac{\overline{Y} - \mu}{\sigma/\sqrt{n}} \leq z(\alpha/2)\right] = 1 - \alpha.$$

This can be algebraically manipulated to obtain

$$1 - \alpha = P\left[-z(\alpha/2) \leq \frac{\overline{Y} - \mu}{\sigma/\sqrt{n}} \leq z(\alpha/2)\right]$$

$$= P\left[-z(\alpha/2)\frac{\sigma}{\sqrt{n}} \leq \overline{Y} - \mu \leq z(\alpha/2)\frac{\sigma}{\sqrt{n}}\right]$$

$$= P\left[-\overline{Y} - z(\alpha/2)\frac{\sigma}{\sqrt{n}} \leq -\mu \leq -\overline{Y} + z(\alpha/2)\frac{\sigma}{\sqrt{n}}\right]$$

$$= P\left[\overline{Y} - z(\alpha/2)\frac{\sigma}{\sqrt{n}} \leq \mu \leq \overline{Y} + z(\alpha/2)\frac{\sigma}{\sqrt{n}}\right] \tag{5.2.1}$$

and the resulting interval

$$\boxed{\overline{y} \pm z(\alpha/2)\frac{\sigma}{\sqrt{n}}} \tag{5.2.2}$$

is a $(1 - \alpha)100\%$ confidence interval for μ. The quantity $(1 - \alpha)100\%$ is called the **confidence coefficient** or **confidence level** for the interval.

Interpretation of the Confidence Coefficient

While the confidence interval (5.2.2) is based on the probability statement (5.2.1), once we have constructed a confidence interval based on a specific value $\overline{y}$, the $1 - \alpha$ confidence coefficient can no longer be interpreted as a probability. The population mean μ is a fixed quantity, and once we have determined specific lower and upper bounds, μ is either in the interval (with probability 1), or it is not in the interval (i.e., is in the interval with probability 0). The correct interpretation of the confidence coefficient is as a long term relative frequency. That is, if the procedure used to obtain the specific interval were repeated a very large number of times, the proportion of intervals containing μ would be $1 - \alpha$.

Large Samples

For large samples the required normality of $\overline{Y}$ is a consequence of the CLT. In addition, for large samples we can deal with the problem that the interval (5.2.2) depends on σ (which is generally unknown) by replacing σ with s.

Example 5.2.1 (*Rail Car Data, p. 112*) For our sample of 43 hold times, $\overline{y} = 5.16$ and $s = 6.44$. Thus, a 95% confidence interval for $\mu =$ the mean hold time is

$$5.16 \pm 2.021(6.44)/\sqrt{43}$$

$$\pm 1.98$$

$$(3.18, 7.14).$$

This is a range of plausible values for μ, at the 95% confidence level.

Small Samples

For small samples the CLT does not imply that $\overline{Y}$ is approximately normal, and one would need the population itself (i.e., the distribution of the individual Y's) to be approximately normal in order for $\overline{Y}$ to be approximately normal. This can be checked using a normal probability plot.

In addition, when n is small, there is greater variability in S, and the confidence interval needs to be adjusted (made slightly larger) to account for this additional variability. Interval (5.2.2) is based on the fact that

$$\frac{\overline{Y} - \mu}{\sigma/\sqrt{n}} \sim N(0, 1)$$

and for small n we need an interval based on the distribution of

$$\frac{\overline{Y} - \mu}{S/\sqrt{n}}. \tag{5.2.3}$$

The t Distribution

When the distribution of the individual Y's is normal, the random variable (5.2.3) has a t **distribution**, whose density function is

$$f(y) = \frac{\Gamma(\frac{\nu+1}{2})}{\Gamma(\frac{\nu}{2})\sqrt{\pi\nu}} \left(1 + \frac{y^2}{\nu}\right)^{-(\nu+1)/2}.$$

The parameter $\nu = n - 1$ is referred to as the **degrees of freedom**.

Notation: We will use $Y \sim t(\nu)$ to denote that the random variable Y has a t distribution with ν df, and $t(\alpha; \nu)$ to denote the upper α quantile of a $t(\nu)$ distribution.

The degrees of freedom (df) are determined by the number of observations (minus one) used to compute s^2. All t distributions are centered at 0 and have a bell shape like the normal distribution, but with "heavier tails" than the normal distribution. That is, more probability is concentrated in the tails and less in the center.

The extent to which the tails are heavier than the normal distribution depends on the df. For small df the heaviness in the tails is very pronounced. For larger df the shape of the t distribution is similar to the shape of the normal distribution (see Figures 5.3–5.5). In fact, it can be shown that a t distribution converges to a $N(0, 1)$ distribution as $n \to \infty$. For practical purposes the t distribution is almost identical to the normal distribution when $n > 30$.

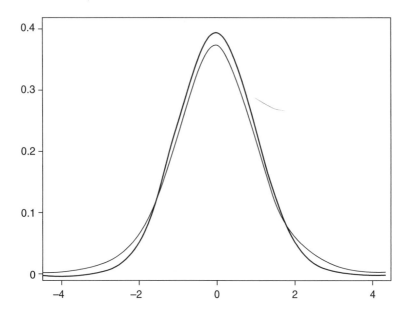

Figure 5.3. A t Distribution with 5 Degrees of Freedom and a $N(0, 1)$ Distribution.

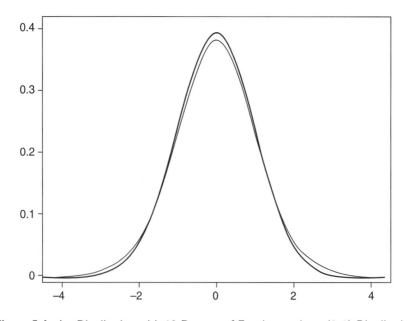

Figure 5.4. A t Distribution with 10 Degrees of Freedom and a $N(0, 1)$ Distribution.

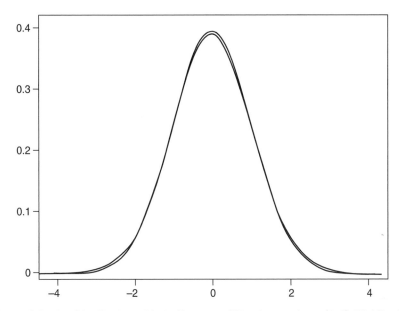

Figure 5.5. A t Distribution with 25 Degrees of Freedom and a $N(0, 1)$ Distribution.

For

$$\frac{\overline{Y} - \mu}{S/\sqrt{n}} \sim t(n - 1)$$

an argument similar to the one leading to the interval (5.2.2) (see Problem 5.2.16) results in the $(1 - \alpha)100\%$ confidence interval

$$\overline{y} \pm t(\alpha/2; n - 1)\frac{s}{\sqrt{n}}. \tag{5.2.4}$$

The upper α quantiles (also called critical values) $t(\alpha; \nu)$ are given in Table B2.

Example 5.2.2 (*Electric Vehicle Data*)   An engineering student is involved with a study of electric vehicles used to monitor parking on campus. In particular, the student needs to estimate how long these vehicles can be operated before they must be recharged. To begin, the student considers one such vehicle and records the amounts of time operated before recharging is needed. These values are given in Table 5.1.

Table 5.1. Times Before Recharging is Required for a Particular Electric Vehicle

5.11 2.10 4.27 5.04 4.47 3.73 5.96 6.21

To calculate a 95% confidence interval for μ = average time for this vehicle, we must first check to see (because of the small sample size) that the data are approximately normally distributed. A normal probability plot of the data is given in Figure 5.6 and looks reasonably linear, so we have no reason to think the times are not normally distributed.

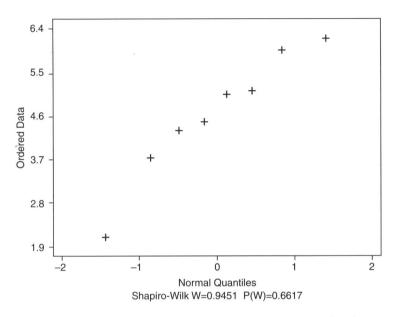

Figure 5.6. A Normal Probability Plot for the Electric Vehicle Data (Example 5.2.2).

The next step is to calculate the sample mean and standard deviation of the data.

$$\overline{y} = \frac{5.11 + 2.10 + \cdots + 6.21}{8} = 4.61$$

$$s^2 = \frac{(5.11 - 4.61)^2 + (2.10 - 4.61)^2 + \cdots + (6.21 - 4.61)^2}{8 - 1} = 1.72.$$

Thus, $s = \sqrt{1.72} = 1.31$, $\nu = 7$, and from Table B2 one finds $t(0.025; 7) = 2.365$. Using formula (5.2.4), a 95% confidence interval for the average time that this vehicle can run on its battery is

$$4.61 \pm 2.365(1.31)/\sqrt{8}$$

$$\pm 1.1$$

$$(3.5, 5.7).$$

We are 95% confident that the average time is somewhere between 3.5 and 5.7 hours.

A confidence interval can be constructed using any confidence level the user requires. (Usually, the confidence level is chosen to be relatively high.) For a fixed sample size increasing the confidence level increases the width of the confidence interval.

Example 5.2.3 (*Rubber-Strip Length Data, p. 138*)   Consider again the random sample of rubber-strip lengths. The normal probability plot was reasonably linear, so we are justified in using formula (5.2.4) data to construct a confidence interval. The summary statistics are

$$\bar{y} = 4.12, \qquad s = 0.092, \qquad n = 12.$$

For a 90% confidence interval we have $\alpha = 1 - 0.9 = 0.1$ and $\alpha/2 = 0.05$, so we need $t(0.05; 11) = 1.796$. A 90% confidence interval for the mean of the population of strip lengths is

$$4.12 \pm 1.796(0.092)/\sqrt{12}$$
$$\pm 0.048$$
$$(4.07, 4.17).$$

We can be 90% confident that the average strip length is between 4.07 and 4.17 inches. Since the target length of the strips was 4 inches, we would conclude that the strips are too long.

Example 5.2.4 (*Clear-Coat Thickness Data, p. 120*)   The clear-coat film build was measured for a sample of 40 cars (`ccthickn.txt`), which had a sample mean of $\bar{y} = 64.27$ and standard deviation $s = 2.73$. We would like to construct a 90% confidence interval for μ. Since $\nu = 39 > 30$, we will use $t(0.05; \infty) = z(0.05) = 1.645$, and our confidence interval is

$$64.27 \pm 1.645(2.73)/\sqrt{40}$$
$$\pm 0.71$$
$$(63.56, 64.98).$$

We can say with 90% confidence that the clear-coat film build population mean is between 63.56 and 64.98. If we wanted an 80% confidence interval ($\alpha = 1 - 0.8 = 0.2$), we would use $z(0.10) = 1.282$ to obtain (63.72, 64.82). If we wanted a 99% confidence interval ($\alpha = 1 - 0.99 = 0.01$), we would use $z(0.005) = 2.576$ and obtain (63.16, 65.38).

Note: In the example above we could not use $t(\alpha; 39)$ because not all large values of ν are tabulated. As ν gets large the t distribution critical values change very little as ν changes. (Recall that the t distribution converges to the normal distribution as ν increases.) If $\nu < 30$, then for values not listed in the table one would use the next smallest value of ν that is listed. If $\nu > 30$, one can use either the closest value of ν or $\nu = \infty$ (which amounts to using the normal distribution).

Sample Sizes

We can exploit our knowledge of confidence intervals to help effectively plan experiments. For example, sample sizes must be sufficiently large in order to obtain reliable, and hence meaningful, estimates. In particular, we have been looking at estimating the mean of a population. One question an experimenter should ask is how large a sample is needed to obtain a reasonable estimate of the mean. The general form of a confidence interval for μ can be written as

$$\widehat{\mu} \pm t \times (\text{estimated standard deviation of } \widehat{\mu})$$

where t is a critical value from the t distribution. The piece that is added and subtracted (i.e., half the interval length) is called the **margin of error** of the interval. We would like to have confidence intervals that have a high level of confidence and are also narrow (i.e., have a small margin of error). The margin of error depends on three things.

The Confidence Level: Increasing the confidence increases the critical value used, and hence increases the margin of error.
The Standard Deviation: When the sample has a large standard deviation (i.e., there's lots of variability in the population), the margin of error will be larger.
The Sample Size: Increasing n decreases the margin of error.

It follows that in order to have a high level of confidence and a small margin of error, one must increase the sample size. The actual sample size needed depends on the confidence level desired, the margin of error desired, and the standard deviation. Using formula (5.2.4), assuming that the sample size will be large enough to use the normal distribution (i.e., $n > 30$), and letting $d =$ margin of error; one obtains

$$d = t(\alpha/2; \infty) \frac{s}{\sqrt{n}}$$

or

$$n = \left[\frac{t(\alpha/2; \infty)s}{d} \right]^2 . \qquad (5.2.5)$$

Before we have taken a sample, the sample standard deviation is unknown. If we have an upper bound on the population standard deviation, then we can calculate an upper bound on the required sample size.

Note: The value of n obtained using equation (5.2.5) will generally not be an integer. Since sample sizes must be integer values, one should round the value obtained for n up to the next integer.

Example 5.2.5   A process engineer notices that some 20 oz. soda bottles coming off a filling line appear to be under filled. From past studies on this filling line

0.4 oz. is a reasonable estimate of the standard deviation for bottle volume. The engineer would like to take as small a sample as possible to construct 95% confidence interval for the mean volume of the bottles with a margin of error of 0.2 oz. or less. How large a sample is needed?

Using equation (5.2.5), one obtains

$$n = \left[\frac{t(0.025; \infty)(0.4)}{0.2} \right]^2 = 15.4 \to 16.$$

Thus, the engineer needs to sample 16 bottles.

When we do not have enough information to determine an upper bound on σ (the more common situation), we can employ *Stein's two-stage procedure* to determine a sample size.

Stein's Two-Stage Procedure

The idea of this procedure is a simple one. We need an estimate of σ in order to find our sample size. If no other information is available, we will perform an initial experiment (of some convenient size n_1), find the standard deviation from this experiment, and call it s_1. Then, using equation (5.2.5), but replacing ∞ with $n_1 - 1$ (the df for s_1), we will calculate n, the sample size needed to obtain the desired confidence. Since we have already gathered n_1 data points, we only need to gather $n_2 = n - n_1$ in the next experiment. By combining the results of the two experiments, we will have the necessary sample size.

More specifically, one would perform the following steps.

1. Choose an initial sample size n_1. This should be some convenient number. One way to determine n_1 would be to guess at σ and use equation (5.2.5).

2. Gather n_1 data points and use them to calculate the sample standard deviation s_1.

3. Compute

$$n_2 = \left[\frac{t(\alpha/2; n_1 - 1) s_1}{d} \right]^2 - n_1. \tag{5.2.6}$$

   If $n_2 > 0$, then we need to gather n_2 more data points. If $n_2 \leq 0$, then the initial sample of size n_1 is large enough.

4. When all the data are gathered, we calculate $\bar{y}$ based on the entire sample. If $n_2 > 0$, then the resulting confidence interval for μ is

$$\bar{y} \pm d. \tag{5.2.7}$$

   If $n_2 \leq 0$ (and one really only has a single sample), then the confidence interval would be constructed using formula (5.2.4).

Example 5.2.6 (*Clear-Coat Thickness Data, p. 120*)   Suppose we are interested in constructing a 99% confidence interval with margin of error no greater than 0.8. An initial sample of size $n_1 = 40$ resulted in $\bar{y}_1 = 64.26$ and $s_1 = 2.72$. How many more observations are needed? Using $t(0.005; 39 \to 40) = 2.704$ and equation (5.2.6), one obtains

$$n_2 = \left[\frac{2.704(2.72)}{0.8} \right]^2 - 40$$
$$= 84.5 - 40 = 44.5 \to 45.$$

Therefore, 45 additional observations are needed. If the additional observations resulted in $\bar{y}_2 = 63.68$, then

$$\bar{y} = \frac{n_1 \bar{y}_1 + n_2 \bar{y}_2}{n_1 + n_2} = \frac{40(64.26) + 45(63.68)}{40 + 45} = 63.95$$

and a 99% confidence interval for the pollution index is 63.95 ± 0.8.

Problems

1. If $z(\alpha) = Q_Z(1 - \alpha)$, then show that $P\left[-z(\alpha/2) \leq Z \leq z(\alpha/2)\right] = 1 - \alpha$.

2. Suppose $n = 20$, $\bar{y} = 3.5$, $s = 1.58$, and the population is approximately normal.

   (a) Find a 95% confidence interval for μ.

   (b) Find a 90% confidence interval for μ.

   (c) Find an 80% confidence interval for μ.

3. Suppose $n = 12$, $\bar{y} = 72$, $s = 10.3$, and the population is approximately normal.

   (a) Find a 90% confidence interval for μ.

   (b) Find a 95% confidence interval for μ.

   (c) Find a 99% confidence interval for μ and explain what it stands for in the context of the problem.

4. A political pollster surveys 30 registered Republicans in Pickens county, South Carolina. Among other things, he asks them their age. The average age in his sample is 44, and the standard deviation is 15.

   (a) If we were to use this information to find a confidence interval, what population parameter would we be estimating? What does this population parameter represent?

   (b) Find a 95% confidence interval for μ and explain what it indicates in the context of the problem.

   (c) Find a 90% confidence interval for μ.

   (d) The pollster would like the 95% confidence interval to be narrower. Explain how this can be achieved.

   (e) Carefully explain what Y is in the context of the problem. Do you think Y is normally distributed? Explain. Does normality or lack of normality have much effect on the confidence intervals you found above? Explain.

   (f) The pollster finds that surveying people takes a lot more time than he had planned. Next time, he might just survey 15 people instead of 30. Do you think this is a good idea? Why or why not?

5. In previous problems (Problem 2.1.1, p. 25; Problem 4.4.3, p. 148) you made graphs of the data in `webvisit.txt` on the number of visits to a web site.

   (a) Do the data appear to be normally distributed? Explain.

   (b) Explain what μ is in the context of this problem.

   (c) This data set has two distinct groups. For the larger group (why this group?)

       i. Construct a 90% confidence interval for μ.

       ii. Construct a 95% confidence interval for μ.

iii. Construct a 98% confidence interval for μ.

6. The data set `surfarea.txt` contains the surface areas for 32 samples of a silica product.

   (a) Construct a 95% confidence interval for the population mean surface area and explain what it stands for.

   (b) You have already graphed this data (Problem 2.1.2, p. 15; Problem 4.4.5, p. 148). Do the surface areas appear to be normally distributed? Why or why not?

   (c) Does your answer to part (b) have any effect on the confidence interval you constructed? Explain.

7. Consider again the data in `viscosity.txt` on failure times of paint that was stored at high temperature to study its stability (see Problem 4.3.7, p. 130). You have already graphed these data (Problem 4.4.1, p. 148). Explain why it would not be a good idea to make a 95% confidence interval for the mean failure time.

8. The data set `absorb.txt` contains the oil absorption values for a silica product from one manufacturing shift.

   (a) Make a normal probability plot of the data. Do the data appear to be normally distributed? Explain your answer.

   (b) Calculate the mean and standard deviation of the sample.

   (c) Construct a 90% confidence interval for the average oil absorption value and explain what it indicates.

   (d) Explain why the normal probability plot was not a necessary step in this problem.

   (e) Construct a histogram for the data. Does looking at the mean of the entire data set seem to be a reasonable thing to do? Explain your answer.

9. A chemist is studying the effects of storage on a monomer used to make optical lenses and wants to estimate the average difference between initial yellowing values and the yellowing values after 2 weeks of storage. If the standard deviation of the differences is 1.02 units and the chemist wants a 95% confidence interval with a margin of error of 0.5 units or less, how many samples should be taken?

10. A motorist wishes to estimate the average gas mileage for her car using a 90% confidence interval with a margin of error of 1 mpg or less. If the standard deviation of mileages is 3 mpg, how many tanks of gas does the motorist need to sample?

11. A civil engineering student needs to estimate the average width of parking spaces at the university for a parking study using a 95% confidence interval with a margin of error of 6 inches or less.

(a) Give a reasonable guess for σ and explain how you arrived at your guess.

(b) Use your guess to calculate how many parking spaces the student needs to sample.

(c) Suppose your guess in part (a) is only half as big as the actual value of σ. Using the actual value of σ, re-calculate how large a sample is needed.

12. One product of a small company is a gel (used in a medical setting) that contains an abrasive material. A key quality measure for this product is the percentage of abrasive material in the product. It can be measured by taking a sample, dissolving the non-abrasive component of the product, and weighing the resulting material. A previous study with $n = 12$ samples found $\bar{y} = 18\%$ and $s = 17\%$. However, there have been some changes made to the formulation so the company wants to verify that the process is still on target. To do this the process engineer wants to construct a 90% confidence interval for μ with a margin of error of 5%, but does not know how many samples to take. Write a short report, suitable for her boss, giving the correct sample size, and explaining how you arrived at the answer.

13. Consider Problem 12 again. The engineer's boss is not sure that a margin of error of 5% is really sufficient. Re-do your calculations for margins of error of 1, 2, 3, 4, and 10%, and report the results in a small table with an explanation of how it can be used.

14. Consider Problem 3, which had $n = 12$, $\bar{y} = 72$, and $s = 10.3$. Suppose this was a preliminary sample, and the experimenters were planning to take a second sample using Stein's procedure. If the experimenters want a 95% confidence interval with a margin of error of 2 or less, how large would the second sample have to be?

15. A political pollster wishes to estimate the average age of the members of the Pickens county Republican Party. If he wants a 99% confidence interval of width 3 years or less, how many party members will he have to survey? Hint: see the information in Problem 4.

16. Using the argument leading to the interval (5.2.2) as a model, provide the argument leading to the interval (5.2.4).

5.3 PREDICTION AND TOLERANCE INTERVALS

A confidence interval makes a statement about where a population parameter (such as μ) is likely to be. Two other important types of intervals are

Prediction Intervals, which indicate where a single future observation is likely to be.
Tolerance Intervals, which indicate where a large proportion of the population is likely to be.

Because a prediction interval is for an individual observation (i.e., a "moving target"), it is necessarily wider than a confidence interval. Similarly, since a tolerance interval is for a proportion of the population, rather than an individual, the tolerance interval will be a little wider than the corresponding prediction interval.

Dependence on the Population Distribution

Unlike confidence intervals, which are estimates of a particular (although unknown) number, prediction and tolerance intervals are descriptions of the entire population. As such, they are more dependent on the shape and variability of the population than confidence intervals are.

Width

In theory, we can make a confidence interval as narrow as we like simply by increasing the sample size. (In practice, this is often difficult to accomplish.) A confidence interval that is "too wide" is an indication that the researcher did not take enough data. However, the width of a prediction or tolerance interval depends mostly on the amount of spread in the population. If the members of the population vary widely, prediction and tolerance intervals will necessarily be wide, no matter how large a sample is taken.

Shape

When n is large, we know that $\overline{Y}$ is approximately normal, even if the underlying population is not. So when constructing a confidence interval, we need not worry too much about whether the underlying population is normally distributed. The same is not true for prediction and tolerance intervals. Since we are trying to pinpoint where individual observations might lie, the shape of the distribution is very important.

Intervals for Normal Populations

The simplest situation for constructing prediction and tolerance intervals is when the underlying population is (at least approximately) normally distributed.

Prediction Intervals

If we knew μ and σ, a $(1-\alpha)100\%$ prediction interval would be

$$\mu \pm z\,(\alpha/2)\,\sigma. \tag{5.3.1}$$

Since we do not know μ, we must estimate it with $\bar{y}$ and the interval becomes

$$\bar{y} \pm z(\alpha/2)\sigma\sqrt{1 + \frac{1}{n}}$$

where the extra $\sigma/\sqrt{n}$ is the additional uncertainty in using $\bar{y}$ to estimate μ (i.e., the standard deviation of $\overline{Y}$). Since we do not know σ either, we must also estimate σ with s and replace the normal critical value with a t critical value (as was done for confidence intervals). The interval then becomes

$$\bar{y} \pm t(\alpha/2; n-1)s\sqrt{1 + \frac{1}{n}}. \tag{5.3.2}$$

Example 5.3.1   Suppose the diameters of computer-mice tracking-balls are normally distributed with a target diameter of 2 cm. A sample of 20 tracking balls has $\bar{y} = 2.01$ cm and $s = 0.0122$ cm. To construct a 95% prediction interval for tracking ball diameters, we find $t(0.025; 19) = 2.093$, and using formula (5.3.2)

$$2.01 \pm 2.093(0.0122)\sqrt{1 + 1/20}$$
$$\pm\, 0.0262$$
$$(1.984, 2.036).$$

A randomly chosen tracking ball will have a diameter between 1.984 and 2.036 cm with confidence 0.95.

Tolerance Intervals

If we knew μ and σ, a tolerance interval covering a proportion p of the population would be

$$\mu \pm z\left(\frac{1+p}{2}\right)\sigma. \tag{5.3.3}$$

Note by comparing formulas (5.3.1) and (5.3.3) that when μ and σ are known, the $1-\alpha$ prediction interval and the tolerance interval covering a proportion p of the population are the same if $1-\alpha = p$. Thus, we see that when μ and σ are estimated, and one compares a prediction interval with the corresponding tolerance interval, it is the coverage of the tolerance interval that must correspond to the confidence of the prediction interval. When the confidence of the prediction interval is the same as the proportion of the population covered by the tolerance interval, the tolerance interval will always be longer.

The theory behind constructing tolerance intervals for normal populations when μ and σ are unknown is beyond the scope of this text. However, given the necessary tables, the actual construction of the intervals is straightforward. A tolerance interval for a normal population has the form

$$\overline{y} \pm rus \qquad (5.3.4)$$

where r and u are factors that take into account the uncertainty in $\overline{y}$ and s, respectively. The factor r depends on the sample size used to compute $\overline{y}$ and the proportion p of the population that one is interested in covering. The factor u depends on the df for s and the desired confidence level γ. Both factors are given in Table B12.

Example 5.3.2   Consider constructing a tolerance interval to cover 95% of the population of computer-mice tracking-balls discussed in Example 5.3.1 with confidence 0.90. Since there were $n = 20$ observations, one finds from Table B12 that if $p = 0.95$, then $r = 2.0080$. With df $= 19$ one finds (also from Table B12) that if $\gamma = 0.90$, then $u = 1.2770$. Using formula (5.3.4), one obtains

$$2.01 \pm (2.0080)(1.2770)(0.0122)$$
$$\pm 0.0313$$
$$(1.979, 2.041).$$

Thus, 95% of the tracking balls will have diameters between 1.979 and 2.041 with 90% confidence. Note that this tolerance interval is slightly larger than the 95% prediction interval in Example 5.3.1.

For values of n and/or df that are not listed in Table B12, one can either construct a conservative interval by using the next smallest value that is listed in the table, or one can use linear interpolation in the table. For values of n not in the table one uses linear interpolation on $1/n$, and for values of df not in the table one uses linear interpolation on $\sqrt{1/\mathrm{df}}$.

Example 5.3.3   A random sample of 58 observation from a normal population results in $\overline{y} = 23.4$ and $s = 3.06$. Suppose one is interested in constructing a 90% tolerance interval with a confidence level 0.99. Linear interpolation for r results in

$$r = 1.6585 + (1.6612 - 1.6585)\left[\frac{1/58 - 1/60}{1/50 - 1/60}\right].$$
$$= 1.6585 + (0.0027)(0.17241)$$
$$= 1.6590$$

and linear interpolation for u results in

$$u = 1.2651 + (1.2973 - 1.2651)\left[\frac{\sqrt{1/57} - \sqrt{1/60}}{\sqrt{1/50} - \sqrt{1/60}}\right]$$

$$= 1.2651 + (0.0322)(0.27218)$$
$$= 1.2739.$$

Using equation (5.3.4), one then obtains

$$23.4 \pm (1.6590)(1.2739)(3.06)$$
$$\pm 6.47$$
$$(16.93, 29.87)$$

A conservative interval using $n = \mathrm{df} = 50$ (see Problem 5.3.5) is $(16.81, 29.99)$.

Intervals for Non-Normal Populations

If the population is not normally distributed, it is difficult to form prediction or tolerance intervals since the formulas would be different for each population. There are three alternatives.

(i) Find a reasonable model for the data, and then find a formula for the prediction or tolerance interval based on that model. A good discussion of this can be found in Hahn and Meeker (1991). This technique is beyond the scope of this text.

(ii) Transform the data to approximate normality, and then find the interval based on the transformed data. If necessary, the endpoints of the resulting interval can be back-transformed to form an interval in the original units.

(iii) Construct **non-parametric** (or **distribution-free**) prediction or tolerance intervals.

Transformations

If a normal probability plot shows curvature, we cannot assume the data come from a normally distributed population, and we cannot use the formulas for normally distributed data. However, we might be able to transform the data so that the transformed values are approximately normally distributed, and construct prediction or tolerance intervals based on the transformed values. Finding an appropriate transformation, when it is possible, involves some trial and error. In order to preserve the relationships among the data points, the transformation must be a one-to-one function and (preferably) monotone increasing. For example, $g(y) = y^2$ is an appropriate transformation if y is strictly positive, but not if the data take on both positive and negative values.

For right-skewed data, such as lifetime data, transformations such as

$$\begin{aligned} g(y) &= \ln(y) \\ g(y) &= y^\gamma \text{ for some } \gamma \in (0, 1) \end{aligned}$$

are often useful because lifetimes are strictly non-negative.

For left-skewed data, transformations such as

$$
\begin{aligned}
g(y) &= e^y \\
g(y) &= y^\gamma \text{ for } \gamma > 1
\end{aligned}
$$

can be tried.

Example 5.3.4 (*Rail Car Data, p. 6*)   Suppose we wanted a prediction interval for the hold times in Example 2.1.1. Such an interval would tell us how long we would predict the next rail car would be held. The distribution of hold times is clearly not normal (see the histogram on p. 7). After a little experimentation, we find that the transformation $g(y) = \ln(y)$ produces approximately normal data, as can be seen from the normal probability plot in Figure 5.7. Therefore, to construct a 95% prediction interval for hold times, we would first transform all the data values and construct a 95% prediction interval for the transformed values. The mean and standard deviation of the transformed values are $\overline{g(y)} = 2.0059$ and $s_{g(y)} = 0.5845$.

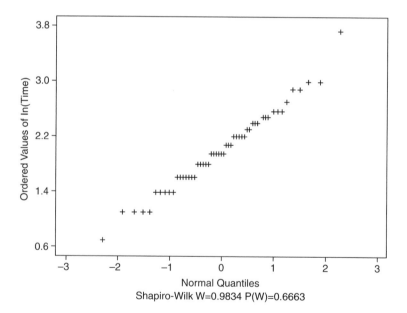

Figure 5.7. A Normal Probability Plot of the Transformed Data (Example 5.3.4).

Using formula (5.3.2), a 95% prediction interval for $g(Y)$ is

$$
2.0059 \pm t(0.025; 40)s\sqrt{1 + \frac{1}{53}}
$$

$$\pm 2.021(0.5845)\sqrt{1 + \frac{1}{53}}$$
$$\pm 1.192$$
$$(0.8139, 3.198).$$

It follows that a 95% prediction interval for Y is

$$\left(e^{0.8139}, e^{3.198}\right) = (2.26, 24.48).$$

That is, with confidence 0.95 a rail car will be held between 2 and 25 days.

Example 5.3.5   One could also use the transformation found in Example 5.3.4 to construct a tolerance interval for the rail car data. For a $p = 0.75$ tolerance interval with confidence $\gamma = 0.95$, one finds from Table B12 (using conservative values $n = 50$ and df $= 50$) that $r = 1.1618$ and $u = 1.1993$. Using formula (5.3.4), the tolerance interval for the transformed data is

$$2.0059 \pm (1.1618)(1.1993)(0.5845)$$
$$\pm 0.814$$
$$(1.192, 2.820)$$

and the corresponding tolerance interval in the original units is

$$\left(e^{1.192}, e^{2.820}\right) = (3.29, 16.78).$$

Note: A tolerance interval with the same confidence as a prediction interval (based on the same data) will not necessarily be larger than the prediction interval. This is the case in Examples 5.3.4 and 5.3.5.

Distribution-Free Intervals

If we cannot find (or do not want to look for) a transformation to approximate normality for the original data, we can use the ordered observations in the data set to form non-parametric prediction and tolerance intervals.

Notation: Recall that $y_{(1)}$ is the smallest observation in the data set, $y_{(2)}$ the next smallest, and $y_{(n)}$ is the largest. The corresponding random variables are $Y_{(1)}, Y_{(2)}, \ldots, Y_{(n)}$.

We know from the probability integral transformation and Example 4.3.3 that

$$P\left(Y \le Y_{(i)}\right) = P\left[F(Y) \le F(Y_{(i)})\right]$$
$$= P\left(U \le U_{(i)}\right) \qquad (5.3.5)$$

where $F(y)$ is the distribution function of Y, and U is a UNIF$[0, 1]$ random variable. Thus, one can study how a new observation Y relates to an order statistic from a previous sample by studying the order statistics from a uniform distribution.

It turns out that choosing n independent observations from a UNIF$[0, 1]$ distribution (i.e., choosing n observations that are uniformly distributed on the interval

$[0, 1])$ is equivalent to choosing $n + 1$ observations uniformly distributed on a circle of circumference one, cutting the circle at the $(n + 1)^{\text{st}}$ point, and straightening the circle into a line. Because of the symmetry on the circle, it does not matter at which point the circle is cut. Further, again because of the symmetry, it is clear that the distribution of the distance between any two consecutive points is the same. More specifically,

$$P\left(U_{(1)} \leq x\right) = P\left(U_{(2)} - U_{(1)} \leq x\right)$$

$$\vdots$$

$$= P\left(U_{(n)} - U_{(n-1)} \leq x\right).$$

Since the n points divide the line into $n + 1$ sub-intervals, and the distribution of the length of each sub-interval is the same, the probability of a new randomly chosen point falling in any particular sub-interval is $1/(n + 1)$. That is,

$$P\left(U \in (U_{(i-1)}, U_{(i)})\right) = \frac{1}{n + 1}.$$

Since the sub-intervals are non-overlapping, a new randomly chosen point can fall in only one sub-interval (i.e., the events $\{$falling in interval $i\}$ are mutually exclusive), and

$$P\left(U \leq U_{(i)}\right) = P\left(U \leq U_{(1)}\right) + P\left(U_{(2)} - U_{(1)} \leq x\right)$$
$$+ \cdots + P\left(U_{(i)} - U_{(i-1)} \leq x\right)$$
$$= \frac{i}{n + 1}. \tag{5.3.6}$$

Similarly,

$$P\left(U > U_{(n+1-i)}\right) = \frac{i}{n + 1}. \tag{5.3.7}$$

It follows from equations (5.3.5)–(5.3.7) that if Y is a randomly chosen new observation,

$$P\left(Y < Y_{(i)}\right) = \frac{i}{n + 1} \quad \text{and} \quad P\left(Y > Y_{(n+1-i)}\right) = \frac{i}{n + 1}.$$

Using $i = 1$ in the above equations, one obtains

$$P\left(Y_{(1)} \leq Y \leq Y_{(n)}\right) = 1 - \frac{1}{n + 1} - \frac{1}{n + 1} = \frac{n - 1}{n + 1}.$$

It follows that the interval

$$\left(Y_{(1)}, Y_{(n)}\right) \tag{5.3.8}$$

is a distribution-free prediction interval for Y with probability $(n - 1)/(n + 1)$. Based on this,

$$\left(y_{(1)}, y_{(n)}\right)$$

is a distribution-free prediction interval for Y with confidence $(n - 1)/(n + 1)$.

Note: This is exactly the same situation as with confidence intervals, and once the random variables $Y_{(1)}$ and $Y_{(n)}$ in the probability statement (5.3.8) are replaced with specific values $y_{(1)}$ and $y_{(n)}$, then the confidence coefficient is no longer a probability, but rather a long term relative frequency.

Using the same line of reasoning with $i = k$, the interval

$$\left(y_{(k)}, y_{(n+1-k)}\right)$$

is a prediction interval for Y with confidence

$$C = 1 - \frac{k}{n+1} - \frac{k}{n+1} = \frac{n+1-2k}{n+1}.$$

Alternatively, one could specify a confidence C and rearrange the above equation to obtain the appropriate value of k. Thus, a **distribution-free prediction interval** for Y with confidence C is

$$\left(y_{(k)}, y_{(n+1-k)}\right)$$

(5.3.9)

$$\text{where } k = [\![(1 - C)(n + 1)/2]\!].$$

Notation: The symbol $[\![x]\!]$ is the greatest integer function, that is, you take x and round the value *down* to an integer.

Note: Since you are rounding down when you calculate k, you must have $(1-C)(n+1)/2 \geq 1$.

Example 5.3.6 (*Rail Car Data, p. 6*)   To form a 95% distribution-free prediction interval for a rail car's hold time, one would compute (using equation (5.3.9))

$$k = [\![(1 - 0.95)(54)/2]\!] = [\![1.35]\!] = 1.$$

Therefore, the interval

$$\left(y_{(1)}, y_{(53)}\right) = (2, 42)$$

is a 95% distribution-free prediction interval for a rail car's hold time. That is, with confidence 0.95 a rail car will be held between 2 and 42 days. Notice that this is slightly wider (i.e., less precise) than the interval we obtained by transforming the data.

Example 5.3.7 (*Rail Car Data, p. 6*)   To form a 90% distribution-free prediction interval for a rail car's hold time, one would compute

$$k = [\![(1 - 0.90)(54)/2]\!] = [\![2.7]\!] = 2$$

and the interval would be

$$\left(y_{(2)}, y_{(52)}\right) = (3, 20).$$

A rail car will be held between 3 and 20 days with confidence 0.9.

The same results used to obtain distribution-free prediction intervals can be used to obtain distribution-free tolerance intervals. They are, however, used in a slightly different fashion due to the fact that for a tolerance interval there is not a future observation that needs to be taken into account. Instead, for a tolerance interval that covers a proportion p of the population with confidence γ, we need to find limits U and L such that

$$P\left[F(U) - F(L) \geq p\right] = \gamma$$

where F is the population distribution function. If the tolerance limits are chosen to be order statistics, say, $L = Y_{(k)}$ and $U = Y_{(n+1-j)}$, then it follows from Example 4.3.3 that

$$P\left[F(Y_{(n+1-j)}) - F(Y_{(k)}) \geq p\right] = P\left[U_{(n+1-j)} - U_{(k)} \geq p\right].$$

Now, since any collection of m consecutive sub-interval distances $U_{(i)} - U_{(i-1)}$ will have the same distribution (due to the symmetry on the circle), it follows that

$$P\left[U_{(n+1-j)} - U_{(k)} \geq p\right] = P\left[U_{(n+1-j-k)} \geq p\right].$$

If the tolerance interval contains $n+1-j-k$ sub-intervals, then $j+k$ sub-intervals have been excluded, and if those included cover at least a proportion p of the population, those excluded must cover no more than a proportion $1 - p$. That is,

$$P\left[U_{(n+1-j-k)} \geq p\right] = P\left[U_{(k+j)} \leq 1 - p\right].$$

Finally, it is possible to show that

$$P\left[U_{(m)} \leq 1 - p\right] = \sum_{i=m}^{n} \binom{n}{i}(1 - p)^i p^{n-i}.$$

Putting all this together, and making the switch from random variables to observed values, a **distribution-free tolerance interval** covering a proportion p of the population with confidence γ is

$$
\left(y_{(k)}, y_{(n+1-j)}\right)
$$
$$
\text{where } \gamma = \sum_{i=j+k}^{n} \binom{n}{i}(1 - p)^i p^{n-i}.
$$

(5.3.10)

Example 5.3.8 (*Rail Car Data, p. 6*)   Choosing $k = j = 1$, it follows from equation (5.3.10) that

$$\left(y_{(1)}, y_{(53)}\right) = (2, 42)$$

is a $p = 0.95$ tolerance interval with confidence

$$\gamma = 1 - p^n - n(1 - p)p^{n-1}$$

$$= 1 - (0.95)^{53} - 53(0.05)(0.95)^{52}$$
$$= 0.75.$$

Ninety-five percent of the rail cars will be held between 2 and 42 days with 75% confidence.

It is clear from equation (5.3.10) that it is only the sum $m = k + j$ that is important, and not the specific values j and k themselves. Rather than specifying m and the coverage p, and computing the resulting confidence using equation (5.3.10), one would really like to be able to specify the confidence and the coverage, and use these to determine m. This is easily done using Figures C1-C3.

Example 5.3.9 (*Rail Car Data, p. 6*) Suppose that one was interseted in obtaining a tolerance interval for rail car holding times with 90% coverage and confidence 0.95. From Figure C2 one finds that for $n = 53$ with $m = 2$ the coverage is about 91%, and with $m = 3$ the coverage is about 89%. Therefore,

$$\left(y_{(1)}, y_{(53)}\right) = (2, 42)$$

is a $p = 0.91$ tolerance interval with confidence 0.95. Alternatively,

$$\left(0, y_{(52)}\right) = (0, 20)$$

is also a $p = 0.91$ tolerance interval with confidence 0.95.

Problems

1. Suppose a sample of $n = 20$ is taken from a normal distribution with $\bar{y} = 3.5$ and $s = 1.58$.

   (a) Find a 95% prediction interval for a new observation.

   (b) Find a 99% prediction interval and explain what it indicates.

   (c) Find a 95% tolerance interval for $p = 0.75$ of the population, and explain what it indicates.

   (d) Find a 90% tolerance interval for $p = 0.95$ of the population, and compare the interval with those in parts (a) and (c).

2. A toothpaste was formulated with a silica content of 3% by weight. The manufacturer randomly samples 33 tubes of toothpaste and finds their mean silica content is 3.4% and the standard deviation is 0.43%.

   (a) Find a 95% prediction interval for the silica content of new randomly chosen tube of toothpaste.

   (b) Find a 99% interval for 95% of the population of valves. Explain what your interval indicates about the population.

   (c) What assumption is necessary for the above intervals to be valid? How could you investigate this assumption (if you had the original data)?

3. In Problem 2.2.5 a set of data was presented on the diameters of tracking balls (for computer mice) from two production lines. The data are reproduced below for convenience.

   Line 1: 2.18 2.12 2.24 2.31 2.02 2.09 2.23 2.02 2.19 2.32
   Line 2: 1.62 2.52 1.69 1.79 2.49 1.67 2.04 1.98 2.66 1.99

   (a) Construct a normal probability plot for each assembly line. Does it appear that the diameters are approximately normally distributed? Explain.

   (b) Construct a 95% prediction interval for a new tracking ball from each assembly line. Assume the observations come from normal distributions.

   (c) Can you construct 95% prediction intervals for a new tracking ball from each assembly line *without* assuming the measurements are normally distributed? Explain. What is the drawback of distribution-free prediction intervals in a situation such as this one?

4. In Problem 5.2.8 (p. 176), you constructed a normal probability plot for the data in `absorb.txt`. Based on that plot:

   (a) Construct a 98% prediction interval for the next shift's average oil absorption. Justify the method you used.

(b) Construct a 98% prediction interval for the next shift's average oil absorption. Justify the method you used.

5. Verify that for Example 5.3.3 a conservative 90% tolerance interval with confidence 0.99 is $(16.81, 29.99)$.

6. A manufacturing process for a medical device uses filtered water that must be strictly monitored for bacteria. The data set `water.txt` contains measurements of the bacteria (in parts per million) from 50 water samples.

(a) Graph the data and convince yourself that it does not follow a normal distribution.

(b) Construct a distribution-free 95% prediction interval for the data.

(c) Construct a distribution-free tolerance interval with 90% coverage and confidence 90%.

5.4 HYPOTHESIS TESTS

A confidence interval is used to find a range of plausible values for a parameter (such as μ). Sometimes, it is necessary to decide if one particular value is plausible for the parameter.

Example 5.4.1   A compact flourecent light bulb is sold with the promise that the average lifetime with normal use is 7 years. Customer complaints have led to a concern that the true average lifetime may be considerably shorter.

Example 5.4.2   Ten years ago, the 5-year survival probability for lung cancer patients was 57%. Today, with improved medical procedures, health care specialists believe the survival probability has increased.

Example 5.4.3   Last year the average number of recordable safety violations for a company with multiple operation sites was 12.3. Has the average changed under a new safety initiative?

In each case a **hypothesis test** (also called a **significance test**) can be used to decide if a particular value is a probable value of the parameter. A hypothesis test consists of choosing between two hypotheses, the **null hypothesis**, which is denoted by H_0, and the **alternative hypothesis**, which is denoted by H_a. The null hypothesis is of the form

$$H_0 : \mu = \mu_0$$

where μ_0 is a specified value, and the alternative hypothesis can be any one of

$$H_a : \mu > \mu_0 \quad \text{a right-tailed test}$$
$$H_a : \mu < \mu_0 \quad \text{a left-tailed test}$$
$$H_a : \mu \neq \mu_0 \quad \text{a two-sided test.}$$

The right-tailed and left-tailed tests are also referred to as one-sided tests. The alternate hypothesis reflects the state of affairs that would cause us to take some action. The null hypothesis, on the other hand, represents the "status quo", or the state of affairs in which no action is needed.

Example 5.4.4 (*Example 5.4.1 continued*)   In this case, μ is the mean lifetime of all lightbulbs in the population. Our hypotheses are

$$H_0 : \mu = 7 \quad \text{versus} \quad H_a : \mu < 7.$$

If the mean lifetime is less than 7 years, the company is making a false claim and will have to either modify the claim or improve the product.

Example 5.4.5   A drum of a chemical product is supposed to weigh 230 kg. If a customer weighs a number of drums and finds the average weight is less than 230 kg, then the customer is likely to file a claim against the supplier. The hypotheses are

$$H_0: \ \mu = 230 \quad \text{versus} \quad H_a: \ \mu < 230.$$

Example 5.4.6   A manufacturing process is supposed to produce bolts of length 5 cm. If the process mean has shifted in either direction, the bolts will not be usable. The hypotheses are

$$H_0: \ \mu = 5 \quad \text{versus} \quad H_a: \ \mu \neq 5.$$

We will use data to choose between H_0 and H_a. If our data seem inconsistent with the null hypothesis, we will *reject the null hypothesis*. Otherwise, we will *fail to reject the null hypothesis*. Thus, we are considering our data as evidence against the null hypothesis. If the evidence is strong enough, we will reject H_0.

Note: You may be wondering why we do not measure how strong the evidence is for the null hypothesis. We cannot do this because the null hypothesis consists of a single number. If the null hypothesis is

$$H_0: \ \mu = 7$$

we cannot possibly prove that the actual value of μ is 7 as opposed to some number arbitrarily close to 7.

How strong must the evidence be, before we will reject H_0? The strength of the evidence is measured in terms of a probability, called the **significance level** of the test. It is the probability of rejecting the null hypothesis when it is correct. Significance levels (also called α-levels) of 0.10, 0.05, or 0.01 are commonly used. The exact level chosen depends on the objectives of the experiment and is chosen by the experimenter.

Performing a Hypothesis Test

To perform a hypothesis test, we must first choose the null and alternate hypotheses. Next, we must choose the α-level for our test. Then we have to take a sample and decide if the observed data are far enough from the value specified in the null hypothesis that the probability of having obtained a value so extreme is less than α when the null hypothesis is true. To make this decision, we first calculate the distance between our estimated parameter and the value specified in the null hypothesis. That distance, after being standardized, is

$$t = \frac{\bar{y} - \mu_0}{s/\sqrt{n}} \tag{5.4.1}$$

which can either be compared with a critical value or used to compute the probability of having gotten a result as far away from μ_0 as the one that was obtained.

Comparing with a Critical Value

When the computed t value (equation (5.4.1)) is compared to a critical value, the particular critical value depends on which alternative hypothesis is of interest. The different possibilities are given in Table 5.2.

Table 5.2. Critical Values for the Three Types of Alternative Hypotheses

Alternate Hypothesis	Rejection Region
$H_a: \mu < \mu_0$	Reject H_0 if $t < -t(\alpha; n-1)$
$H_a: \mu > \mu_0$	Reject H_0 if $t > t(\alpha; n-1)$
$H_a: \mu \neq \mu_0$	Reject H_0 if $|t| > t(\alpha/2; n-1)$

Note: Similarly to the situation for confidence intervals, the above critical values are only appropriate if $\overline{Y}$ is at least approximately normally distributed. Thus, for small sample sizes (i.e., $n \leq 30$) where the CLT is not applicable, one must check the normality of the individual Y's. A normal probability plot is one way to do this.

Example 5.4.7   Consider the drums of material in Example 5.4.5 for which a one-sided (left-tailed) test was of interest. Suppose we take a random sample of 25 drums and calculate $\overline{y} = 229$. This is below the target mean of 230, but is it far enough below 230 to believe that the supplier is short filling the drums? If $s = 4$, then using equation (5.4.1), one obtains

$$t = \frac{229 - 230}{4/\sqrt{25}} = -1.25.$$

For $\alpha = 0.05$ the critical value is $-t(0.05; 24) = -1.711$. Since -1.25 is not less than -1.711, we would not reject the null hypothesis. There is not enough evidence to conclude that $\mu < 230$.

Example 5.4.8 (*Temperature Data*)   With the rise in cost of home heating oil government agencies try to anticipate requests for heating bill assistance by monitoring the average low temperature for the month and comparing it to historical averages. For one western US city the historical average low temperature for March is 29 degrees. If the most recent month of March has been colder than normal then the number of assistance requests will be greater than "normal". The hypotheses of interest are

$$H_0: \mu = 29 \quad \text{versus} \quad H_a: \mu < 29.$$

The newspaper reports for the month of March that $\overline{y} = 26$ and $s = 5$. Using $\alpha = 0.05$, we will reject H_0 if $t < -t(0.05; 30) = -1.697$. The value of the test statistic is

$$t = \frac{26 - 29}{5/\sqrt{31}} = -3.34$$

so we reject H_0. There is evidence that the average temperature was lower than normal, hence, the government agencies should expect an increase in requests.

Two-Sided Tests and Confidence Intervals

Performing a two-sided hypothesis test on μ with level of significance α is equivalent to constructing a $(1 - \alpha)100\%$ confidence interval for μ and rejecting $H_0: \mu = \mu_0$ if μ_0 does not fall in the interval.

Example 5.4.9   Consider Example 5.4.6 where the bolts have a target length of 5 cm. Suppose we have chosen $\alpha = 0.01$ and from a random sample of 20 bolts we calculate $\overline{y} = 5.2$ cm and $s = 0.07$ cm. A normal probability plot indicates the data are at least approximately normally distributed. Since we are interested in a two-sided alternative, the appropriate critical value is $t(0.005; 19) = 2.861$, and we would reject H_0 if $|t| > 2.861$. Since

$$t = \frac{\overline{y} - \mu}{s/\sqrt{n}} = \frac{5.2 - 5}{0.07/\sqrt{20}} = 12.78$$

we reject H_0. Our evidence shows that the length of the bolts is not on target. The shipment should be rejected.

Alternatively, we could have constructed the 99% confidence interval

$$5.2 \pm t(0.005; 19)\frac{0.07}{\sqrt{20}}$$
$$\pm 2.861(0.0157)$$
$$\pm 0.045$$
$$(5.155, 5.245)$$

and rejected H_0 because $5 \notin (5.155, 5.245)$.

p -Values

An alternative to comparing the computed t value to a critical value is to compute a ***p*-value**. A p-value, sometimes referred to as the descriptive level of significance, is the probability (when H_0 is true) of obtaining a value as (or more) extreme than the one actually obtained. The specific computation (i.e., exactly what constitutes more extreme) depends on the alternative hypothesis. The three possibilities are given in Table 5.3, where $X \sim t(n - 1)$.

Note: For small samples the (approximate) normality of the underlying population is necessary for computing p-values using the t distribution, just as it was for using critical values from the t distribution.

Table 5.3. p-Values for the Three Types of Alternative Hypotheses

Alternative Hypothesis	p-Value
$H_a: \mu < \mu_0$	$P[X < t]$
$H_a: \mu > \mu_0$	$P[X > t]$
$H_a: \mu \neq \mu_0$	$2P[X > |t|]$

Unless n is large enough to use the standard normal distribution, p-values will have to be obtained by using a computer. Having obtained a p-value, one would

$$\text{reject } H_0 \text{ if } p\text{-value} < \alpha. \tag{5.4.2}$$

This has the advantage over comparing with a critical value that the test can easily be carried out for any α value without having to look up additional critical values.

Example 5.4.10   In Example 5.4.7, we were interested in a left-tailed test. Therefore, one could have computed (here $X \sim t(24)$)

$$p\text{-value} = P[X < -1.25] = 0.1117.$$

Since 0.1117 is not less than 0.05, we would not reject H_0.

Example 5.4.11 (*Clear-Coat Thickness Data, p. 120*)   Suppose that one was interested in testing whether the average clear-coat film build was 64 microns. Since no particular direction for the alternative is specified (e.g., we are not asked to test if the index is less than 64), a two-sided test is appropriate. From the sample of $n = 40$ cars we found $\bar{y} = 64.26$ and $s = 2.72$. Therefore (equation (5.4.1)),

$$t = \frac{64.26 - 64}{2.72/\sqrt{40}} = 0.604.$$

Since $n > 30$, we can use the standard normal distribution ($Z \sim N(0,1)$) to compute a p-value, and one finds (Table 5.3)

$$p\text{-value} = 2P[Z > |0.60|] = 2[1 - \Phi(0.60)] = 2[1 - 0.7257] = 0.549.$$

This p-value is large enough that it is larger than any reasonable α level (e.g., $0.549 > 0.1$), therefore one would not reject H_0.

Sample Sizes

In any hypothesis test there are two errors that can arise. Neither of these errors is a "mistake" in the sense of an arithmetic mistake. These are incorrect conclusions that we might arrive at because we were misled by the data. We might reject H_0 when H_0 is true, or we might fail to reject H_0 when H_0 is false. Usually, both errors have serious consequences. In terms of the drum example, we would not want to accuse the supplier of short filling the drums when they are not (this might strain relations with the supplier, lead to material cost increases, etc.), but we also do not want the supplier to get away with anything. These two kinds of errors are called

$$
\begin{array}{ll}
\textbf{Type I} & \text{rejecting } H_0 \text{ when } H_0 \text{ is true} \\
\textbf{Type II} & \text{failing to reject } H_0 \text{ when } H_0 \text{ is false.}
\end{array}
\tag{5.4.3}
$$

When we choose the α-level of the test, we are choosing a bound on the $P(\text{Type I error})$. The $P(\text{Type II error})$ is harder to quantify, because it depends on the distance between the actual value of μ and μ_0. However, we can say a few things about the $P(\text{Type II error})$.

- $P(\text{Type II error})$ increases as α decreases. So it is a good idea not to choose α *too* small.
- $P(\text{Type II error})$ decreases as n increases. So it is a good idea to take as large a sample as possible.

The **power** of a hypothesis test is the probability that the test will reject H_0 when the true mean is actually μ_1 rather than μ_0. Thus, the power is a function of μ_1, and

$$
\text{Power}(\mu_1) = 1 - P(\text{Type II error} \mid \mu = \mu_1).
\tag{5.4.4}
$$

Notation: The $\mid$ in a probability statement is read as "given", and indicates one is interested in the probability conditional on whatever follows the $\mid$.

Specifying both the significance level and the power of the test at μ_1 will determine the required sample size.

If the Variance is Known

If the variance is known and our observations come from a normal population (i.e., $Y_i \sim N(\mu, \sigma^2)$, where σ^2 is known), then it is possible to derive an explicit formula for the sample size n needed to obtain a power of γ for detecting a mean of μ_1 when testing $H_0 \colon \mu = \mu_0$ using a level of significance of α.

Note: When one is sampling from a normal population, σ^2 will generally be unknown. However, the formulas obtained here will also be useful when using the normal approximation for inference with binomial distributions.

Example 5.4.12   Let $\mu = $ traffic volume (per week) at a web site. We want to test

$$
H_0 \colon \mu = 22{,}000 \qquad \text{versus} \qquad H_a \colon \mu > 22{,}000.
$$

If it is known that traffic volume is normally distributed and the variance of traffic volume is 1000, how large a sample would be needed to detect a volume of 22,500 with power (probability) 0.8 when using $\alpha = 0.05$?

If σ^2 is known, then one would test

$$H_0: \ \mu = \mu_0 \quad \text{versus} \quad H_a: \ \mu > \mu_0$$

by computing

$$z = \frac{\overline{y} - \mu_0}{\sigma/\sqrt{n}}$$

and rejecting H_0 if $z > z(\alpha)$. (One could think of this as using the rejection region from Table 5.2, but with infinite df since σ is known.) Specifying the power at μ_1 to be γ, it follows from equation (5.4.4) and definition (5.4.3) that

$$\begin{aligned}
\gamma &= P\left[\text{rejecting } H_0 \mid \mu = \mu_1\right] \\
&= P\left[\frac{\overline{Y} - \mu_0}{\sigma/\sqrt{n}} > z(\alpha) \ \Big| \ \mu = \mu_1\right] \\
&= P\left[\frac{\overline{Y} - \mu_0}{\sigma/\sqrt{n}} - \frac{\mu_1 - \mu_0}{\sigma/\sqrt{n}} > z(\alpha) - \frac{\mu_1 - \mu_0}{\sigma/\sqrt{n}} \ \Big| \ \mu = \mu_1\right] \\
&= P\left[\frac{\overline{Y} - \mu_1}{\sigma/\sqrt{n}} > z(\alpha) - \frac{\mu_1 - \mu_0}{\sigma\sqrt{n}} \ \Big| \ \mu = \mu_1\right] \\
&= P\left[Z > z(\alpha) - \frac{\mu_1 - \mu_0}{\sigma/\sqrt{n}}\right] \quad (5.4.5)
\end{aligned}$$

where $Z \sim N(0,1)$. Therefore,

$$z(\gamma) = z(\alpha) - \frac{\mu_1 - \mu_0}{\sigma/\sqrt{n}}$$

and solving for n results in

$$n = \frac{[z(\alpha) - z(\gamma)]^2}{(\mu_1 - \mu_0)^2}\sigma^2. \quad (5.4.6)$$

Example 5.4.13   A hypothesis test is to be carried out to see if the average potency of a batch of a drug is at its specified value of 5 mg of drug per tablet, or if it is too potent. The variation in the amount of drug between tablets is known to be $\sigma^2 = 1 \ (\text{mg})^2$. The level of significance of the test is specified to be 0.05, and one is interested in being able to detect a batch having potency of 6 mg per tablet with probability 0.8 (i.e., at least 80% of the time). What size sample is necessary? Using Table B1 for the normal distribution and equation (5.4.6) with $\alpha = 0.05, \gamma = 0.8, \mu_0 = 5, \mu_1 = 6$, and $\sigma^2 = 1$, one obtains

$$n = \frac{[z(0.05) - z(0.8)]^2}{(5-6)^2}(1) = (1.645 + 0.842)^2 = 6.19.$$

Since sample sizes must be integer values, one would use $n = 7$ tablets in order to be sure that the power is at least as large as was specified.

Sample size equation (5.4.6) also holds for the one-sided alternative $H_a: \ \mu < \mu_0$ (see Problem 5.4.11).

Example 5.4.14   Suppose that in Example 5.4.13 one were interested in testing whether the batch of drug was at its specified potency against the alternative that it was not potent enough. The test is to be carried out at level of significance 0.05, and one is interested in being able to detect a batch whose actual potency is only 3.5 mg with probability 0.9. Using equation (5.4.6) with $\alpha = 0.05, \gamma = 0.9, \mu_0 = 5, \mu_1 = 3.5$, and $\sigma^2 = 1$, one obtains

$$n = \frac{[z(0.05) - z(0.9)]^2}{(5-3.5)^2}(1) = \frac{(1.645 + 1.282)^2}{2.25} = 3.81$$

and would, therefore, sample $n = 4$ tablets.

For a two-sided test one would reject H_0 if

$$|z| = \left| \frac{\overline{y} - \mu_0}{\sigma/\sqrt{n}} \right| > z(\alpha/2).$$

Obtaining a specified power γ for detecting a mean of μ_1 would require

$$\gamma = P\left[\left| \frac{\overline{Y} - \mu_0}{\sigma/\sqrt{n}} \right| > z(\alpha/2) \ \middle| \ \mu = \mu_1 \right]$$

$$= P\left[\frac{\overline{Y} - \mu_0}{\sigma/\sqrt{n}} < -z(\alpha/2) \ \text{or} \ \frac{\overline{Y} - \mu_0}{\sigma/\sqrt{n}} > z(\alpha/2) \ \middle| \ \mu = \mu_1 \right].$$

Since the events

$$\left\{ \frac{\overline{Y} - \mu_0}{\sigma/\sqrt{n}} < -z(\alpha/2) \right\} \ \text{and} \ \left\{ \frac{\overline{Y} - \mu_0}{\sigma/\sqrt{n}} > z(\alpha/2) \right\}$$

are mutually exclusive ($\overline{Y}$ cannot be both large and small at the same time), the probability of one or the other occurring is the sum of their individual probabilities of occurrence. Therefore, an argument similar to the one leading to equation (5.4.5) results in

$$\gamma = P\left[Z < -z(\alpha/2) - \frac{\mu_1 - \mu_0}{\sigma/\sqrt{n}}\right] + P\left[Z > z(\alpha/2) - \frac{\mu_1 - \mu_0}{\sigma/\sqrt{n}}\right]. \qquad (5.4.7)$$

One of the two probabilities in equation (5.4.7) will always be fairly small. When $\mu_1 > \mu_0$, then the two probabilities are as shown in Figure 5.8, and

$$\gamma \doteq P\left[Z > z(\alpha/2) - \frac{\mu_1 - \mu_0}{\sigma/\sqrt{n}}\right]. \qquad (5.4.8)$$

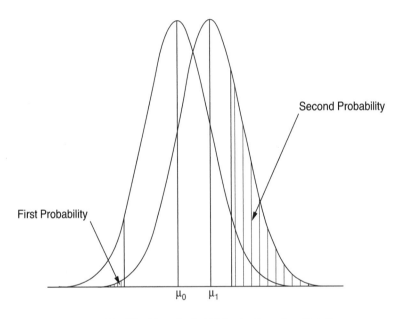

Figure 5.8. The Two Probabilities in Equation (5.4.7).

Since equation (5.4.8) is the same as equation (5.4.5) except that α in equation (5.4.5) has been replaced by $\alpha/2$ in equation (5.4.8), one immediately obtains

$$n = \frac{[z(\alpha/2) - z(\gamma)]^2}{(\mu_1 - \mu_0)^2}\sigma^2 \qquad (5.4.9)$$

which is the counterpart to equation (5.4.6). Equation (5.4.9) also holds when $\mu_1 < \mu_0$ (see Problem 5.4.12).

Example 5.4.15   A new process to manufacture resistors is being considered. Before investigating the economic ramifications of switching to the new process, the company is interested in determining if the new process significantly changes the characteristics of the resistors. The average lifetime of the resistors is of particular importance. The company is willing to assume that the lifetimes are normally distributed and that the variability of the new process is the same as the old process, 100 (hours)^2. If a test is to be conducted at level of significance 0.05 and the company is interested in being able to detect a difference in lifetime of five hours with probability 0.95, how large a sample size is necessary? Using equation (5.4.9), one obtains

$$n = \frac{[z(0.025) - z(0.95)]^2}{5^2}(100) = \frac{(1.96 + 1.645)^2}{25}(100) = 51.98$$

and a sample of size $n = 52$ resistors is needed.

If the Variance is Unknown

When the variance is unknown (and the observations are normally distributed), then tests on μ are based on the t distribution (see Table 5.2). In that case, having power γ to detect H_a: $\mu = \mu_1 > \mu_0$ would require

$$\gamma = P[t > t(\alpha; n - 1) \mid \mu = \mu_1]$$
$$= P\left[\frac{\overline{Y} - \mu_0}{S/\sqrt{n}} > t(\alpha; n - 1) \;\middle|\; \mu = \mu_1\right]$$
$$= P\left[\frac{\frac{\overline{Y} - \mu_1}{\sigma/\sqrt{n}} + \frac{\mu_1 - \mu_0}{\sigma/\sqrt{n}}}{S/\sqrt{n}} > t(\alpha; n - 1) \;\middle|\; \mu = \mu_1\right]$$
$$= P\left[\frac{Z + \frac{\mu_1 - \mu_0}{\sigma/\sqrt{n}}}{S/\sqrt{n}} > t(\alpha; n - 1)\right] \qquad (5.4.10)$$

where $Z \sim N(0, 1)$. Equation (5.4.10) differs from equation (5.4.5) in two important ways. First, the random variable

$$\frac{Z + \frac{\mu_1 - \mu_0}{\sigma/\sqrt{n}}}{S/\sqrt{n}} \sim t'\left(\frac{\mu_1 - \mu_0}{\sigma/\sqrt{n}}, n - 1\right)$$

depends on both n and $(\mu_1 - \mu_0)/\sigma$; and second, the critical value $t(\alpha; n - 1)$ is a function of n. Thus, it is not possible to solve analytically for n as was done to obtain the sample size formula (5.4.6). Instead, it is necessary to solve equation (5.4.10) numerically.

Note: The random varianble $t'(\lambda, \nu)$ is said to have a **non-central t distribution** *with non-centrality parameter λ and df ν. We will not need to worry about the specifics of this distribution since solutions to equation (5.4.10) are provided in Table B3.*

Because of its dependence on the unknown σ, solutions to equation (5.4.10) are based on

$$d = \frac{\mu - \mu_0}{\sigma}.$$

That is, the difference $\mu_1 - \mu_0$ that one is interested in detecting is expressed in units of σ. For specified power, α, and d; the necessary sample size n can be found in Table B3.

Example 5.4.16   Let μ = traffic volume (per week) at a web site. We want to test

$$H_0: \ \mu = 22,000 \quad \text{versus} \quad H_a: \ \mu > 22,000.$$

If the difference of interest is $d = 0.40$, and we want the test to have 90% power with $\alpha = 0.05$, then the required sample size is found from Table B3 to be $n = 55$ weeks.

Recall that in the σ known case approximate sample sizes for two-sided tests were obtained using the same formula as for one-sided tests, but with α replaced by $\alpha/2$. Exactly the same situation occurs when σ is unknown. Therefore, Table B3 can also be used for two-sided tests. Note that since Table B3 in indexed by the error probability α in a single tail of the distribution, using these values for a two-sided test corresponds to a test of level 2α.

Example 5.4.17   Suppose that for the new resistor manufacturing process discussed in Example 5.4.15 the company has no information on what the variability will be. If a test is to be conducted at level of significance 0.05 and the company is interested in being able to detect a difference in lifetime of 0.6σ hours with probability 0.95, how large a sample size is necessary? From Table B3 with $2\alpha = 0.1$, power $= 0.95$, and $d = 0.6$, one obtains $n = 32$.

Given α, a difference d, and a sample size n, Table B3 can also be used to obtain the power of the test.

Example 5.4.18   (*Temperature Data, p. 192*)   Suppose the government agencies were concerned about an average temperature of 0.4σ below the historical average. What is the power of the test in Example 5.4.8? Recall we used $\alpha = 0.05$. From Table B3, the power with a sample size of $n = 31$ is 0.70. Suppose the agencies needed the power to be 0.9. Then a sample of at least $n = 55$ days is needed.

Example 5.4.19   Suppose that for the new resistor manufacturing process discussed in Example 5.4.17 the company was only able to sample 16 resistors. What would the power of an $\alpha = 0.05$ test be for detecting a difference in lifetime of 0.6σ hours? From Table B3 with $2\alpha = 0.1$ and $d = 0.6$, one finds a sample of size $n = 15$ results in a power of 0.7, and a sample of size $n = 17$ results in a power of 0.75. Therefore, the power with a sample of size $n = 16$ is somewhere between 0.7 and 0.75.

Problems

1. A sample of size 45 is taken from a particular population and results in $\bar{y} = 122$ and $s = 33.7$. Test

$$H_0: \ \mu = 100 \quad \text{versus} \quad H_a: \ \mu > 100$$

using $\alpha = 0.05$. Explain what your results indicate.

2. A sample of size 20 is taken from a normal distribution and results in $\bar{y} = 3.5$ and $s = 1.58$. Test the hypothesis that the population mean is 3 using $\alpha = 0.10$. Explain what your results indicate.

3. A technician often places samples in a temperature-controlled cold room in which a constant temperature of $30°F$ is supposed to be maintained. The laboratory technician thinks the room is overheating. He obtains a sensitive thermometer and measures the room temperature once an hour for 8 hours with the following results: $34, 35, 41, 34, 34, 27, 31, 37$.

   (a) Explain why the technician should not simply adjust the thermostat if the room feels warm to him.

   (b) Make a normal probability plot of the data and comment on it.

   (c) Calculate the sample mean and standard deviation.

   (d) Test the technician's hypothesis using $\alpha = 0.01$ and explain what the results indicate.

4. A city near a government weapons clean-up site is concerned about the level of radioactive byproducts in their water supply. The EPA limit on naturally occurring radiation is 5 picocuries per liter of water. A random sample of 26 water specimens from the city's water supply gives $\bar{y} = 4.82$ and $s = 0.79$. Does this provide sufficient evidence that the water supply is safe (i.e., below the maximum level allowed)? Test, using $\alpha = 0.01$, and explain what your results mean.

5. A textile mill weaves cloth that is used to make sheets. The top grade of sheets has at least 200 threads per inch. (Because higher thread count indicates higher quality, it will not matter if the thread count is above 200.) Random samples from 20 areas of a piece of cloth result in an average thread count of $\bar{y} = 190$ with $s = 50$. Assuming normality, test the appropriate hypothesis using $\alpha = 0.05$, and explain what your results mean.

6. A toothpaste was formulated with a silica content of 3% by weight. The manufacturer randomly samples 33 tubes of toothpaste and finds their mean silica content is 3.4% and the standard deviation is 0.43%. Test the hypothesis that the mean silica content is not on target using a p-value and $\alpha = 0.05$. Explain what the results of your test indicate in terms the manufacturer can understand.

7. Consider again the rail-car hold-time data in `railcars2.txt`. Is there sufficient evidence to conclude that the average hold time is more than 7 days? Test, using a p-value and $\alpha = 0.01$, and explain what your results indicate.

8. Consider again the oil absorption data in `absorb.txt`. Can we conclude that the average oil absorption is less than 100 units? Test, using a p-value and $\alpha = 0.05$, and explain what your results mean.

9. Consider Problem 3 again. Explain in words what Type I and Type II errors would be, in the context of the problem. What are the consequences of both types of errors? Which error has more serious consequences?

10. Consider Problem 4 again. Explain in words what Type I and Type II errors would be, in the context of the problem. What are the consequences of both types of errors? Which error has more serious consequences?

11. Show that equation (5.4.6) also applies when one is testing against the alternative H_a: $\mu < \mu_0$.

12. Show that equation (5.4.9) also applies when $\mu_1 < \mu_0$.

13. Suppose it is desired to test the hypothesis that the melting point of an alloy is $1200°C$. If the melting point differs by more than $20°C$, it is desired to change the composition.

   (a) Assuming normality and that σ is approximately $15°C$, how many determinations should be made so that when using $\alpha = 0.05$ the $20°$ difference can be detected with power 0.95?

   (b) Compare this with the situation where σ is unknown.

   (c) With 20 observations what would be the power of an $\alpha = 0.05$ test to detect:

      i.. A $20°$ difference if σ is known to be $15°$?

      ii.. A difference of 1.4σ if σ is unknown?

14. Consider Problem 5 again.

   (a) Estimate how large a difference from the null hypothesis would be important.

   (b) Estimate how large a sample would be needed to detect this size difference with 90% power, using the α level given.

   (c) Discuss the consequences of Type I and Type II errors in this problem. Which seems more serious?

   (d) Considering your answer to part (c), do you think the chosen α level is too large, or too small? Explain.

15. Consider Problem 6 again. Estimate how large a difference from the null hypothesis would be important. Estimate how large a sample would be needed to detect this size difference with a power of 0.8, using the α level given.

5.5 INFERENCE FOR BINOMIAL POPULATIONS

Up to now, the material in this chapter has focused on the study of the mean μ of a continuous population. These methods can be extended to other types of populations and other parameters. Consider, for example, the proportion of defectives from a binomial population. Recall that in a binomial population each member can be put in one of two categories. For example, we might inspect a shipment of parts and categorize each part as defective or nondefective. What can we infer about the proportion of defectives in the population?

We have already shown that a binomial random variable can be thought of as a sum of independent Bernoulli random variables, and therefore, the sample proportion of defectives

$$\widehat{p} = \frac{\text{\# of defectives in sample}}{n}$$

is a sample mean, and the CLT applies. Hence, for large samples

$$\widehat{p} \sim N\left(p, \frac{p(1-p)}{n}\right).$$

Note: The sample size needed to ensure the normality of $\widehat{p}$ depends on p. Since p is unknown, a reasonable condition is n should be large enough that $n\widehat{p} > 5$ and $n(1-\widehat{p}) > 5$.

Confidence Intervals

We can use the above information to make inferences about p. A $(1 - \alpha)100\%$ confidence interval for p would be

$$\widehat{p} \pm z(\alpha/2)\sqrt{\frac{p(1-p)}{n}}. \tag{5.5.1}$$

In practice, however, we do not know p (in fact, it is what we are trying to estimate). There are two solutions to this problem, an approximate solution and a conservative solution.

An Approximate Solution

An approximate $(1 - \alpha)100\%$ confidence interval is obtained by simply replacing the unknown value p in formula (5.5.1) with $\widehat{p}$. This results in

$$\widehat{p} \pm z(\alpha/2)\sqrt{\frac{\widehat{p}(1-\widehat{p})}{n}}. \tag{5.5.2}$$

Example 5.5.1   Suppose we sample 200 items and find 27 that are defective. What can we say about p = proportion of defectives in the population? We know

$$\widehat{p} = \frac{27}{200} = 0.135$$

and an approximately 95% confidence interval for p is (formula (5.5.2))

$$0.135 \pm 1.96\sqrt{\frac{(0.135)(0.865)}{200}} = 0.135 \pm 0.0474 = (0.09, 0.18).$$

That is, we estimate that 9% to 18% of the population is defective.

A Conservative Solution

Since $p \in [0,1]$, it follows that $p(1-p) \le 0.25$. So a conservative confidence interval (i.e., worst case) is obtained by replacing $p(1-p)$ in formula (5.5.1) with 0.25. This results in

$$\widehat{p} \pm z(\alpha/2)\sqrt{\frac{0.25}{n}} = \widehat{p} \pm z(\alpha/2)\frac{1}{2\sqrt{n}} \,. \tag{5.5.3}$$

Example 5.5.2   In Example 5.5.1, a conservative 95% confidence interval for p is (formula (5.5.3))

$$0.135 \pm 1.96\sqrt{\frac{0.25}{200}} = 0.135 \pm 0.0693 = (0.07, 0.20).$$

Sample Sizes

As before, we can find the sample size needed to achieve a specified margin of error d by rearranging formula (5.5.3) to obtain

$$n = \left[\frac{z(\alpha/2)}{2d}\right]^2 . \tag{5.5.4}$$

Note: Since one would generally determine a sample size before taking a sample, it is unlikely that an estimate $\widehat{p}$ would be available, and hence formula (5.5.2) would be of no use.

Example 5.5.3   Suppose a process engineer needs to report to management the proportion of defective lenses produced on a particular production line. The engineer feels that a 95% confidence interval with a margin of error of 0.05 or less would give an appropriate estimate. Using equation (5.5.4),

$$n = \left[\frac{1.96}{2(0.05)}\right]^2 = 384.16 \to 385.$$

The engineer would need to sample 385 lenses to achieve the desired precision.

Hypothesis Tests

We can test a hypothesis about p in much the same way we tested hypotheses about μ. In this situation, the test statistic is

$$z = \frac{\widehat{p} - p_0}{\sqrt{\frac{p_0(1-p_0)}{n}}} \tag{5.5.5}$$

where p_0 is the value specified in H_0. Under H_0 (if n is large enough) z has a $N(0,1)$ distribution. The three possible alternative hypotheses and their corresponding rejection regions (similar to Table 5.2) are given in Table 5.4. The p-values for the three possible alternative hypotheses (similar to Table 5.3) are given in Table 5.5.

Table 5.4. Critical Regions for the Three Possible Alternative Hypotheses

Alternate Hypothesis	Rejection Region
$H_a:\ p < p_0$	Reject H_0 if $z < -z(\alpha)$
$H_a:\ p > p_0$	Reject H_0 if $z > z(\alpha)$
$H_a:\ p \neq p_0$	Reject H_0 if $|z| > z(\alpha/2)$

Table 5.5. The p-Values for the Three Possible Alternative Hypotheses

Alternative Hypothesis	p-Value
$H_a:\ p < p_0$	$\Phi(z)$
$H_a:\ p > p_0$	$1 - \Phi(z)$
$H_a:\ p \neq p_0$	$2[1 - \Phi(|z|)]$

Example 5.5.4 (*Defective Rates*) Suppose that 1000 items are sampled and 18 of them are defective. Can we conclude that the defective rate is above 1% at a significance level of 0.05? We are interested in testing

$$H_0:\ p = 0.01 \quad \text{versus} \quad H_a:\ p > 0.01$$

and

$$\widehat{p} = \frac{18}{1000} = 0.018.$$

Using equation (5.5.5), one obtains

$$z = \frac{0.018 - 0.01}{\sqrt{\frac{(0.01)(0.99)}{1000}}} = 2.54.$$

With $\alpha = 0.05$ we should reject H_0 if $z > z(0.05) = t(0.05; \infty) = 1.645$. Since $2.54 > 1.645$, we reject H_0 and conclude that the defective rate is above 1%.

Example 5.5.5 (*Defective Rates, p. 205*)   Suppose one wanted to conduct the test in the previous example by computing a p-value. Since we have a right-tailed test (see Table (5.5)),

$$p\text{-value} = 1 - \Phi(2.54) = 1 - 0.9945 = 0.0055$$

which is less than 0.05, so (see condition (5.4.2)) H_0 should be rejected.

Example 5.5.6   Let p be the proportion of bottles that overflow in a filling process. In a sample of 72 bottles, eight overflowed (which means they have to be pulled from the line). The filling line typically runs with 5% or fewer overflows. We wish to test

$$H_0: \; p = 0.05 \qquad \text{versus} \qquad H_a: \; p > 0.05$$

using $\alpha = 0.10$ and by computing a p-value. Since $\widehat{p} = 8/72 = 0.111$ and

$$z = \frac{0.111 - 0.05}{\sqrt{\frac{0.05(0.95)}{72}}} = 2.37$$

one obtains

$$p\text{-value} = 1 - \Phi(2.37) = 1 - 0.9911 = 0.0089.$$

Since $0.0089 < \alpha$, we reject H_0. The proportion of over-full bottles appears higher than desired, and the process engineer should investigate for possible causes.

Example 5.5.7   Let p be the proportion of students who lose their computer passwords every semester. From a sample of 72 students 15 had lost their passwords. We wish to test

$$H_0: \; p = 0.25 \quad \text{versus} \quad H_a: \; p \neq 0.25$$

using $\alpha = 0.10$ and by computing a p-value. Since $\widehat{p} = 15/72 = 0.208$ and

$$z = \frac{0.208 - 0.25}{\sqrt{\frac{0.25(0.75)}{72}}} = -0.82$$

one obtains

$$p\text{-value} = 2(1 - \Phi(0.82)) = 2(0.2061) = 0.4122.$$

Since $0.4122 \geq \alpha$, we fail to reject H_0. There is not enough evidence to say that the proportion of students who lose their passwords every semester is different from 0.25.

Sample Sizes

The sample size derivation for normal populations when σ is known can be easily modified for testing hypotheses on proportions. For a one-sided test with specified power of γ at $p = p_1 > p_0$, one has

$$\gamma = P\left[\text{rejecting } H_0 \mid p = p_1\right]$$

$$= P\left[\frac{\hat{p} - p_0}{\sqrt{\frac{p_0(1-p_0)}{n}}} > z(\alpha) \mid p = p_1\right]$$

$$= P\left[\frac{\hat{p} - p_0}{\sqrt{\frac{p_1(1-p_1)}{n}}} > z(\alpha)\sqrt{\frac{p_0(1-p_0)}{p_1(1-p_1)}} \mid p = p_1\right]$$

$$= P\left[\frac{\hat{p} - p_0}{\sqrt{\frac{p_1(1-p_1)}{n}}} - \frac{p_1 - p_0}{\sqrt{\frac{p_1(1-p_1)}{n}}} > z(\alpha)\sqrt{\frac{p_0(1-p_0)}{p_1(1-p_1)}} - \frac{p_1 - p_0}{\sqrt{\frac{p_1(1-p_1)}{n}}} \mid p = p_1\right]$$

$$= P\left[\frac{\hat{p} - p_1}{\sqrt{\frac{p_1(1-p_1)}{n}}} > z(\alpha)\sqrt{\frac{p_0(1-p_0)}{p_1(1-p_1)}} - \frac{p_1 - p_0}{\sqrt{\frac{p_1(1-p_1)}{n}}} \mid p = p_1\right]$$

$$= P\left[Z > z(\alpha)\sqrt{\frac{p_0(1-p_0)}{p_1(1-p_1)}} - \frac{p_1 - p_0}{\sqrt{\frac{p_1(1-p_1)}{n}}}\right] \qquad (5.5.6)$$

where $Z \sim N(0,1)$. Therefore,

$$z(\gamma) = z(\alpha)\sqrt{\frac{p_0(1-p_0)}{p_1(1-p_1)}} - \frac{p_1 - p_0}{\sqrt{\frac{p_1(1-p_1)}{n}}}$$

and solving for n results in

$$n = \left[\frac{z(\alpha)\sqrt{p_0(1-p_0)} - z(\gamma)\sqrt{p_1(1-p_1)}}{p_1 - p_0}\right]^2. \qquad (5.5.7)$$

From the result in the σ known normal case and equation (5.5.7), it follow immediately that for a two-sided test on a proportion the approximate sample size needed for a power of γ at p_1 is

$$n = \left[\frac{z(\alpha/2)\sqrt{p_0(1-p_0)} - z(\gamma)\sqrt{p_1(1-p_1)}}{p_1 - p_0}\right]^2. \qquad (5.5.8)$$

Example 5.5.8   A bottle-filling process currently has an overflow rate of 5%. It is hoped that an adjustment to the process will decrease the overflow rate. How many bottles would have to be sampled from the adjusted process to be able to detect a decrease in the overflow rate to 3% with probability 0.9 when testing at an α level of 0.05? Using equation (5.5.7) with $\alpha = 0.05$, $\gamma = 0.9$, $p_0 = 0.05$, and $p_1 = 0.03$, one obtains

$$n = \left[\frac{1.645\sqrt{0.05(1-0.05)} + 1.28\sqrt{0.03(1-0.03)}}{0.05 - 0.03} \right]^2 = 831.9$$

and a sample of 832 bottles would be needed.

Problems

1. A production engineer monitoring a process that makes magic markers samples 500 markers at random and tests to see how many are defective. He finds 50 defective markers.

   (a) Estimate the proportion of defectives in the population.

   (b) Construct a 95% confidence interval for the proportion of defectives.

   (c) Construct a 99% confidence interval for the proportion of defectives.

   (d) How many markers must be tested to form a 99% confidence interval with a margin of error of 1% or less?

   (e) How many markers must be tested to form a 90% confidence interval with a margin of error of 1% or less?

   (f) The engineer assumes the population proportion of defectives is 10%. Test the hypothesis that the proportion of defectives has risen above 10%, using $\alpha = 0.05$. Explain what your results mean.

   (g) Calculate the p-value for the hypothesis test you just performed, and explain what it indicates about your test. How confident do you feel in your results?

   (h) The engineer would like the hypothesis test above to have 95% power for detecting a difference of 1% in the defective rate. How large a sample is needed to achieve this?

2. A consumer products company surveys 30 shoppers at a mall and asks them if they like the packaging of a new soap product. The results of the survey are:

$$\text{Yes: 10; No: 19; No opinion: 1.}$$

   (a) You have already used these data to estimate the proportion of shoppers who would like the package for the soap (see Problem 4.2.9, p. 104). Construct a 95% confidence interval for your estimate.

   (b) Prior to conducting this survey it was believed that only 25% of shoppers would like this package. Does the poll described above provide evidence that the percentage is different? Test, using $\alpha = 0.10$, and explain what your result means. Also report the p-value and explain what it means.

3. Suppose you want to estimate the proportion of customers who would buy a soap product based on its package. You devise a small survey, to be administered at a mall by volunteers. How many customers should you survey to get an estimate of this proportion with 95% confidence and margin of error of 2% or less?

4. An e-commerce business wishes to estimate the proportion of its web site users who entered false phone numbers when registering. A sample of 250 users is

chosen at random. Of these, 37 are discarded because of database errors. Of the remaining users, 19 had registered with false phone numbers.

(a) Find a 99% confidence interval for the proportion of registered users with false phone numbers and explain what it indicates. Use language that a manager could easily understand.

(b) The manager wants the margin of error to be 2% or less. How large a sample should be taken?

(c) The manager decided that a margin of error of 3% will be sufficient. How large a sample must be taken now?

(d) The business is interested in trying to redesign its web site to cut down on the number of registrations with false phone numbers. How large a sample of registrations would be needed if they wanted to test their success using $\alpha = 0.1$ and wanted to be able to detect a decrease from 10 to 5% with probability 0.8?

5. One goal of diagnostic medicine is to develop non-invasive tests to replace, or at least precede, invasive tests. For example, one might want to develop a blood test for liver disease, which alternatively can be tested with a liver biopsy. Two critical measures of goodness for a diagnostic test are the **sensitivity** of the test, that is, the proportion of truly positive subjects that test positive; and the **specificity** of the test, the proportion of truly negative subjects that test negative.

Suppose researchers have just discovered a new version of a diagnostic blood test. Data from 100 subjects were collected. Sixty of the subjects were known healthy patients and 40 had liver disease. The test results are given below.

		Test Result		
		Pos	Neg	
Known Result	Pos	38	2	40
	Neg	8	52	60
		46	54	100

(a) Estimate the sensitivity and specificity of the test using 95% confidence intervals.

(b) How large a sample of healthy subjects is needed to estimate specificity within 0.05 units with confidence of 95%?

(c) What is the power of this test to detect an increase of 0.03?

(d) A previous version of this diagnostic test had sensitivity 0.90. Using $\alpha = 0.05$, test to see if the sensitivity of the new improved test is greater.

6. A manufacturer of an electrical component claims to have made changes in their manufacturing process to improve the quality of the component. Previously the component had a nonconforming rate of 10%. The supplier would like to provide you with 8 units for destructive testing to demonstrate that the nonconforming rate is now below 10%.

   (a) If the nonconforming rate has dropped to say 8%, what is the probability that you would see no non-conforming items in a sample of size 8?

   (b) If the non-conforming rate has increased to say 12% what is the probability that you would see no non-conforming items in a sample of size 8?

   (c) How large of a sample would you need to estimate the non-conforming rate to within 5% with 95% confidence?

5.6 CHAPTER PROBLEMS

1. Consider the set of data $\{18, 7, 10, 21, 14, 19\}$.

   (a) Construct a normal probability plot for this data set. What do you conclude?

   (b) Construct a 95% confidence interval for the unknown mean of the population from which the above sample was obtained.

   (c) Construct an interval that with 95% confidence will contain 99% of the population values.

2. The October 19, 1990 edition of the Clemson University *Tiger* contained the information that in the last 86 World Series 46 times the series has been won by the team that began the series at home. Let p denote the probability of the World Series being won by the team that begins the series at home.

   (a) Compute a point estimate of p.

   (b) Construct a 90% confidence interval for p.

   (c) Use the interval in part (b) to test whether starting the series at home has any effect on winning the series. State your null and alternative hypotheses.

   (d) If one was interested in constructing a 75% confidence interval for p with interval length of no more than 0.4, how large a sample would be required? Comment on your result.

3. A pharmaceutical firm produces capsules that are supposed to contain 50 mg of a certain drug. The precise weights vary from capsule to capsule but can be assumed to have a normal distribution.

   (a) Suppose that one is interested in constructing a 95% confidence interval that would estimate the mean weight of capsules to within 2 mg, and you suspect that σ is no larger than 3 mg. How large a sample size is necessary?

   (b) A random sample of 16 capsules results in the statistics $\bar{x} = 47$ mg and $s = 4$ mg. Construct a 90% confidence interval for the mean weight of the capsules produced by this process.

   (c) How large should a second sample be in order to estimate the mean weight of capsules to within 1.5 mg with 95% confidence?

   (d) If the second sample resulted in $\bar{x} = 44$ and $s = 4.5$ mg, what would you obtain for the 95% confidence interval?

4. In order to determine the average amount of unburned hydrocarbons produced by a particular brand of carburetor (adjusted in what is believed to be the optimal way), a random sample of 35 carburetors was taken, and the amount of unburned hydrocarbons recorded. The three largest and three smallest values were 1.7, 2, 4.1 and 7.9, 8.2, 9.3. The average amount was

computed to be 5.7 with a sample standard deviation of 0.9. The manufacturer suspects that the distribution of unburned hydrocarbons is not normal.

(a) Construct an 88% confidence interval for the average amount of unburned hydrocarbons for the entire population of this brand of carburetor.

(b) Construct an 88% prediction interval for the amount of unburned hydrocarbons produced by the next carburetor tested.

(c) Construct an interval that with 90% confidence will cover 90% of the population.

5. Given below are data on the diameters of steel ball bearings from a particular production process.

$$1.72 \quad 1.62 \quad 1.69 \quad 0.79 \quad 1.79 \quad 0.77$$

(a) Construct a normal probability plot for these values. What do you conclude?

(b) Construct a 70% prediction interval for the diameter of a new randomly chosen ball bearing.

6. A particular population is known to have an exponential distribution with mean of 5.

(a) If 40 items are selected at random from this population, what is the probability that their average will exceed 5.5?

(b) If one item is selected at random from this population, what is the probability that its value will exceed 5.5?

7. A new process is being proposed for the manufacture of automobile tires. The manufacturing company is interested in constructing a 90% confidence interval for the rate of tread wear for tires manufactured using the new process.

(a) An initial sample of size 25 results in $\bar{x} = 8$ mm/10,000 miles and $s = 2.5$ mm/10,000 miles. Construct the desired interval.

(b) If the company is interested in obtaining a 90% confidence interval with margin of error 0.5 mm/10,000 miles, how many additional observations would be needed?

8. Below are values for the resistances of 20 randomly selected resistors from a particular production process.

$$4.239 \quad 4.398 \quad 4.419 \quad 4.444 \quad 4.448 \quad 4.561 \quad 4.744 \quad 4.803 \quad 4.912 \quad 4.916$$
$$4.967 \quad 5.037 \quad 5.110 \quad 5.164 \quad 5.285 \quad 5.309 \quad 5.309 \quad 5.548 \quad 6.043 \quad 7.160$$

(a) Construct a normal probability plot for these data.

(b) Construct an 80% prediction interval for the resistance of a new randomly selected resistor from this process.

   (c) Construct an interval that will cover 80% of the population with 90% confidence.

9. A new model of Sun workstation is being considered for purchase. The company is willing to consider purchasing the new model only if it results in faster computing times. The average time for a particular kind of program on the current computer is 35 seconds.

   (a) A random sample of 38 runs with the new model results in an average time of 33 seconds with a standard deviation of 5 seconds. Perform the desired hypothesis test by computing a p-value.

   (b) How large a sample size would be needed to conduct the test in part (a) at level $\alpha = 0.05$ and have a power of 0.8 for detecting a difference from the null value (35 seconds) of 0.4σ?

10. Suppose a certain oil producing country exports $(1 + X)$ million barrels of oil per day, where X is a random variable having a normal distribution with mean zero and variance $1/25$.

   (a) Find the exact probability that in 16 days more than 17 million barrels of oil will be exported.

   (b) If the exact distribution of X is unknown, but one knows that $E[X] = 0$ and $\text{Var}[X] = 1/6$, then find an approximation for the probability that in 36 days less than 35 million barrels of oil will be exported.

   (c) Suppose that not only is the exact distribution of X unknown, but also $E[X]$ and $\text{Var}[X]$ are unknown. If a random sample of 36 days' productions results in $\bar{x} = 1$ million barrels of oil per day and $s = 1/10$ million barrels of oil per day, construct a 95% confidence interval for $E[X]$.

11. A textile mill weaves cloth that is used to make sheets. The top grade of sheets has at least 200 threads per inch. (Because higher thread count indicates higher quality, it will not matter if the thread count is above 200.) Random samples from 35 areas of a piece of cloth result in an average thread count of $\bar{y} = 190$ with $s = 50$.

   (a) Test the appropriate hypothesis by comparing with a critical value using $\alpha = 0.05$, and explain what your results mean.

   (b) Compute the p-value for the test in part (a) and use it to perform the hypothesis test.

12. Atlas Fishing line produces 10-lb test line. Twelve randomly selected spools are subjected to tensile-strength tests, and the results are given below.

9.8 10.2 9.8 9.4 9.7 9.7 10.1 10.1 9.8 9.6 9.1 9.7

(a) Is it misleading to advertise this as "10-lb test line"? Explain what your results mean.

(b) Construct a 95% prediction interval for the tensile strength of the next randomly selected spool.

(c) Construct an interval that with 90% confidence will include the tensile strength of 95% of the spools.

(d) How large a second sample would be needed in order to construct a 99% confidence interval with length no more than 0.5?

13. A particular production process produces bolts whose diameter is a $N(3, 16)$ random variable.

(a) Find the probability that a randomly selected bolt will have a diameter less than 3.2.

(b) Find the probability that the average diameter from a sample of 35 bolts will be less than 3.2.

14. Customers arrive at a counter in accordance with a Poisson process of density three arrivals per minute.

(a) What is the expected number of arrivals per minute?

(b) What is the standard deviation of the number of arrivals per minute?

(c) Use the CLT to approximate the probability that the average number of arrivals in 75 1-minute periods is between 2.6 and 3.4.

15. You are interested in testing whether a particular brand of floppy disks has more than 5% defectives. A random sample of $n = 325$ disks contains 18 defectives.

(a) State the null and the alternative hypotheses.

(b) Conduct the test by computing a p-value.

16. The time between telephone calls at a company switchboard has an exponential distribution with mean of 3 minutes.

(a) Find the probability that the average of 45 of these times is less than 2.5 minutes.

(b) Find the 0.15 quantile for the average of 45 of these times.

17. Suppose one performs the following acceptance sampling scheme. From a lot of N light bulbs (N is a large number), 100 bulbs are sampled at random without replacement and tested. The probability that a bulb is defective is 0.1, and the lot is accepted if no more than 15 defective bulbs are found in the sample. Find an approximation (using the continuity correction) for the probability that the lot is accepted.

18. Customers arrive at a queue according to a Poisson process with an average arrival rate of 4 per hour.

(a) Find the probability of exactly three arrivals in 1 hour.

(b) Find the probability of at least three arrivals in 2 hours.

(c) Use the normal distribution to approximate the probability in part (a).

19. A random sample of size 100 is taken from a population with density function $f(x) = \frac{3}{4}(x^2 + 1)$ for $0 < x < 1$. Approximate the probability that the sum of the values is greater than 50.

6

COMPARING TWO POPULATIONS

In the last chapter, we looked at inference (i.e., confidence intervals, prediction intervals, and hypothesis tests) based on a sample coming from a single population. Frequently, scientific experiments are performed to compare different **treatments**. These treatments might be different techniques, different processes, different suppliers, and so forth. In that case the data will consist of samples from different populations, and the experimenter will be interested in comparing the responses from one treatment to the responses from another. In this chapter, we will consider comparing responses from two experimental groups. Chapter 7 expands these ideas to three or more experimental groups.

The methods for comparing two groups depend on the type of populations (continuous or discrete) as well as how the experiment was performed. With continuous data we will want to make inferences about $\mu_1 - \mu_2$, whereas with binomial data we will want to make inferences about $p_1 - p_2$.

6.1 PAIRED SAMPLES

A paired-samples design requires a careful experimental setup, but when appropriate, it provides an efficient method for comparing two population means. The experimenter starts with pairs of experimental units. In each pair one unit is randomly assigned to one treatment and the other unit is assigned to the other treatment.

Notation: The paired responses are denoted y_{1j} and y_{2j} where $j = 1, \ldots, n$; and n is the number of pairs.

Then the difference $d_j = y_{1j} - y_{2j}$ is calculated for each pair. If most of the differences are positive, we would conclude that the first treatment has a higher mean response. Conversely, if most of the differences are negative, we would conclude that the second treatment has a higher mean response. If the differences are randomly distributed around 0, we would conclude that the two treatments have similar mean responses. We have, in essence, reduced the problem to the type of one-sample problem discussed in the last chapter.

Not all paired-sample experiments are performed on pairs of identical experimental units, as in the description above. A paired-sample experiment can also be performed by applying both treatments to the *same* experimental unit, provided there is no "carry-over" effect from one treatment to the other. This is often done in clinical trials where one is interested in comparing two drugs. Both drugs are given to each participant (in a random order), but with a "wash out" period between the two so that the first drug does not influence the effect of the second drug.

Example 6.1.1 (*UV Coating Data*)   Scientists developing a new UV coating for eyeglasses felt they were ready to field test the coating (as opposed to running simulated tests in the laboratory). Ten volunteers were to be used in the study in which the scientists wanted to be able to compare the durability of the experimental coating to a current commercial coating. The durability of the coatings would be evaluated by taking an initial haze measurement on each lens and then repeating that measurement after the glasses had been used by the volunteers for 3 months. The change in the haze value (3 month − initial) would be the reported durability number. Smaller values would indicate better durability.

The scientists were concerned that the way the volunteers used their glasses could influence the outcome of the study. For instance, one volunteer was an avid mountain biker, and they expected her lenses to see more extreme conditions than those of the other volunteers. Many factors, such as the care of the lenses, the amount of use, and type of use could all influence the durability measure of the coating. To account for these differences a paired-differences study was designed. Each volunteer in the study would have one lens with the experimental coating and one lens with the commercial coating. Then the differences in the durability of each pair of lenses would be used to ultimately judge the durability of the new coating. The assignment of the coatings to the lenses (i.e., right or left) was done at random.

For the jth pair of lenses let

$$y_{1j} = \text{durability for the commercial coating}$$
$$y_{2j} = \text{durability for the experimental coating.}$$

We can compare the coatings by calculating $d_j = y_{1j} - y_{2j}$ for each pair. The data are given in Table 6.1 and in uvcoatin.txt. To test whether or not the experimental coating is more durable than the commercial coating, we would test

$$H_0: \ \mu_1 = \mu_2 \quad \text{versus} \quad H_a: \ \mu_1 > \mu_2$$

where μ_1 is the mean change in haze over the three month trial period for the commercial product (μ_2 for the experimental) and smaller values are desired since

Table 6.1. Comparisons of Two Coatings

Pair	y_1	y_2	d
1	8.9	8.5	0.4
2	9.4	9.3	0.1
3	11.2	10.8	0.4
4	11.4	11.6	−0.2
5	13.0	12.9	0.1
6	6.4	6.5	−0.1
7	13.4	13.1	0.3
8	5.6	5.1	0.5
9	4.8	4.3	0.5
10	15.8	15.6	0.2

they indicate less change in haze (i.e., greater durability). In terms of the differences d_j the hypothesis test would be written as

$$H_0: \; \mu_d = 0 \quad \text{versus} \quad H_a: \; \mu_d > 0.$$

We can perform this test by computing the t statistic (equation (5.4.1)) and comparing it with $t(\alpha; 9)$. If we use $\alpha = 0.1$, then $t(0.1; 9) = 1.383$. Since $\bar{d} = 0.22$ and $s_d = 0.244$, it follows that

$$t = \frac{0.22 - 0}{0.244/\sqrt{10}} = 2.85$$

and since $2.85 > 1.383$, we would reject H_0. The experimental coating appears more durable than the commercial product.

Example 6.1.2 (*Continuation of Example 6.1.1*) The scientists also wanted to know *how much* more durable the experimental coating was than the commercial coating. (This information can be used to help determine if this product is ready to be commercialized.) A 95% confidence interval for μ_d is

$$\bar{d} \pm t(0.025; 9) s_d/\sqrt{n}$$
$$0.22 \pm 2.262(0.244)/\sqrt{10}$$
$$\pm 0.17$$
$$(0.05, 0.39).$$

Note: The above analyses depend on the d_j's being at least approximately normal since we are dealing with a sample size of only 10. This can be checked using a normal probability plot of the d_j's, which is given in Figure 6.1. Considering the small sample size, the plot looks extremely linear.

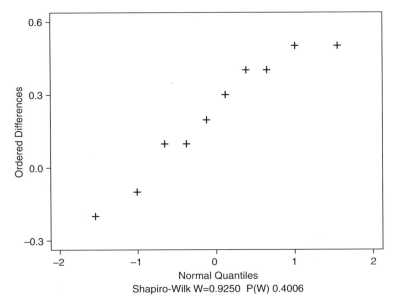

Figure 6.1. A Normal Probability Plot of the Differences in Durability of Two UV Coatings.

Problems

1. Use the data below to do the following.

Pair	Before (I)	After (II)
1	12	15
2	3	2
3	9	14
4	2	6
5	7	10
6	8	5

  (a) Calculate the pairwise differences.

  (b) Make a normal probability plot of the pairwise differences.

  (c) Calculate the mean and standard deviation of the differences.

  (d) Test the hypothesis that $\mu_2 > \mu_1$ using $\alpha = 0.10$. Explain what your results mean.

  (e) Construct a 90% confidence interval for $\mu_2 - \mu_1$. Explain what it indicates.

2. The data set `calcium.txt` contains the calcium levels of eleven test subjects at zero hours and three hours after taking a multi-vitamin containing calcium.

  (a) Make the appropriate normal probability plot and comment on it.

  (b) Is there any evidence that calcium levels are affected by this vitamin? Test, using $\alpha = 0.10$.

  (c) Construct a 90% confidence interval for the mean difference in calcium levels.

  (d) Write a short report on the experiment and its results.

3. The data set `yellow.txt` contains yellowing data (b is a color measure) for 23 samples of monomer used in the casting of optical lenses. The data include an initial measurement and a measurement after 1 month of storage. It is hoped that the material is stable and that the yellowing value does not change over time.

  (a) Calculate the pairwise differences (initial − 1 month) and find the mean and standard deviation of the differences.

  (b) Make a normal probability plot of the differences and explain what it indicates.

  (c) Find a 95% confidence interval for the average difference in yellowing values. Explain what the confidence interval means.

4. The data set `wash.txt` contains data from a study of a test procedure used for automotive paint. The test consists of applying drops of acid onto a painted steel panel at 1 minute intervals for 30 minutes. The panel is then rinsed and later rated by a person on how well the paint withstood the acid. It was thought that over time the ratings of the panels changed due to further degradation in the paint from residual acid on the panel. It was also thought that rinsing the panel with a neutralizing wash after the test might solve this degradation problem. A test of 36 panels (18 with the wash, 18 without) was conducted. The panels were rated initially and then again after being stored for 2 weeks.

   (a) Calculate the pairwise differences and make a normal probability plot of them for each test procedure (wash/no wash).

   (b) For each of the two test methods test to see if there is a change in ratings over time. Test, using $\alpha = 0.05$, and explain what your results mean. Do these results support the thoughts of the lab personnel?

   (c) Write a report explaining the experiment and its results in your own words.

5. A diagnostic kit (for measuring a particular protein level in blood) can be used to test 16 blood samples. To check for kit-to-kit consistency the manufacturer tested the blood from 16 patients with two different kits. The data are recorded in `diagnostic.txt`. What is your assessment of kit-to-kit consistency.

6. The diameters of 14 rods are measured twice, once with each of two different calipers. Is there a difference between the two types of caliper? Write a brief report addressing this question based on the data in `caliper.txt`.

6.2 INDEPENDENT SAMPLES

There are many situations in which the pairing of observations is either not reasonable or simply not possible. Thus, we are often interested in comparing two populations by considering *independent* (i.e., unrelated) random samples from the two.

Such an experiment is performed by starting with a single pool of experimental units and randomly assigning units to one treatment group or the other. This random assignment is important because it ensures that any initial differences among the experimental units will be randomly scattered over both treatment groups, thus avoiding bias.

As a general rule, we would plan to have equal-size samples in the two treatment groups. However, this is not strictly necessary, and even when we plan for equal sample sizes, an experimental unit may be lost or destroyed, upsetting this balance. In the formulas below, we let n_1 denote the sample size from the first population or treatment group, and n_2 the sample size from the second.

The purpose of the experiment is to make inferences about $\mu_1 - \mu_2$, which would be estimated by $\overline{y}_{1\bullet} - \overline{y}_{2\bullet}$. Thus, in order to make inferences it is necessary to know the distribution of $\overline{Y}_{1\bullet} - \overline{Y}_{2\bullet}$. We know that if either (a) the data are approximately normal or (b) n_1 and n_2 are large, then (at least approximately)

$$\overline{Y}_{1\bullet} \sim N(\mu_1, \sigma_1^2/n_1)$$

and

$$\overline{Y}_{2\bullet} \sim N(\mu_2, \sigma_2^2/n_2).$$

In order to combine these distributions to obtain the distribution of $\overline{Y}_{1\bullet} - \overline{Y}_{2\bullet}$, we will use the fact that the sum of two independent normal random variables is also normal, and the properties of expected values and variances. Since

$$E(\overline{Y}_{1\bullet} - \overline{Y}_{2\bullet}) = E(\overline{Y}_{1\bullet}) + E(-\overline{Y}_{2\bullet}) = \mu_1 - \mu_2$$

and

$$\begin{aligned}
\mathrm{Var}(\overline{Y}_{1\bullet} - \overline{Y}_{2\bullet}) &= \mathrm{Var}(\overline{Y}_{1\bullet}) + \mathrm{Var}(-\overline{Y}_{2\bullet}) \\
&= \mathrm{Var}(\overline{Y}_{1\bullet}) + \mathrm{Var}(\overline{Y}_{2\bullet}) \\
&= \frac{\sigma_1^2}{n_1} + \frac{\sigma_2^2}{n_2}
\end{aligned} \tag{6.2.1}$$

it follows that

$$\overline{Y}_{1\bullet} - \overline{Y}_{2\bullet} \sim N\left(\mu_1 - \mu_2, \frac{\sigma_1^2}{n_1} + \frac{\sigma_2^2}{n_2}\right). \tag{6.2.2}$$

In order to do any inference on $\mu_1 - \mu_2$, we need to estimate the variance (6.2.1). We will consider two cases: (i) the variances σ_1^2 and σ_2^2 are equal, and (ii) they are not equal.

Equal Variances

Often it is the case that a treatment affects only the mean of the population and does not affect the variance. In such cases, random assignment allows us to assume the two populations have the same variance. We will use σ^2 to denote the common variance of the two populations.

If the sample sizes are equal, the best estimate of σ^2 is simply the average of the sample variances (see equation (3.1.4)). If the sample sizes are unequal, then the sample variance from the larger sample is actually a better estimate than the sample variance from the smaller sample, and the best estimate of σ^2 is a *weighted average* of the two, namely,

$$s_p^2 = \frac{(n_1 - 1)s_1^2 + (n_2 - 1)s_2^2}{n_1 + n_2 - 2}. \tag{6.2.3}$$

The weights used are the df (see p. 167). This is called a **pooled sample variance** because it pools together information from both samples. The df for s_p^2 are the sum of the df's for s_1^2 and s_2^2, namely,

$$n_1 - 1 + n_2 - 1 = n_1 + n_2 - 2.$$

In this case it follows from formula (6.2.2) that

$$\frac{\overline{Y}_{1\bullet} - \overline{Y}_{2\bullet} - (\mu_1 - \mu_2)}{s_p\sqrt{\frac{1}{n_1} + \frac{1}{n_2}}} \sim t(n_1 + n_2 - 2). \tag{6.2.4}$$

Confidence Intervals

Using formula (6.2.4) and an argument similar to that for the one-sample case given in Section 5.2, a $(1 - \alpha)100\%$ confidence interval for $\mu_1 - \mu_2$ is

$$\overline{y}_{1\bullet} - \overline{y}_{2\bullet} \pm t(\alpha/2; n_1 + n_2 - 2)s_p\sqrt{\frac{1}{n_1} + \frac{1}{n_2}}. \tag{6.2.5}$$

Example 6.2.1   A company suspects that the configuration of the rail cars used to ship a product may have an effect on the moisture level of the product when it reaches its destination. Data from a customer on the product's moisture level when it arrived was collected for two types of rail cars and is stored in `railcar3.txt`. Side-by-side box plots (Figure 6.2) suggest a difference between the two types of rail cars.

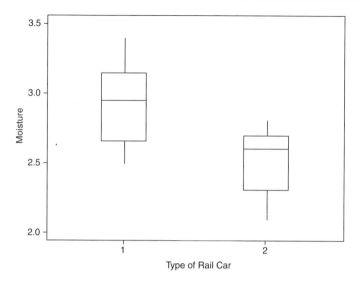

Figure 6.2. Moisture Levels for Product Stored in Two Types of Rail Cars.

Some summary statistics are given in Table 6.2.

Table 6.2. Summary Statistics for the Moisture Levels of a Product Stored in Two Types of Rail Cars

Type 1	Type 2
$n_1 = 8$	$n_2 = 9$
$\bar{y}_{1\bullet} = 2.93$	$\bar{y}_{2\bullet} = 2.52$
$s_1^2 = 0.0964$	$s_2^2 = 0.0694$

From the values in Table 6.2 one can compute (equation (6.2.3))

$$s_p^2 = \frac{7(0.0964) + 8(0.0694)}{15} = 0.082.$$

To find out if the moisture values differ due to rail car type, we can form a confidence interval for $\mu_1 - \mu_2$. Using 95% confidence and formula (6.2.5), one obtains

$$\bar{y}_{1\bullet} - \bar{y}_{2\bullet} \pm t(0.025; 15)\sqrt{0.082}\sqrt{\frac{1}{8} + \frac{1}{9}}$$

$$0.403 \pm 2.131\sqrt{0.082}\sqrt{\frac{1}{8} + \frac{1}{9}}$$

$$\pm 0.296$$

$$(0.107, 0.699).$$

This means that on average the moisture level of the product shipped in the type 1 rail cars is at least 0.107 units higher than for the type 2 rail cars, and perhaps as much as 0.699 units higher.

Hypothesis Tests

The hypothesis

$$H_0: \ \mu_1 = \mu_2 \tag{6.2.6}$$

is equivalent to $H_0: \ \mu_1 - \mu_2 = 0$, and using formula (6.2.4), one finds that the appropriate test statistic for hypothesis (6.2.6) is

$$t = \frac{\bar{y}_{1\bullet} - \bar{y}_{2\bullet}}{s_p \sqrt{\frac{1}{n_1} + \frac{1}{n_2}}} . \tag{6.2.7}$$

The computed value of t (equation (6.2.7)) would be compared to a critical value from a $t(n_1 + n_2 - 2)$ distribution as shown in Table 6.3.

Table 6.3. Critical Values for Comparing the Computed Value of t from Two Independent Samples

Alternate Hypothesis	Rejection Region
$H_a: \ \mu_1 < \mu_2$	Reject H_0 if $t < -t(\alpha; n_1 + n_2 - 2)$
$H_a: \ \mu_1 > \mu_2$	Reject H_0 if $t > t(\alpha; n_1 + n_2 - 2)$
$H_a: \ \mu_1 \neq \mu_2$	Reject H_0 if $|t| > t(\alpha/2; n_1 + n_2 - 2)$

Example 6.2.2   An equipment upgrade was made to only one of two conveyor belts leading to a packing room. The plant manager wanted to know if the upgrade had decreased down time on that conveyor. Over a one week period the plant personnel were supposed to study the conveyors during 10 randomly chosen one-hour time periods and keep track of the number of minutes in the hour that they were stopped. Summary statistics are given in Table 6.4. Note that one of the sampling times was missed for the "new" conveyor belt. In order to determine if

Table 6.4. Summary Statistics for Conveyor Belt's Down Times

Old	New
$n_1 = 10$	$n_2 = 9$
$\bar{y}_{1\bullet} = 8.8$	$\bar{y}_{2\bullet} = 3.9$
$s_1^2 = 2.25$	$s_2^2 = 3.24$

the old conveyor belt has more down time one would test (denoting old as 1 and new as 2)

$$H_0: \ \mu_1 = \mu_2 \quad \text{versus} \quad H_a: \ \mu_1 > \mu_2.$$

We will use $\alpha = 0.05$. That is, we will reject H_0 if $t > t(0.05; 17) = 1.740$. An estimate of the variance of down time per hour is

$$s_p^2 = \frac{9(2.25) + 8(3.24)}{17} = 2.72$$

and our test statistic (equation (6.2.7)) is

$$t = \frac{\bar{y}_{1\bullet} - \bar{y}_{2\bullet}}{s_p \sqrt{\frac{1}{n_1} + \frac{1}{n_2}}} = \frac{8.8 - 3.9}{\sqrt{2.72}\sqrt{\frac{1}{10} + \frac{1}{9}}} = 6.47.$$

Since $6.47 > 1.740$, we would reject H_0. The data demonstrate that the old conveyor has more down time than the new one. The plant manager now has the data needed to convince management to upgrade the second conveyor belt.

Note: When the sample sizes are small, it is necessary that the data come from populations that are normally distributed. You can check this by making a normal probability plot of the residuals.

Sample Sizes

Table B4 gives the necessary sample sizes for two-sample t tests. It is used in exactly the same way as Table B3 for one-sample t tests, which was described in Chapter 5. Table B4 is based on the assumption that the two populations being compared have equal variances.

Example 6.2.3 Suppose that in Example 6.2.1 the company is interested in detecting a difference in the two means of 0.4σ with probability 0.95 when testing using $\alpha = 0.01$. How large a sample size is needed?

From Table B4 with $\alpha = 0.01, d = 0.4$, and power $= 0.95$; one finds $n = 227$. Thus, one would need 227 observations with each type of rail car.

Unequal Variances

When the variances of the two populations are either clearly not equal or if one is not willing to assume that they are equal, the variance (6.2.1) must be estimated in a different way. Rather than combining the sample variances to estimate a common variance, one uses s_1^2 to estimate σ_1^2 and s_2^2 to estimate σ_2^2. This results in the variance (6.2.1) being estimated as

$$\frac{s_1^2}{n_1} + \frac{s_2^2}{n_2}$$

and the statistic for testing hypothesis (6.2.6) becomes

$$\boxed{\frac{\overline{y}_{1\bullet} - \overline{y}_{2\bullet}}{\sqrt{\frac{s_1^2}{n_1} + \frac{s_2^2}{n_2}}}} \cdot \qquad (6.2.8)$$

Estimating the variance in this fashion creates an additional problem. Namely,

$$\frac{\overline{Y}_{1\bullet} - \overline{Y}_{2\bullet}}{\sqrt{\frac{s_1^2}{n_1} + \frac{s_2^2}{n_2}}}$$

does not have an exact t distribution. One can approximate its distribution with a t distribution, but the df have to be adjusted. There are two choices for adjusting the df:

1. Use a conservative df

$$\mathrm{df} = \min\{n_1 - 1, n_2 - 1\}. \qquad (6.2.9)$$

   This produces a conservative test (or confidence interval) in the sense that the α level is guaranteed to be no larger than the α value used.

2. Use an approximate df

$$\mathrm{df} = \frac{\left(\frac{s_1^2}{n_1} + \frac{s_2^2}{n_2}\right)^2}{\frac{1}{n_1-1}\left(\frac{s_1^2}{n_1}\right)^2 + \frac{1}{n_2-1}\left(\frac{s_2^2}{n_2}\right)^2}. \qquad (6.2.10)$$

   This approximation is due to Satterthwaite (1946) and provides approximate tests and confidence intervals.

The df computed using equation (6.2.10) are always between $\min\{n_1 - 1, n_2 - 1\}$ and $n_1 + n_2 - 2$. Computer programs generally provide tests and confidence intervals using the approximate df (6.2.10) for the unequal variances situation. However, for computation by hand using the conservative df (6.2.9) is much easier and has the advantage of a guaranteed α level.

Example 6.2.4 Two different chemical processes are being considered for producing a resin needed in the manufacture of paint. Four batches are run with the first process and five batches are run with the second process. Each batch is then assayed to determine the yield. The results are given Table 6.5 (and in `assay.txt`), together with some summary statistics.

Table 6.5. Yields for Batches of Paint from Two Different Processes

	Process I	Process II
	36.5	94.9
	39.1	74.8
	44.0	68.3
	38.2	75.0
		85.7
$\overline{y}_{i\bullet}$	39.45	79.74
s_i^2	10.36	110.8

One is interested in testing to see if the two processes are different with respect to their average yields, but there is concern (particularly if the yields are different) that the variances may not be the same. Thus, using the statistic (6.2.8), one would compute

$$t = \frac{\overline{y}_{1\bullet} - \overline{y}_{2\bullet}}{\sqrt{\frac{s_1^2}{n_1} + \frac{s_2^2}{n_2}}}$$

$$= \frac{39.45 - 79.74}{\sqrt{\frac{10.36}{4} + \frac{110.8}{5}}}$$

$$= -8.10$$

which at $\alpha = 0.01$ and with the conservative df (equation (6.2.9)) would be compared with $t(0.005; 3) = 5.841$. Since $|-8.10| > 5.841$, one would reject H_0 and conclude that Process II has a higher yield. Alternatively, one could compute a p-value (using a computer) as

$$p\text{-value} = 2P[t(3) > 8.10] = 2(0.002) = 0.004$$

and one would reject H_0 since $0.004 < 0.01$.

Example 6.2.5 (*Continuation of Example 6.2.4*) If one wanted to use the approximate df (equation (6.2.10)) for comparing the two processes, one would compute

$$df = \frac{\left(\frac{10.36}{4} + \frac{110.8}{5}\right)^2}{\frac{1}{3}\left(\frac{10.36}{4}\right)^2 + \frac{1}{4}\left(\frac{110.8}{5}\right)^2} = 4.9$$

and the corresponding p-value (again obtained using a computer) would be

$$p\text{-value} = 2P[t(4.9) > 8.10] = 2(0.00063) = 0.0013.$$

Note that $\min\{4, 5\} = 4 \le 4.9 \le 7 = n_1 + n_2 - 2$.

Problems

1. Samples are taken under two sets of experimental conditions. The summary statistics for each are given below.

Condition 1	Condition 2
$\bar{y}_{1\bullet} = 121$	$\bar{y}_{2\bullet} = 89$
$s_1 = 25$	$s_2 = 22$
$n_1 = 21$	$n_2 = 24$

   (a) Test to see if the two experimental conditions have different average responses using $\alpha = 0.01$. Explain what your results mean.

   (b) Construct a 99% confidence interval for the average difference in response.

   (c) What assumption (other than normality) was made in the previous parts of the problem? What aspect of the data makes that assumption reasonable?

2. A QC technician measures the quality for two lots of product. In the first lot, 25 units are tested and the average measurement is 75 with a standard deviation of 15. In the second lot, 32 units are tested with an average measurement of 69 and a standard deviation of 12. Given that higher is better for this quality measurement, test the hypothesis that the quality of the second lot is lower than that of the first lot using $\alpha = 0.05$. You may assume the two lots come from populations with the same variance.

3. The percentage of "white area" in a sample of a dark product is used to evaluate how well a white raw material has been mixed into the product. Using the data from whitearea.txt, which contains samples from two different mixing processes, do the following.

   (a) Make a normal probability plot of the residuals and comment on it.

   (b) Test to see if white area percentages differ for the two groups using $\alpha = 0.10$. Explain what your results mean.

   (c) Construct a 90% confidence interval for the mean difference in white area percentages and explain what it indicates.

4. The data set labcomp.txt contains data on four laboratories and two types of surface area measurements, 1 point and 5 point (see Problem 2.3.2., p. 31). Only two of the labs are equipped to run both tests. For these two labs do the following.

   (a) Calculate the mean and standard deviation of each measurement for each lab.

   (b) Test to see whether the two labs get different results for the 1-point test using $\alpha = 0.01$. Be sure to check the normality assumption before performing the hypothesis test.

(c) Repeat the procedure in part (b) for the 5-point test.

(d) The data were of interest because it was suspected that the two labs might be running these tests differently. Do your hypothesis tests provide any evidence of this? Explain.

5. Refer to the data in Problem 2.2.5. (p. 25). Check to see if you can reasonably assume the data came from two normal populations. If so, test for differences between the two assembly lines using $\alpha = 0.05$. If a difference is found, construct a confidence interval for it. Write a summary of your findings.

6. A lens casting facility tested the UV absorbance for lenses cured in one of two different ovens, and the data are stored in `uvoven.txt`.

(a) Plot the data using side-by-side box plots. What do the plots illustrate about the group variances?

(b) Test to see if the measurements from each oven are equal using a conservative test.

(c) Construct an approximate 95% confidence interval for the mean difference in the results from the two ovens.

6.3 COMPARING TWO BINOMIAL POPULATIONS

If we want to compare proportions from two different populations, such as the proportion of defectives produced by two assembly lines (or assembly processes), we need to make some inference about the difference in proportions $p_1 - p_2$. This difference would be estimated by $\widehat{p}_1 - \widehat{p}_2$, and if n_1 and n_2 are large enough,

$$\widehat{p}_1 \sim N\left(p_1, \frac{p_1(1-p_1)}{n_1}\right)$$

and

$$\widehat{p}_2 \sim N\left(p_2, \frac{p_2(1-p_2)}{n_2}\right).$$

Using the same argument as was used to obtain the distribution of $\overline{Y}_{1\bullet} - \overline{Y}_{2\bullet}$ (formula (6.2.4)), it follows that

$$\widehat{p}_1 - \widehat{p}_2 \sim N\left(p_1 - p_2, \frac{p_1(1-p_1)}{n_1} + \frac{p_2(1-p_2)}{n_2}\right). \tag{6.3.1}$$

Note: A good rule of thumb for when n_1 and n_2 are large enough is $n_1\widehat{p}_1 > 5, n_1(1-\widehat{p}_1) > 5, n_2\widehat{p}_2 > 5$, and $n_2(1-\widehat{p}_2) > 5$.

Confidence Intervals

As with a single proportion there are two possibilities for constructing confidence intervals: approximate intervals and conservative intervals.

Approximate Intervals

A $(1-\alpha)100\%$ approximate confidence interval for $p_1 - p_2$ is

$$\widehat{p}_1 - \widehat{p}_2 \pm z(\alpha/2)\sqrt{\frac{\widehat{p}_1(1-\widehat{p}_1)}{n_1} + \frac{\widehat{p}_2(1-\widehat{p}_2)}{n_2}}. \tag{6.3.2}$$

Example 6.3.1   Given two suppliers of a chip for a cell phone, an engineer must decide which company to use as the primary supplier. There is pressure to use supplier 2 as its chip is priced well below the chip from supplier 1. Inspection of a single shipment of 500 chips ordered from each supplier shows eight defective chips from supplier 1, and 14 defective chips from supplier 2. Does one supplier seem to be better than the other?

An approximate 95% confidence interval for $p_1 - p_2$ is

$$0.016 - 0.028 \pm 1.96\sqrt{\frac{0.016(0.984)}{500} + \frac{0.028(0.972)}{500}}$$

$$-0.012 \pm 0.0182$$
$$(-0.0302, 0.0062).$$

The fact that this interval contains zero indicates that there is no evidence of a difference between the two suppliers.

Conservative Intervals

Replacing $p_i(1 - p_i)$ with the upper bound of 0.25, one obtains the conservative $(1 - \alpha)100\%$ confidence interval

$$\widehat{p}_1 - \widehat{p}_2 \pm z(\alpha/2)\frac{1}{2}\sqrt{\frac{1}{n_1} + \frac{1}{n_2}}. \tag{6.3.3}$$

Example 6.3.2 (*Continuation of Example 6.3.1*)   A conservative 95% confidence interval for the difference in proportions defective is (formula (6.3.3))

$$0.016 - 0.028 \pm (1.96)\frac{1}{2}\sqrt{\frac{1}{500} + \frac{1}{500}}$$
$$-0.012 \pm 0.062$$
$$(-0.074, 0.05).$$

Sample Sizes

The sample size necessary to achieve a specified margin of error d is obtained by rearranging formula (6.3.3) (with $n_1 = n_2 = n$) to obtain

$$n = \frac{1}{2}\left[\frac{z(\alpha/2)}{d}\right]^2. \tag{6.3.4}$$

Example 6.3.3   How large a sample would be needed to compare two proportions using a 99% confidence interval with a margin of error of 0.05?

From equation (6.3.4) one obtains

$$n = \frac{1}{2}\left[\frac{2.58}{0.05}\right]^2 = 1331.3 \rightarrow 1332$$

and 1332 samples are needed from each population.

Hypothesis Tests

When testing a hypothesis about the difference between two proportions, we do not use either

$$\frac{\widehat{p}_1(1 - \widehat{p}_1)}{n_1} + \frac{\widehat{p}_2(1 - \widehat{p}_2)}{n_2} \quad \text{or} \quad \frac{1}{4}\left(\frac{1}{n_1} + \frac{1}{n_2}\right)$$

to estimate the variance. Because our null hypothesis says that the two proportions are the same, we can get a slightly more powerful test by calculating

$$\widehat{p} = \frac{\text{total number of defectives in both samples}}{n_1 + n_2} \tag{6.3.5}$$

and using

$$\widehat{p}(1 - \widehat{p})\left(\frac{1}{n_1} + \frac{1}{n_2}\right)$$

to estimate the variance. This leads to the test statistic

$$z = \frac{\widehat{p}_1 - \widehat{p}_2}{\sqrt{\widehat{p}(1 - \widehat{p})}\sqrt{\frac{1}{n_1} + \frac{1}{n_2}}} \tag{6.3.6}$$

which would be compared to one of the critical values in Table 6.6.

Table 6.6. Critical Values for Comparing Two Proportions

Alternate Hypothesis	Rejection Region
$H_a\colon p_1 < p_2$	Reject H_0 if $z < -z(\alpha)$
$H_a\colon p_1 > p_2$	Reject H_0 if $z > z(\alpha)$
$H_a\colon p_1 \neq p_2$	Reject H_0 if $|z| > z(\alpha/2)$

Alternatively, one could compute a p-value (as shown in Table 6.7) and compare it with α.

Table 6.7. p-Values for Comparing Two Proportions

Alternative Hypothesis	p-Value
$H_a\colon p_1 < p_2$	$\Phi(z)$
$H_a\colon p_1 > p_2$	$1 - \Phi(z)$
$H_a\colon p_1 \neq p_2$	$2[1 - \Phi(|z|)]$

Example 6.3.4   A plant runs two shifts a day. There has been a long-standing argument between the shifts on how the process should be run. The plant manager

decides to monitor the defective product from each shift to determine if one shift is "better" than the other at running the process. Of 312 units produced by shift 1, 26 are defective and of 329 units produced by shift 2, 17 are defective.

Using $\alpha = 0.10$ to test the hypotheses

$$H_0: \; p_1 - p_2 = 0 \quad \text{versus} \quad H_a: \; p_1 - p_2 \neq 0$$

that the two shifts have different proportions of defectives, we would calculate

$$\widehat{p}_1 = 26/312 = 0.083$$
$$\widehat{p}_2 = 17/329 = 0.052$$
$$\widehat{p} = \frac{26 + 17}{312 + 329} = \frac{43}{641} = 0.067$$

and

$$z = \frac{0.083 - 0.052}{\sqrt{0.067(0.933)}\sqrt{\frac{1}{312} + \frac{1}{329}}} = \frac{0.031}{0.0198} = 1.57.$$

Since $|1.57| \not> z(0.05) = 1.645$, we would not reject H_0. There is no evidence to support a difference in shifts. Alternatively, we could have computed (Table 6.7)

$$p\text{-value} = 2[1 - \Phi(1.57)] = 2(1 - 0.9418) = 0.1164$$

and since $0.1164 \not< 0.10$, we would not reject H_0.

Sample Sizes

Exact sample sizes are not available for a test of equal proportions. However, one can compute conservative sample sizes for any combination of α and power γ when $p_1 - p_2 = d$. Alternatively, one can obtain approximate sample sizes for $\alpha = \beta = 0.1$ or $\alpha = \beta = 0.05$ from Figure C4 in Appendix C. Figure C4 also contains scales that show how β changes when α's other than 0.1 and 0.05 are used.

Computing Conservative Sample Sizes

For a test of equal proportions against the one-sided alternative $H_a: \; p_1 > p_2$, the sample size necessary to obtain a power of γ when $p_1 - p_2 = d$ would be obtained as the value of n satisfying

$$\gamma = P\left[\frac{\widehat{p}_1 - \widehat{p}_2}{\sqrt{\widehat{p}(1 - \widehat{p})}\sqrt{\frac{2}{n}}} > z(\alpha) \;\middle|\; p_1 - p_2 = d \right].$$

Unfortunately, when $p_1 \neq p_2$, we do not know exactly what the distribution is of

$$\frac{\widehat{p}_1 - \widehat{p}_2}{\sqrt{\widehat{p}(1 - \widehat{p})}\sqrt{\frac{2}{n}}} . \tag{6.3.7}$$

However, in order to obtain some idea of the necessary sample size, we can consider the statistic

$$\frac{\widehat{p}_1 - \widehat{p}_2}{\sqrt{p_1(1 - p_1) + p_2(1 - p_2)}\sqrt{\frac{1}{n}}}$$

from which the statistic (6.3.7) was obtained (by estimating p_1 and p_2 under H_0). Letting

$$c = \sqrt{p_1(1 - p_1) + p_2(1 - p_2)}\sqrt{\frac{1}{n}}$$

for convenience of notation, we need to find the value of n satisfying

$$\gamma = P\left[\frac{\widehat{p}_1 - \widehat{p}_2}{c} > z(\alpha) \,\middle|\, p_1 - p_2 = d\right]$$

$$= P\left[\frac{\widehat{p}_1 - \widehat{p}_2 - d}{c} > z(\alpha) - \frac{d}{c} \,\middle|\, p_1 - p_2 = d\right]$$

$$= P\left[Z > z(\alpha) - \frac{d}{c}\right].$$

Thus,

$$z(\gamma) = z(\alpha) - \frac{d}{\sqrt{p_1(1 - p_1) + p_2(1 - p_2)}\sqrt{\frac{1}{n}}}$$

and solving for n results in

$$n = \left[\frac{z(\alpha) - z(\gamma)}{d}\right]^2 [p_1(1 - p_1) + p_2(1 - p_2)].$$

We still cannot compute a value for n since p_1 and p_2 are unknown, but a conservative value of n can be obtained by replacing

$$p_1(1 - p_1) + p_2(1 - p_2)$$

with an upper bound subject to the restriction that $p_1 - p_2 = d$. That is, we need to find the value of p_1 that maximizes

$$\begin{aligned}
g(p_1) &= p_1(1 - p_1) + p_2(1 - p_2) \\
&= p_1(1 - p_1) + (p_1 - d)(1 - p_1 + d) \\
&= 2p_1(1 - p_1) + 2dp_1 - d(1 - d).
\end{aligned}$$

Differentiating with respect to p_1 and setting the result equal to zero results in

$$g'(p_1) = 2(1 - 2p_1 + d) = 0$$

and for a fixed value of d, $p_1(1 - p_1) + p_2(1 - p_2)$ is maximized when

$$p_1 = 0.5 + d/2$$

$$p_2 = 0.5 - d/2 \tag{6.3.8}$$

and that maximum is

$$p_1(1 - p_1) + p_2(1 - p_2) \le 2(0.5 + d/2)(0.5 - d/2) = 0.5 - d^2/2.$$

Therefore, a conservative value of n for a one-sided alternative is

$$n = \left[\frac{z(\alpha) - z(\gamma)}{d} \right]^2 (0.5 - d^2/2). \tag{6.3.9}$$

Example 6.3.5   An engineer wishes to compare the proportions of sub-assemblies from Companies A and B that can withstand a simulated-use resistance test. The comparison is to be done at level $\alpha = 0.1$, and it is desired to be able to detect when the proportion from Company A is 0.1 less than the proportion from Company B with probability 0.9. That is, the engineer is interested in an $\alpha = 0.1$ level test of

$$H_0 \colon p_A = p_B \quad \text{versus} \quad H_a \colon p_A < p_B$$

with power 0.9 when $p_B - p_A = 0.1$. Using equation (6.3.9), one finds that a sample size of

$$n = \left[\frac{1.282 + 1.282}{0.1} \right]^2 [0.5 - (0.1)^2/2] = 325.4 \to 326$$

would be required from each company.

As with all the previous sample size formulas, sample sizes for a two-sided alternative are obtained by simply replacing α in the equation for a one-sided alternative with $\alpha/2$. Thus, the sample size needed to obtain a power of γ when $p_1 - p_2 = d$ for a two-sided comparison of two proportions at level α is

$$n = \left[\frac{z(\alpha/2) - z(\gamma)}{d} \right]^2 (0.5 - d^2/2). \tag{6.3.10}$$

Example 6.3.6   In Example 6.3.4, how large a sample would have been required in order to detect a difference in defective rates of 0.05 with probability 0.8? We have $\alpha = 0.1$, $\gamma = 0.8$, and $d = 0.05$. Thus, using equation (6.3.10), one would need

$$n = \left[\frac{1.645 + 0.84}{0.05} \right]^2 [0.5 - (0.05)^2/2] = 1232$$

units from each shift.

Example 6.3.7   In Example 6.3.4, what would the power be for detecting a difference in proportions defective of 0.05 when using samples of only 320 (essentially what was used)? Rearranging equation (6.3.10), one obtains

$$z(\gamma) = z(\alpha/2) - \frac{d\sqrt{n}}{\sqrt{0.5 - d^2/2}}$$

$$= z(0.05) - \frac{0.05\sqrt{320}}{\sqrt{0.5 - (0.05)^2/2}}$$

$$= 1.645 - 1.266$$

$$= 0.379$$

and from Table B1

$$\gamma = 1 - \Phi(0.379) \doteq 1 - 0.6480 = 0.352.$$

Thus, the test would only have a power of approximately 35%.

Obtaining Approximate Sample Sizes

For $\alpha = \beta = 0.1$ or $\alpha = \beta = 0.05$ one can read the approximate sample size for a test of equal proportions from Figure C4 in Appendix C.

Example 6.3.8   Previous experience has shown that approximately 35% ($p_2 = 0.35$) of all cell phone calls from a particular provider suffer from extensive static. It is hoped that a new design will decrease this by 20% ($p_1 = 0.15$). Using an $\alpha = 0.05$ significance test, how large a sample size would be needed to detect a 20% decrease with probability 0.95? From the lower chart in Figure C4 one finds a sample size of $n = 110$ is required for each proportion.

Example 6.3.9   If one had used a significance level of $\alpha = 0.01$ in Example 6.3.8, what would the power of the test be with $n = 110$? This can be determined using the horizontal scales at the bottom of Figure C4. Drawing a line through $\alpha = 0.01$ and pivot point 2 (pivot point 1 would be used with the $\alpha = \beta = 0.1$ chart), one finds $\beta = 0.2$, which corresponds to a power of 0.8.

Example 6.3.10   Consider again the situation described in Example 6.3.5. That problem specified $\alpha = 0.1$, power $= 0.9$ (or $\beta = 0.1$), and $p_B - p_A = 0.1$. From the $\alpha = \beta = 0.1$ chart in Figure C4 one finds that for any proportion p_A between 0.1 and 0.75, a sample size of more than 200 is required. In order to obtain a specific sample size from the curves one would need to know either the smaller proportion p_1 (or the larger proportion p_2 from which p_1 can be obtained). Alternatively, one can use the conservative value of $p_1 = 0.5 - d^2/2$ (equation 6.3.8) as was done when computing the sample size in Example 6.3.5. With $p_1 = 0.5 - (0.1)^2/2 = 0.495$, one finds a sample size of approximately $n = 400$.

Note: The proportion p_2 in equation 6.3.8 was relabeled as p_1 in Example 6.3.10 because in Figure C4 the smaller proportion is labeled as p_1.

Problems

1. Samples of size 100 are taken from two binomial populations. In the first population, $\hat{p} = 0.41$; in the second population, $\hat{p} = 0.69$. Construct a 90% confidence interval for the difference in population proportions.

2. Samples of lenses coated with two different anti-reflective coatings are tested for defectives. A lens is counted as "defective" if it has visible scratches. The data are given below.

Type I	Type II
$n = 263$	$n = 219$
defective $= 19$	defective $= 14$

   (a) Test the hypothesis that one type of coating results in more defectives than the other using $\alpha = 0.05$. Explain what your results mean.

   (b) Construct a 95% confidence interval for the difference in proportions of defectives and explain what it indicates.

   (c) Write a brief memo summarizing the study and its results.

3. Molded fiberglass parts are to be tested for defectives. Two different grades of fiberglass were used for the parts, and one is interested in testing for a difference in the proportions defective using $\alpha = 0.05$.

   (a) How many parts of each grade must be sampled in order to detect a difference in proportions defective of 0.1 with probability 0.8?

   (b) The examination of 550 parts from the first grade of fiberglass resulted in 14 defectives. From 400 parts of the second grade of fiberglass there were 11 defectives. Conduct the test for differences, and explain what your results mean.

4. A production engineer in a toothpaste plant decides to study how often the toothpaste tubes are "over-filled". An "over-filled" tube "oozes" toothpaste from the tube when the end is being crimped and requires a quick slowdown of the line to remove the excess toothpaste from the crimping machine. The engineer monitors two different assembly lines for 1 hour each. On the first line, 523 tubes are filled in the hour, and the engineer records 72 over-filled tubes. On the second line, 483 tubes are filled, with 41 over-filled tubes.

   (a) Test to see if the second line has a lower "over-fill" rate using $\alpha = 0.05$, and explain what your results mean.

   (b) Construct a 95% confidence interval for the difference in "over-fill" rates between the two assembly lines.

   (c) How large a sample would be required for the test in part (a) to have a power of 0.9 for detecting a difference of 0.05?

(d) How large a sample would be required for the confidence interval in part (b) to have a margin of error of 0.05?

5. Two non-invasive diagnostic test kits are available to test for a disease. In one study, 90 patients known to have the disease (as determined from a biopsy, the "gold standard" for this disease) were tested with kit A, 85 tested positive. In a second study of patients known to have the disease 93 of 100 tested positive with kit B. The producers of kit B need to know if there is a difference in the two kits. They need a 95% confidence interval for the difference in sensitivity (sensitivity is the percent of positive patients who test positive), and they need the sample size required to detect a difference of 2.5% in sensitivity between the two kits at a 95% level with probability 0.8. Write a brief report for the company summarizing the results of your analyses.

6. A validation study on the strength of a seal of a plastic tube containing a medical gel material was performed. Part of the study was to compare failure rates at the extremes of the allowable temperature range of the sealing machine. At the low temperature setting, 50 tubes were sealed and tested, two failed. At the high end of the temperature setting, 53 tubes were sealed and tested, three failed. Is there a difference in failure rates at the two ends of the temperature range?

7. Re-do Example 6.3.8 using equation (6.3.9).

6.4 CHAPTER PROBLEMS

1. The times (in seconds) to run six randomly selected computer jobs on a new CPU are $\{1.2, 1.8, 2.1, 1.7, 3.1, 2.5\}$. You are interested in purchasing this unit only if the average job time is less than that of your current unit, which is 2.5 seconds.

   (a) Construct a normal probability plot for these data.

   (b) Test the hypotheses of interest at the 0.05 level.

   (c) Construct an interval that with 95% confidence will contain 99% of the population (time) values.

   (d) If one were interested in being able to detect a difference in speed (between the current unit and the new unit) of 2σ with probability 0.9, how large a sample size is necessary?

   (e) Suppose the new CPU unit (call it unit A) were to be compared with a new CPU made by a different manufacturer (call it unit B). The times for six randomly selected computer jobs on unit B are $\{1.4, 2.0, 2.1, 1.9, 2.7, 3.0\}$. Test if there any difference between the two companies at level $\alpha = 0.05$ by constructing the appropriate interval.

   (f) Suppose that the two CPU units were tested by selecting the computer jobs from six classes of jobs (spreadsheet applications, Fortran programs, multimedia applications, etc.) with the first values (1.2 and 1.4) being from the first class, the second values (1.8 and 2.0) being from the second class, etc. Is there any difference between the two companies at level $\alpha = 0.05$?

2. Suppose that one is interested in finding out if one's ability to shoot free throws in basketball is a function of gender. For random samples of 25 males and 20 females, one records the proportion of participants who make more than 30% of their free throws. These proportions are 0.3 and 0.35 for males and females, respectively.

   (a) Construct a 90% confidence interval for the difference in the two proportions.

   (b) Use the interval in part (a) to test the hypothesis of interest.

3. In order to compare the average amount of unburned hydrocarbons produced by a particular brand of carburetor adjusted in two different ways (A and B), a random sample of 64 carburetors was taken. Of the 64 carburetors 31 were subjected to adjustment A, 33 were subjected to adjustment B, and the amounts of unburned hydrocarbons produced were recorded. The average amount of unburned hydrocarbons using adjustment A was computed to be 5.7 with a sample standard deviation of 0.9. The average amount using adjustment B was computed to be 4.3 with a sample standard deviation of 1.7.

(a) Test to see if there is any difference in the two adjustments with regard to unburned hydrocarbons at level $\alpha = 0.1$ by comparing with the appropriate critical value.

(b) Compute the p-value for the test in part (a).

4. Two different sorting algorithms were compared on unsorted arrays and the sorting times in seconds are recorded below.

$$\text{Algorithm A: } 6 \quad 4 \quad 8 \quad 6$$
$$\text{Algorithm B: } 7 \quad 3 \quad 9 \quad 9$$

(a) Suppose that you are interested in testing to see if Algorithm A sorts faster. State the null and alternative hypotheses.

(b) How the data should be analyzed depends on exactly how the experiment was conducted. Using $\alpha = 0.1$, test the hypotheses in part (a) if eight unsorted arrays were assigned at random to the two algorithms.

(c) Using $\alpha = 0.1$, test the hypotheses in part (a) if the same four unsorted arrays were used with each algorithm. The four columns of data represent the four arrays (e.g., the (6, 7) pair is for the first array).

(d) Which design would you prefer, and why?

(e) What assumptions are necessary for part (b), and what assumptions are necessary for part (c)?

5. One is interested in comparing two brands of dog food: Purina and Alpo. As a first attempt, five randomly selected dogs are fed one cup of Purina and five different randomly selected dogs are fed one cup of Alpo. The time it takes the dogs to finish the food is recorded (in minutes) below.

Purina	0.5	0.9	2.2	1.4	5.0
Alpo	1.1	0.2	3.0	0.8	2.4

(a) Test to see if there is any difference in the two brands by constructing a confidence interval. Use $\alpha = 0.05$.

(b) Since there was such wide variability in the times for different dogs, a second experiment was run where only five dogs were used and each dog received (at different times) both brands of food. The results are given below.

		Dog			
	1	2	3	4	5
Purina	0.5	0.9	2.2	1.4	5.0
Alpo	0.2	0.8	2.4	1.1	3.0

Test to see if Alpo is preferred using $\alpha = 0.05$.

6. As an alternative to the experiments in Problem 5, it was decided to record simply whether a dog took longer than 1 minute to eat the cup of food. This time 50 dogs were used with half of them getting Purina and the other half getting Alpo. The results are given below.

Time	Purina	Alpo
> 1 min.	10	14
≤ 1 min.	15	11

(a) Test to see if there is an difference in the two brands by computing a p-value and using $\alpha = 0.05$.

(b) Check the assumptions required to perform the test in part (a).

7. A new genetically engineered sleeping pill is to be compared with a sleeping pill made from only natural ingredients (a paired test).

(a) Suppose the comparison is to be done by constructing a 90% confidence interval and one desires an interval with length no more than 2. If an initial sample of size 10 results in an average difference of 15 minutes until sleep with $s_d^2 = 25$, what sample size is needed at the second stage?

(b) If the average difference for the observations at the second stage is 13 minutes with $s_d^2 = 16$, construct the 90% confidence interval.

7

ONE-FACTOR MULTI-SAMPLE EXPERIMENTS

Up to this point we have seen how to estimate means, find confidence intervals, and perform hypothesis tests for one- and two-sample experiments. In this chapter, we will look at the more general case of a multi-sample experiment. Generally, one is interested in comparing I **treatments**, which are also referred to as I levels of a **factor**.

Example 7.0.1 (*Reaction Yield Data*)   A chemist wanted to compare the yield of a reaction using three different catalysts, which are labeled "A", "B", and "C". The three types are three levels of the factor catalyst. The dependent variable is the product yield obtained from a chemical reaction in a particular size reactor (measured in lbs.). In total 12 observations were made, but the experiment is **unbalanced**: only three replicates were made for "A", four for "B", and five for "C".

Notation: Recall the notation we have already used to keep track of the observations in a multi-sample experiment. The observations are denoted y_{ij} where i indicates the level of the factor (i.e., which treatment) and j indexes the replicate observations for the treatment. In addition, $\bar{y}_{i\bullet}$ is the sample mean of the ith treatment, and n_i is the sample size of the ith treatment. The total sample size is denoted by N, and the number of treatments is I.

To illustrate this notation, each observation and the sample means and standard deviations are labeled in Table 7.1.

Table 7.1. Notation for Observations, Sample Means, and Sample Standard Deviations

A	B	C
$y_{11} = 3.3$	$y_{21} = 4.0$	$y_{31} = 5.3$
$y_{12} = 4.0$	$y_{22} = 5.0$	$y_{32} = 6.5$
$y_{13} = 4.7$	$y_{23} = 4.3$	$y_{33} = 6.4$
	$y_{24} = 5.5$	$y_{34} = 7.0$
		$y_{35} = 7.7$
$\bar{y}_{1\bullet} = 4.0$	$\bar{y}_{2\bullet} = 4.7$	$\bar{y}_{3\bullet} = 6.58$
$s_1 = 0.7$	$s_2 = 0.68$	$s_3 = 0.88$
$n_1 = 3$	$n_2 = 4$	$n_3 = 5$

7.1 BASIC INFERENCE

One of the first things one would want to do with a multi-sample set of data is find a useful way to graph it. Some possibilities are: a scatter plot with the treatment on the horizontal axis, side-by-side box plots, or an error-bar chart. For the example above, an error-bar chart was chosen because the sample sizes were small. (Box plots are not very meaningful with only three or four observations.) The bars on the error-bar chart in Figure 7.1 are one standard deviation (i.e., for "A" the bars are of length 0.7).

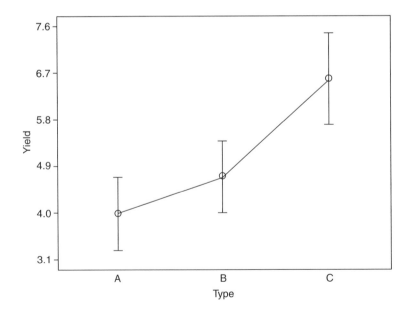

Figure 7.1. Error-Bar Chart for the Reaction Yield Data in Example 7.0.1.

We can see from the error-bar chart that catalyst "C" results in a higher yield than the other two catalysts. The mean for "B" is slightly higher than for "A", but there is so much overlap in the samples that it is difficult to say whether "B" is actually better than "A".

Recall from Section 3.1 that a mathematical model for such an experiment is given by equation (3.1.1), which is

$$Y_{ij} = \mu_i + \epsilon_{ij}.$$

As before, μ_i represents the population mean of the ith factor level (e.g., the ith catalyst), and ϵ_{ij} represents the random variation of an individual experimental unit (in the reaction yield example the random variation in yield of an individual reaction). The ϵ_{ij}'s will be assumed to be (at least approximately) $N(0, \sigma^2)$ and independent. An estimate of μ_i is (equation (3.1.2))

$$\widehat{\mu}_i = \overline{y}_{i\bullet}$$

and the residuals for model (3.1.1) are (equation (3.1.5))

$$\widehat{\epsilon}_{ij} = y_{ij} - \overline{y}_{i\bullet}.$$

The common variance of the ϵ_{ij}'s (i.e., σ^2) is estimated with the pooled sample variance

$$s_p^2 = \frac{(n_1 - 1)s_1^2 + (n_2 - 1)s_2^2 + \cdots + (n_I - 1)s_I^2}{(n_1 - 1) + (n_2 - 1) + \cdots + (n_I - 1)}.$$

The pooled sample variance is also referred to as the mean square error (MS_e) and reduces to

$$\text{MS}_e = s_p^2 = \frac{(n_1 - 1)s_1^2 + (n_2 - 1)s_2^2 + \cdots + (n_I - 1)s_I^2}{N - I} . \tag{7.1.1}$$

In the equal sample size case it reduces still further to

$$\text{MS}_e = s_p^2 = \frac{s_1^2 + \cdots + s_I^2}{I} . \tag{7.1.2}$$

Example 7.1.1   (*Reaction Yield Data, p. 245*)   The pooled estimate of variance is (equation (7.1.1))

$$\text{MS}_e = s_p^2 = \frac{2(0.7)^2 + 3(0.68)^2 + 4(0.88)^2}{12 - 3} = 0.607.$$

A $(1 - \alpha)100\%$ confidence interval for any μ_i is given by

$$\bar{y}_{i\bullet} \pm t(\alpha/2; N - I)\sqrt{\frac{\text{MS}_e}{n_i}}\ .$$

Notice that this is just like a confidence interval for a one-sample problem, except that s^2 has been replaced by MS_e, which has $N - I$ degrees of freedom. This is done because MS_e, which is computed using all the samples, is a better estimate of σ^2 than the estimate from any one sample.

Prediction intervals for multi-sample data are also similar to those for a single sample. Only the variance estimate and the corresponding degrees of freedom are different.

Example 7.1.2   (*Reaction Yield Data, p. 245*)   In this example μ_1 represents the mean yield for catalyst "A", which is the catalyst used in the current commercial product. A 95% confidence interval for μ_1 is

$$4 \pm t(0.025; 9)\sqrt{\frac{0.607}{3}}$$
$$\pm 2.262(0.4498)$$
$$\pm 1.02$$
$$(3, 5).$$

So, the average yield obtained with catalyst "A" is between 3 and 5 lbs., at least for the size reactor used in the test.

Similarly, a 95% prediction interval is

$$4 \pm 2.262\sqrt{0.607\left(1 + \frac{1}{3}\right)}$$
$$\pm 2.03$$
$$(2, 6).$$

This means there is a 95% probability that the yield on a single batch of this chemical will be between 2 and 6 lbs. for the type of reactor used in the test.

Problems

1. A paint company wanted to compare the effect of three different "UV packages" (used to improve paint durability with respect to UV exposure) on the appearance of an automotive paint. Each UV package was added to five paint samples that were then sprayed onto test panels and cured. Gloss measurements for each test panel are given below.

UV Package

UV1	UV2	UV3
88	85	82
87	87	83
88	88	85
86	86	82
89	85	84

   (a) Graph the data.

   (b) Calculate a 95% confidence interval for each mean.

   (c) What is a range of values that have a 95% probability of containing the gloss of a panel that has paint containing UV package #2?

2. Four catalysts were studied for their effect on the hardness of an optical lens coating. The hardness measurements for each catalyst are listed below. Write a report (including graphs) detailing the results of this experiments. Note that larger measurement values are more desirable.

Catalyst

CAT1	CAT2	CAT3	CAT4
87	104	68	74
90	104	71	73
102	98	73	79
102			8
7			85

   Is there a best catalyst? What hardness might one expect to achieve if a lens were coated using that catalyst in the formulation? Substantiate your answers.

3. An environmental consulting firm collects fish from sites along a stream that runs near abandoned lead mines. The fish are then tested for levels of various toxins to monitor a stream reclamation project. The file `fish.txt` contains the values of a toxin in fish from four sites along this stream. Analyze the data and write a report for the consulting firm.

7.2 THE ANALYSIS OF MEANS

In a one-factor multi-sample experiment the first question one is interested in answering is whether the factor has any effect. Since we have assumed the different levels of the factor have the same variance, in order to check if the factor has an effect we need only check to see if the means for the different levels are all the same. That is, one would test

$$H_0: \ \mu_1 = \mu_2 = \cdots = \mu_I \tag{7.2.1}$$

versus the alternative hypothesis that at least one of the μ_i's is different.

A good way to do this is using the **analysis of means** (ANOM). This test not only answers the question of whether or not there are any differences among the factor levels, but as we will see below, when there are differences, it also tells us which levels are better and which are worse.

The idea behind the analysis of means is that if H_0 is true, the I factor levels all have the same population mean. Therefore, all the $\overline{y}_{i\bullet}$'s should be close to the overall mean $\overline{y}_{\bullet\bullet}$. So we will reject H_0 if any one of the $\overline{y}_{i\bullet}$'s is too far away from $\overline{y}_{\bullet\bullet}$. The difference

$$\boxed{\widehat{\alpha}_i = \overline{y}_{i\bullet} - \overline{y}_{\bullet\bullet}} \tag{7.2.2}$$

is an estimate of α_i, the effect of the ith level of the factor, so the ANOM is actually testing to see if any level has an effect.

In the discussion of the ANOM that follows, we will assume that all the factor levels have equal sample sizes. The case where the sample sizes are unequal is slightly more complicated, and is covered in the next section.

The ANOM belongs to the class of **multiple comparison procedures**. It tests hypothesis (7.2.1) by constructing multiple confidence intervals, one interval for each α_i. The general form for these intervals is

$$\widehat{\alpha}_i \pm \text{ critical value } \times \text{ (estimated standard deviation of } \widehat{\alpha}_i).$$

With a little algebra one can show that (see Problem 7.2.6)

$$\text{Var}(\widehat{\alpha}_i) = \sigma^2 \frac{I-1}{N}$$

and therefore, the intervals are

$$\widehat{\alpha}_i \pm h\sqrt{\text{MS}_e}\sqrt{\frac{I-1}{N}} \tag{7.2.3}$$

where h is a critical value that accounts for the fact that multiple intervals are being constructed.

Rather than computing all of these intervals, the procedure can be simplified by noting that one would reject H_0 if any of the α_i's is significantly different from zero, which corresponds to the interval (7.2.3) not containing zero for some i. The interval (7.2.3) would not contain zero if either all the values in the interval were negative (corresponding to the upper confidence limit being negative) or if all the values in the interval were positive (corresponding to the lower confidence limit being positive). The lower limit would be positive if

$$0 < \widehat{\alpha}_i - h\sqrt{\mathrm{MS}_e}\sqrt{\frac{I-1}{N}}$$

$$= \overline{y}_{i\bullet} - \overline{y}_{\bullet\bullet} - h\sqrt{\mathrm{MS}_e}\sqrt{\frac{I-1}{N}}$$

which is equivalent to

$$\overline{y}_{i\bullet} > \overline{y}_{\bullet\bullet} + h\sqrt{\mathrm{MS}_e}\sqrt{\frac{I-1}{N}}.$$

A similar argument for a negative upper confidence limit would lead to a lower limit for $\overline{y}_{i\bullet}$, and the ANOM is performed by computing upper and lower decision limits

$$\mathrm{UDL} = \overline{y}_{\bullet\bullet} + h(\alpha; I, N-I)\sqrt{\mathrm{MS}_e}\sqrt{\frac{I-1}{N}}$$

$$\mathrm{LDL} = \overline{y}_{\bullet\bullet} - h(\alpha; I, N-I)\sqrt{\mathrm{MS}_e}\sqrt{\frac{I-1}{N}}$$

$$(7.2.4)$$

and checking to see if any of the treatment means (i.e., factor level means) fall outside these limits. This is generally done graphically by plotting the treatment means together with lines for the decision limits (decision lines), and rejecting H_0 if any of the treatment means plot outside the decision lines. The critical values $h(\alpha; I, \nu)$ depend on

$$\alpha = \text{the level of significance desired}$$
$$I = \text{the number of means being compared}$$
$$\nu = \text{the degrees of freedom for } \mathrm{MS}_e$$

and are given in Table B5.

Example 7.2.1 (*Optical Lens Data, p. 26*)   Consider the four optical coatings again. The four treatment averages were (Example 3.1.2, p. 39)

$$\overline{y}_{1\bullet} = 9.45, \qquad \overline{y}_{2\bullet} = 12.61, \qquad \overline{y}_{3\bullet} = 12.19, \qquad \overline{y}_{4\bullet} = 10.05$$

$\overline{y}_{\bullet\bullet} = 11.08$, $\mathrm{MS}_e = 1.68$, and $N - I = 28 - 4 = 24$. Using $\alpha = 0.01$, one finds $h(0.01; 4, 24) = 3.35$ and

$$h(0.01; 4, 24)\sqrt{MS_e}\sqrt{\frac{I-1}{N}} = 3.35\sqrt{1.68}\sqrt{\frac{3}{28}} = 1.42.$$

Thus,

$$UDL = 11.08 + 1.42 = 12.5$$
$$LDL = 11.08 - 1.42 = 9.66$$

and one obtains the ANOM chart in Figure 7.2. The decision lines make it clear that there is a significant effect (at $\alpha = 0.01$) due to Coating 2 being significantly bad and Coating 1 being significantly good.

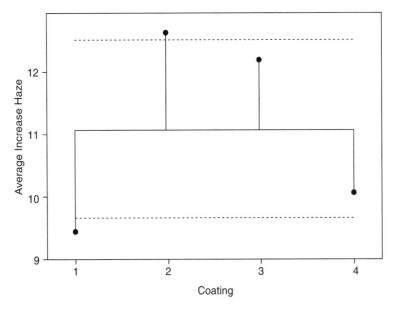

Figure 7.2. ANOM Chart for Example 7.2.1 with $\alpha = 0.01$.

A report of the experiment would look something like the following.

An experiment was performed to compare the durability of four coatings for plastic optical lenses. The four coatings were each applied to seven different lenses. The lenses were then abraded a standard amount, after which the increase in haze was recorded for each lens. Statistical analysis using $\alpha = 0.01$ showed that Coating 1 was the best, Coatings 3 and 4 were in between, and Coating 2 was the worst. On average Coating 1 was better than the overall average by 1.63 (11.08 − 9.45) units.

Assumptions: Recall that the assumptions behind the ANOM are that the ϵ_{ij}'s are approximately normally distributed and independent, and that the populations all have approximately the same variance. Therefore, before using the ANOM to analyze a set of data, one should check the following.

1. Is it reasonable to assume the population variances are equal?

    (a) Plot the residuals versus the sample means. Do the residuals seem to fall within a horizontal band?

    (b) Test H_0: $\sigma_1^2 = \cdots = \sigma_I^2$ using the test procedures given at the end of this chapter.

2. Is it reasonable to assume the errors (i.e., the ϵ_{ij}'s) are normally distributed?

    (a) Make a normal probability plot of the residuals. Does the plot indicate serious non-normality?

    (b) Note that non-constant (i.e., unequal) variances will show up as a non-linear normal probability plot, so this plot is also useful for checking the assumption of equal variances.

Example 7.2.2 (*Tennis Ball Data*)   A new fiber is being studied for use in tennis balls. The new fiber is supposed to improve the durability of the tennis ball cover (and hence, increase the lifetime of the tennis ball). Researchers have developed two possible coverings for tennis balls from this new fiber. To gauge their progress they run a durability test with two commercial products (A and B) and the two candidate coverings (C and D). Five balls of each type are randomly chosen and marked with a stripe to gauge wear. Each ball is put into a machine that simulates a serve. The response variable is the number of serves a ball can withstand before the wear stripe is worn. The results are given in Table 7.2 and in `tennis.txt`.

Table 7.2. Results of the Tennis Ball Study (Example 7.2.2)

Type			
A	B	C	D
151	149	98	239
113	200	123	182
154	185	105	205
121	170	146	215
135	211	129	235

$\bar{y}_{A\bullet} = 126.6 \quad \bar{y}_{B\bullet} = 183.0 \quad \bar{y}_{C\bullet} = 120.2 \quad \bar{y}_{D\bullet} = 215.2$
$s_A^2 = 266.8 \quad s_B^2 = 600.5 \quad s_C^2 = 368.7 \quad s_D^2 = 541.2$

Before performing an ANOM on the data, we must first check the assumptions of approximate normality of the errors and approximately equal population variances. We will check the normality assumption by making a normal probability plot of the residuals. The residuals are (equation (3.1.5))

$$\hat{\epsilon}_{11} = 151 - 126.6 = 24.4$$
$$\hat{\epsilon}_{12} = 149 - 183.0 = -34.0$$
$$\vdots$$
$$\hat{\epsilon}_{45} = 235 - 215.2 = 19.8.$$

These residuals are then ranked from smallest to largest and plotted versus the corresponding normal quantiles, resulting in the normal probability plot given in Figure 7.3. A few points at each end of the plot do not seem to fall along the same line as the rest of the residuals, but there is nothing to indicate serious non-normality or unequal variances. As a double check on the variances one can look at a plot of the residuals versus the sample means. That plot is given in Figure 7.4 and also indicates (because the residuals all fall within a horizontal band) that the assumption of equal variances is reasonable.

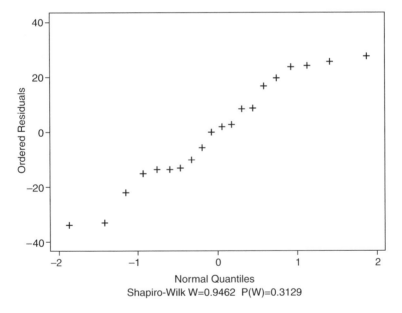

Shapiro-Wilk W=0.9462 P(W)=0.3129

Figure 7.3. A Normal Probability Plot of the Residuals in Example 7.2.2.

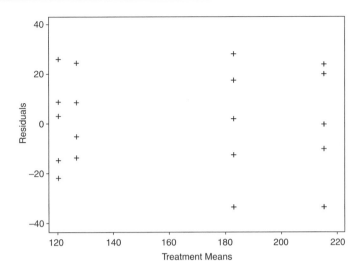

Figure 7.4. A Plot of the Residuals Versus the Sample Means Used to Check for Unequal Variances.

Now, since the assumptions are reasonable, one can proceed with the ANOM. Using the summary statistics, one can compute

$$\bar{y}_{\bullet\bullet} = \frac{126.6 + \cdots + 215.2}{4} = 161.25$$

$$MS_e = \frac{266.8 + \cdots + 541.2}{4} = 444.3$$

where MS_e has $N - I = 20 - 4 = 16$ degrees of freedom. The ANOM decision lines using $\alpha = 0.01$ are

$$161.25 \pm h(0.01; 4, 16)\sqrt{444.3}\sqrt{\frac{3}{20}}$$

$$\pm 3.54(8.164)$$

$$\pm 28.9$$

$$(132.3, 190.2).$$

The ANOM chart given in Figure 7.5 shows that there are significant differences due to types A and C being below average and type D being above average. Therefore, the four types of tennis balls can be divided into three categories: type D (an experimental) is best, type B (a commercial) is average, and types A and C are worst. So one of the experimental coverings holds great promise as a commercial product, while the other should not be pursued further. The ANOM chart also provides an easy way to assess **practical significance**. That is, once we have found

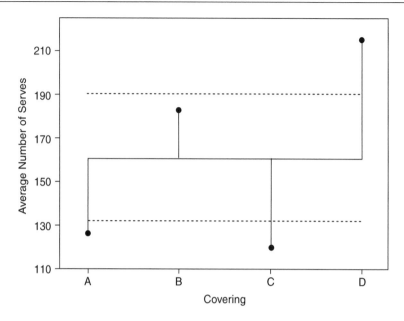

Figure 7.5. An ANOM Chart for the Tennis Ball Data in Example 7.2.2 with $\alpha = 0.01$.

that the treatments are statistically significantly different, one can ask if there are really any practical differences. In the case of the tennis balls, on the average type D withstands $215.2 - 161.25 = 53.95$ more serves than the average of all types. Thus, the difference appears to be of practical significance. Note that practical significance is not a statistical issue, and it only makes sense to ask about practical significance if one first finds statistical significance since without statistical significance one cannot distinguish between the treatments.

Example 7.2.3   An experiment was performed to study how three different types of filaments affect the lifetime of light bulbs. Measurements were taken on 10 bulbs for each of the different filament types, and the data are recorded in Table 7.3.

Table 7.3. The Lifetimes of Light Bulbs with Three Different Types of Filaments

Type 1	Type 2	Type 3
0.6	3.3	1.0
1.7	31.9	1400.0
1.3	8.5	10.2
1.3	52.5	1017.0
9.0	3.6	700.5
0.6	1.1	156.6
1.8	26.4	34.9
1.4	1.1	54.8
8.6	4.5	3.5
12.6	2.5	8.1

Filament

A normal probability plot to check the assumption of normality is given in Figure 7.6. From the normal probability plot it is clear that the residuals are not

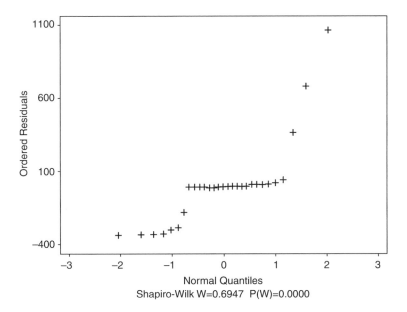

Shapiro-Wilk W=0.6947 P(W)=0.0000

Figure 7.6. A Normal Probability Plot for the Filament Data in Example 7.2.3.

even approximately normally distributed, and it would therefore not be appropriate to apply the ANOM to these data. However (see Problem 7.2.4), if one transforms the data using a log transformation (i.e., $x \rightarrow \ln(x)$) the data then look at least approximately normal. Note also that if one applies the ANOM with $\alpha = 0.01$ to the untransformed data, no significant differences are found, but for the transformed data there are significant differences.

Example 7.2.4 The data in Table 7.4 are random samples from three normal populations. Population 1 is $N(25, 0.01)$, Population 2 is $N(5, 25)$, and Population 3 is $N(0, 100)$. The vast differences in the sample variances would suggest that the variances are not all equal. This is confirmed by the normal probability plot given in Figure 7.7 where one sees lines with three different slopes, which correspond to the three different variances.

Table 7.4. Random Samples From Three Normal Populations

	Population	
1	2	3
24.92	8.518	−7.096
25.01	8.678	4.808
25.12	5.442	25.86
25.07	5.436	14.68
25.21	0.736	−11.62
25.11	−2.260	−0.150
24.95	0.342	−12.92
25.01	6.814	−4.502
$s_1^2 = 0.009$	$s_2^2 = 16.33$	$s_3^2 = 181.4$

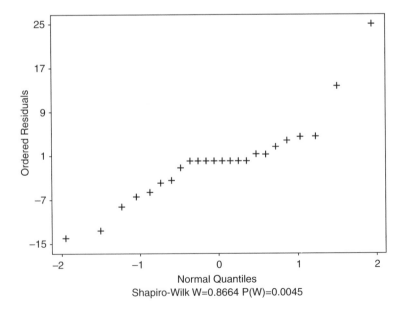

Shapiro-Wilk W=0.8664 P(W)=0.0045

Figure 7.7. A Normal Probability Plot of the Data in Example 7.2.4.

Problems

1. An experiment has five levels of treatment. A summary of the results is given below.

Treatment	$\bar{y}_{i\bullet}$	s	n
1	15.5	3.1	6
2	16.3	2.8	6
3	21.4	2.6	6
4	15.8	3.2	6
5	24.3	3.3	6

(a). Calculate the MS_e.

(b). Perform an ANOM using $\alpha = 0.05$, and explain what your results mean.

2. Using the adhesion data set `adhesion.txt` and $\alpha = 0.10$, perform an ANOM and explain what the results indicate.

3. The paint on a car consists of multiple layers (an undercoat, a primer, a base coat, and a clear coat). In the development of a clear coat, the chemists became concerned that their formulation did not perform the same over all colors (color is a component of the base coat layer). To study the situation they tested their clear coat formulation over four base coats and measured a number of properties, including gloss. The gloss measurements are in `gloss.txt`.

(a) Perform an ANOM on the data using $\alpha = 0.05$.

(b) Write a summary of your results (high gloss values are desired).

4. Refer to Example 7.2.3 (p. 256).

(a) Make a normal probability plot of the residuals for the transformed data.

(b) Construct an ANOM chart for the untransformed data using $\alpha = 0.01$.

(c) Construct an ANOM chart for the transformed data using $\alpha = 0.01$.

5. An experiment was performed to study how three processors affect the lifetime of a battery used in notebook computers. Measurements were taken on ten batteries with each of the different processor types, and the data are recorded in `battery.txt`.

(a) Make a normal probability plot of the residuals for these data.

(b) Transform the data using a log transformation, and make a normal probability plot of the residuals for the transformed data.

(c) Construct an ANOM chart for the untransformed data using $\alpha = 0.01$.

(d) Construct an ANOM chart for the transformed data using $\alpha = 0.01$.

6. Use the fact that $\widehat{\alpha}_i$ can be written as

$$\widehat{\alpha}_i = \left(1 - \frac{1}{I}\right)\overline{y}_{i\bullet} - \frac{1}{I}\sum_{j \neq i}\overline{y}_{j\bullet}$$

and the properties of variances to show that $\text{Var}(\widehat{\alpha}_i) = \sigma^2(I-1)/N$.

7. Using the data in `fish.txt` first introduced in Problem 7.1.3, construct an ANOM chart with decisions lines for both $\alpha = 0.01$ and $\alpha = 0.10$.

8. Three batches of swimming pool chlorine tablets were tested in an accelerated test. One hour in the test pool is equivalent to 1 day in a home swimming pool. Analyze the data in `tablets.txt` and write a brief report.

7.3 ANOM WITH UNEQUAL SAMPLE SIZES

The ANOM as described in the previous section assumes a **balanced design**. That is, it assumes the different factor levels all have the same number of observations. When that is not the case, the test procedure is similar, but slightly more complicated. The complication is due to the fact that the decision lines around $\bar{y}_{\bullet\bullet}$ will be tighter for the larger samples and wider for the smaller samples. The decision lines are

$$\bar{y}_{\bullet\bullet} \pm m(\alpha; I, N - I)\sqrt{MS_e}\sqrt{\frac{N - n_i}{Nn_i}} \qquad (7.3.1)$$

where the critical values $m(\alpha; I, \nu)$ are given in Table B6 and similarly to the $h(\alpha; I, \nu)$ critical values depend on

α = the level of significance desired

I = the number of means being compared

ν = the degrees of freedom for MS_e.

These critical values are slightly larger than the corresponding h critical values to allow for any possible set of sample sizes. The quantity

$$\frac{N - n_i}{Nn_i}$$

is a generalization of $(I - 1)/N$ (from the equal sample size case), and reduces to that when all the n_i's are equal (see Problem 7.3.4).

Assumptions: The assumptions for the ANOM are the same whether the sample sizes are balanced or unbalanced, namely, the ϵ_{ij}'s are approximately normally distributed and independent with approximately equal variances.

Example 7.3.1 (*Reaction Yield Data, p. 245*)   To compare the three types of catalyst one would compute

$$\bar{y}_{\bullet\bullet} = \frac{3.3 + 4.0 + \cdots + 7.7}{12} = 5.31$$

and construct three different sets of decision lines around $\bar{y}_{\bullet\bullet}$ using formula (7.3.1) and $MS_e = 0.607$ from Example 7.1.1 (p. 245). The ANOM decision lines for $\alpha = 0.05$ are

$$A: 5.31 \pm m(0.05; 3, 9)\sqrt{0.607}\sqrt{\tfrac{12-3}{12(3)}}$$
$$\pm 2.89(0.3896)$$
$$\pm 1.126$$

$$B: 5.31 \pm 2.89\sqrt{0.607}\sqrt{\tfrac{12-4}{12(4)}}$$
$$\pm 0.919$$

$$C: 5.31 \pm 2.89\sqrt{0.607}\sqrt{\tfrac{12-5}{12(5)}}$$
$$\pm 0.769.$$

The ANOM chart is given in Figure 7.8, from which it is clear that there are differences in the three catalysts due to catalyst A producing significantly low yield and catalyst C producing significantly high yield.

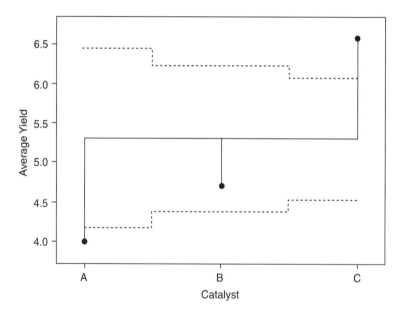

Figure 7.8. ANOM Chart for the Reaction Yield Data (Example 7.3.1) with $\alpha = 0.05$.

Example 7.3.2 (*Soap Packaging Data*) A small company's most popular soap product must be repackaged because the current package type is no longer available. A small consumer study had been performed to determine the best new package for the soap (Box, Foil, or Shrink Wrap). The study had shown that the foil wrapper was the most popular wrapper, however, the director of this project wanted more information. An in-store test was run where all three packages were available and

Y = number of boxes sold in a week was recorded. This was intended to be a balanced experiment, run for 5 weeks, however, after 3 weeks the company had supplier problems and did not have any more boxes to use in the study. The resulting data is given in Table 7.5 and in `soap.txt`.

Table 7.5. Average Number of Boxes Sold in a Week for Each Type of Packaging (Example 7.3.2)

	Package Type	
Box	Foil	Shrink Wrap
38.1	52.8	33.1
32.6	57.3	29.7
35.4	56.6	38.3
	58.0	32.6
	55.7	33.7
$\bar{y}_{B\bullet} = 35.4$	$\bar{y}_{F\bullet} = 56.1$	$\bar{y}_{SW\bullet} = 33.5$
$s_B = 2.75$	$s_F = 2.02$	$s_{SW} = 3.10$
$n_B = 3$	$n_F = 5$	$n_{SW} = 5$

Thus,

$$\bar{y}_{\bullet\bullet} = \frac{3(35.4) + 5(56.1) + 5(33.5)}{13} = 42.6$$

and using equation (7.1.1),

$$MS_e = \frac{2(2.75)^2 + 4(2.02)^2 + 4(3.10)^2}{10} = 6.99.$$

Also, $\nu = 13 - 3 = 10$ and $m(0.05; 3, 10) = 2.83$. So the decision lines are

$$\text{Box:} \quad 42.6 \ \pm \ 2.83\sqrt{6.99}\sqrt{\tfrac{13-3}{13(3)}}$$
$$\pm \ 3.79$$
$$(38.81, 46.39)$$

$$\text{Foil and Shrink Wrap:} \quad 42.6 \ \pm \ 2.83\sqrt{6.99}\sqrt{\tfrac{13-5}{13(5)}}$$
$$\pm \ 2.62$$
$$(39.98, 45.22)$$

and the ANOM chart is given in Figure 7.9.

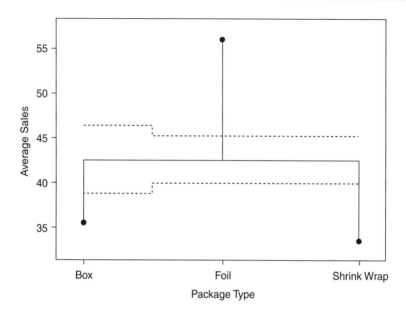

Figure 7.9. ANOM Chart for the Soap Packaging Data (Example 7.3.2) with $\alpha = 0.05$.

From the ANOM chart it is clear that there are significant differences in the way the soap packaging affects sales. The foil package results in significantly higher sales, while the shrink wrap results in significantly lower sales. This study confirms the previous consumer study, which indicated that the foil wrapper would be the best packaging for this soap product.

Problems

1. Fluoride levels were measured in ppm for samples drawn from three separate municipal water sources (see `fluoride.txt`). The ideal level of fluoride is 1.0 ppm.

   (a) Use a computer to make side-by-side box plots and error-bar charts for these data.

   (b) Use a computer to calculate the mean, standard deviation, and sample size for each water source.

   (c) Find the MS_e.

   (d) Compare mean fluoride levels of the three water sources using $\alpha = 0.10$ and the ANOM.

   (e) Write a summary of the results.

2. For the vitamin data set `vitamin.txt` consider the increase in calcium level for each vitamin studied.

   (a) Calculate the mean and standard deviation of the calcium increase for each vitamin.

   (b) Calculate the MS_e and explain what it estimates.

   (c) Compare the three treatments using the ANOM and $\alpha = 0.05$. Explain what your results mean.

3. A contracting firm wishes to compare the drying times (in hours) of various cements used in sidewalks. Type A is a standard concrete, while Types B and C are quick drying compounds (more expensive than the standard compound). The test squares of cement were poured and then tested every 15 minutes until they were dry enough to be walked on. The data are in `cement.txt`.

   (a) Suggest a reason why the experimental design is unbalanced.

   (b) Although it is not stated, what is probably the engineer's main concern in doing this experiment?

   (c) Perform an ANOM for these data using $\alpha = 0.05$ and explain your results.

   (d) Discuss practical versus statistical difference in the framework of this problem.

4. Show that $(N - n_i)/(Nn_i)$ reduces to $(I - 1)/N$ when all the n_i's are equal.

5. An environmental consulting firm collected fish from four sites around a pond feed by the stream discussed in Problem 7.1.3. The scientists collected five fish from each site but in testing some samples were destroyed by accident resulting in an uneven number of readings from each site. The data are stored in `fish2.txt`. Analyze the data using ANOM.

7.4 ANOM FOR PROPORTIONS

The ANOM procedures described in last Sections 7.2 and 7.3 are based on the assumption that the observations are normally distributed. Those procedures are easily modified to allow comparisons of several proportions when the sample sizes are large enough to use the normal approximation for the binomial distribution. To compare several proportions we will test the hypothesis

$$H_0\colon\ p_1 = \cdots = p_I$$

against the alternative that at least one of the p_i's is different.

Equal Sample Sizes

When the proportions are all based on equal sample sizes, then one modifies the decision lines (7.2.4). Recall from Section 4.2 that one can think of the sample proportion $\widehat{p}_i$ from a sample of size n as being the average of n Bernoulli random variables. That is,

$$\widehat{p}_i = \frac{1}{n}\sum_{j=1}^{n} X_{ij}$$

where

$$X_{ij} = 1 \text{ if the } j^{\text{th}} \text{ item from the } i^{\text{th}} \text{ group is a success}$$
$$= 0 \text{ otherwise.}$$

The X_{ij}'s are the counterparts to the Y_{ij}'s used in the decision lines (7.2.4). The variance of X_{ij} is $p_i(1 - p_i)$ (see equation (4.2.3), p. 93), which under the null hypothesis does not depend on i. Thus, $p(1 - p)$ is the counterpart to σ^2, and the best estimate of that quantity (assuming the null hypothesis is true) is $\bar{p}(1 - \bar{p})$, where

$$\bar{p} = \frac{\widehat{p}_1 + \cdots + \widehat{p}_I}{I}.\tag{7.4.1}$$

The degrees of freedom associated with this estimate are ∞ since we are using the normal approximation. Taking all this into account results in the decision lines

$$\text{UDL} = \bar{p} + h(\alpha; I, \infty)\sqrt{\bar{p}(1 - \bar{p})}\sqrt{\tfrac{I-1}{N}}$$

$$\tag{7.4.2}$$

$$\text{LDL} = \bar{p} - h(\alpha; I, \infty)\sqrt{\bar{p}(1 - \bar{p})}\sqrt{\tfrac{I-1}{N}}.$$

Example 7.4.1   A national HMO (Curem, Inc.) has contracts with thousands of physicians' practices. Curem, Inc. has just instituted a new billing procedure. The new procedure is described on the Curem, Inc. web site. A letter was sent to each practice announcing the web site. The HMO wants to know whether the physicians' practices are knowledgeable about the new billing procedure. They decide to conduct a random phone survey of the practices. A question regarding the new billing procedure is posed to each interviewee. The HMO is interested in comparing six types of practices: small urban, medium urban, large urban, small suburban, medium suburban, and large suburban. Any difference among these types of practice will be used to help design training for the new billing procedure. It was decided to randomly select 70 practices of each type to survey. The numbers of practices that were able to correctly answer the question are given in Table 7.6. From the

Table 7.6. The Number of Medical Practices Able to Correctly Answer a Question on the New Billing Procedure

Practice Type	Small Urban	Medium Urban	Large Urban	Small Suburban	Medium Suburban	Large Suburban
Number Correct	9	44	36	7	33	41

data in Table 7.6 one can compute

$$\widehat{p}_1 = 0.129, \quad \widehat{p}_2 = 0.629, \quad \widehat{p}_3 = 0.514, \quad \widehat{p}_4 = 0.100, \quad \widehat{p}_5 = 0.471 , \quad \widehat{p}_6 = 0.586,$$

and

$$\bar{p} = \frac{0.129 + \cdots + 0.586}{6} = 0.405.$$

In order to be sure the sample size is large enough to use the normal approximation to the binomial (and, therefore, the decision lines (7.4.2)), one would check that $n\widehat{p}_i > 5$ and $n(1 - \widehat{p}_i) > 5$ for all i. It suffices to check these conditions for just

$$\widehat{p}_{\min} = \text{the smallest } \widehat{p}_i$$

and

$$\widehat{p}_{\max} = \text{the largest } \widehat{p}_i.$$

In fact, one only really needs to check that

$$n\widehat{p}_{\min} > 5 \quad \text{and} \quad n(1 - \widehat{p}_{\max}) > 5 . \qquad (7.4.3)$$

For this example we have

$$n\widehat{p}_{\min} = 70(0.100) = 7 > 5$$
$$n(1 - \widehat{p}_{\max}) = 70(1 - 0.629) = 70(0.371) = 25.97 > 5.$$

Thus, the sample size is large enough, and the $\alpha = 0.01$ decision lines (7.4.2) are

$$0.405 \pm h(0.01; 6, \infty)\sqrt{0.405(1 - 0.405)}\sqrt{\frac{5}{420}}$$
$$\pm 3.14(0.0536)$$
$$\pm 0.168$$
$$(0.237, 0.573).$$

The ANOM chart is given in Figure 7.10 from which one sees that small practices did significantly poorly, while medium urban and large suburban practices did significantly well.

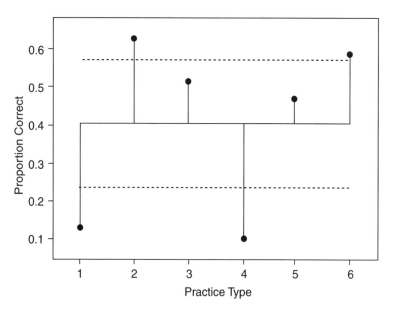

Figure 7.10. ANOM Chart for the HMO Data (Example 7.4.1) with $\alpha = 0.01$.

Unequal Sample Sizes

When the proportions are based on unequal sample sizes, then one modifies the decision lines (7.3.1). The best estimator of $p(1 - p)$ is still $\bar{p}(1 - \bar{p})$, but now it is computed using

$$\bar{p} = \frac{n_1 \hat{p}_1 + \cdots + n_I \hat{p}_I}{N} \,. \qquad (7.4.4)$$

Note that $\bar{p}$ is really nothing more than the overall proportion of defectives in the combined samples. The degrees of freedom associated with this estimator are again ∞ since we are using the normal approximation. Taking all this into account results in the decision lines

$$\text{UDL} = \bar{p} + m(\alpha; I, \infty)\sqrt{\bar{p}(1-\bar{p})}\sqrt{\frac{N-n_i}{Nn_i}}$$

$$\text{LDL} = \bar{p} - m(\alpha; I, \infty)\sqrt{\bar{p}(1-\bar{p})}\sqrt{\frac{N-n_i}{Nn_i}} .$$

$(7.4.5)$

Again, before computing these decision lines, one would want to verify that the sample sizes are large enough to use the normal approximation by checking

$$n_i\hat{p}_i > 5 \quad \text{and} \quad n_i(1-\hat{p}_i) > 5 \quad \text{for } i = 1, \ldots, I.$$

$(7.4.6)$

Example 7.4.2   The purchasing department must choose between three competing suppliers for a component. Inspection of shipments ordered from each supplier showed 15 defective components out of 1000 from Supplier 1, 26 defective components out of 1500 from Supplier 2, and 20 defective components out of 500 from Supplier 3. Are there any differences in the suppliers' defective rates at the 0.05 significance level?

The three sample proportions are

$$\hat{p}_1 = \tfrac{15}{1000} = 0.015, \quad \hat{p}_2 = \tfrac{26}{1500} = 0.0173, \quad \hat{p}_3 = \tfrac{20}{500} = 0.04,$$

and (equation $(7.4.4)$)

$$\bar{p} = \frac{15 + 26 + 20}{1000 + 1500 + 500} = \frac{61}{3000} = 0.020.$$

Checking condition $(7.4.6)$ one would compute

$$n_1\hat{p}_1 = 15, \quad n_1(1-\hat{p}_1) = 985;$$
$$n_2\hat{p}_2 = 26, \quad n_2(1-\hat{p}_2) = 1474;$$
$$n_3\hat{p}_3 = 20, \quad n_3(1-\hat{p}_3) = 480.$$

Since these quantities are all greater than 5, we can proceed with the ANOM. For $\alpha = 0.05$ the decision lines are

Supplier 1: $0.020 \pm m(0.05; 3, \infty)\sqrt{0.020(1-0.020)}\sqrt{\frac{3000-1000}{3000(1000)}}$

$ \pm 2.39(0.00361)$

$ \pm 0.00863$

$ (0.0114, 0.0286)$

Supplier 2: $0.020 \pm m(0.05; 3, \infty)\sqrt{0.020(1-0.020)}\sqrt{\frac{3000-1500}{3000(1500)}}$

$ \pm 2.39(0.00256)$

$ \pm 0.00611$

$ (0.0139, 0.0261)$

Supplier 3: $0.020 \pm m(0.05; 3, \infty)\sqrt{0.020(1-0.020)}\sqrt{\frac{3000-500}{3000(500)}}$

$ \pm 2.39(0.00572)$

$ \pm 0.01370$

$ (0.0063, 0.0337).$

The ANOM chart is given in Figure 7.11 from which one sees that there are significant differences due to Supplier 3 producing significantly more defectives.

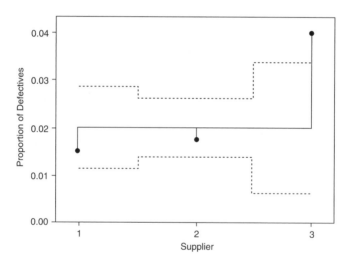

Figure 7.11. ANOM Chart for Defectives From Three Suppliers with $\alpha = 0.05$.

Problems

1. A study of tinted lenses was conducted to see if tinted lenses were preferred over non-tinted lenses, and which colors were preferred. For each of four colors 36 people were asked to choose between non-tinted lenses and tinted lenses. The results are given in the table below.

Tint	Selected as Better	Sample Proportion
Blue	27	0.75
Green	20	0.56
Purple	16	0.44
Red	10	0.28

   (a) Construct an ANOM chart to test for significant differences between tints at the $\alpha = 0.05$ level, and write a brief report to accompany the chart.

   (b) Test at the $\alpha = 0.05$ level to see if tinted lenses are preferred.

2. Four treatments were studied in the development of a fiberglass with an end use in the molded fiberglass market. The choice of resin and the use (or non-use) of a vacuum in the process were studied. Thirty-two test panels were molded for each treatment and the number of defective panels was counted. Unfortunately, before all the panels could be evaluated, three of the panels produced with resin 2 and the vacuum off were damaged. The numbers of undamaged defective panels are given below.

Resin	Vacuum	Defectives	# of Panels
1	Off	8	32
1	On	6	32
2	Off	11	29
2	On	10	32

   Use the ANOM to evaluate this study.

3. Using the safety data in `safety.txt`, calculate the number of incidents per employee at each location. Use the ANOM to compare the safety records at the various work locations.

4. A plant manager felt that the third shift was producing more non-conforming items than the other two shifts. Before implementing an improvement initiative the manager collected data for a week on the number of non-conforming items per shift. Shift 1 produced 963 out of 10,024 units; Shift 2 produced 873 out of 9553 units; and Shift 3 produced 1076 out of 13,244 units. Is there a difference between shifts? Create an ANOM chart that will help the plant manager understand the data that has been collected.

5. In the validation study on tube seals (see Problem 6.3.6, p. 241) 2 of 50 tubes failed at the low temperature setting, and 3 of 53 tubes failed at the high temperature setting. An additional sample of 56 tubes was taken at the center point of the temperature range. Only one failure was recorded from this sample. Use the ANOM to test for differences in failure rates due to temperature.

7.5 THE ANALYSIS OF VARIANCE

The **analysis of variance** (ANOVA) is an alternative to the ANOM for testing

$$H_0: \ \mu_1 = \cdots = \mu_I$$

against the alternative that at least one of the μ_i's is different. The assumptions underlying the ANOVA are the same as those for the ANOM. That is, the ϵ_{ij} are all at least approximately normally distributed and independent, and the I populations all have equal variances. These assumptions can be checked in the same way as for the ANOM (see p. 253).

In the one-factor setting the ANOM is generally preferred over the ANOVA since it reveals which treatments are responsible for any significant difference found and provides a means for assessing practical significance. The ANOVA, on the other hand, only tells us whether or not the means look different. As we will see in the next chapter, however, for multi-factor studies the ANOVA provides information that the ANOM does not. We have opted to introduce the ANOVA in its simplest form, namely, the one-factor setting.

Equal Sample Sizes

The basic idea behind the ANOVA is fairly simple. If H_0 is true, then the sample means all come from the same population, and the sample variance of the $\overline{y}_{i\bullet}$'s is an estimate of $\mathrm{Var}(\overline{Y}_{i\bullet}) = \sigma^2/n$. Thus, if H_0 is true, the **treatment mean square**

$$\mathrm{MS}_{\text{treatment}} = n(\text{variance of the } \overline{y}_i\text{'s}) = \frac{n\sum(\overline{y}_i - \overline{y})^2}{I-1} \qquad (7.5.1)$$

is an estimate of σ^2. When H_0 is false, $\mathrm{MS}_{\text{treatment}}$ will tend to be larger than σ^2.

Since MS_e is an estimate of σ^2 regardless of whether H_0 is true, the ANOVA tests H_0 by computing

$$F = \frac{\mathrm{MS}_{\text{treatment}}}{\mathrm{MS}_e} \qquad (7.5.2)$$

and rejecting H_0 if F is too large. Too large is defined as greater than $F(\alpha; \nu_1, \nu_2)$, where these critical values are found in Table B7 and depend on

$$\alpha = \text{the level of significance desired}$$
$$\nu_1 = \text{the numerator degrees of freedom}$$
$$= I - 1 \text{ for the one-way layout}$$
$$\nu_2 = \text{the denominator degrees of freedom}$$

$= N - I$ for the one-way layout.

In intuitive terms, the test statistic F compares the variation among the treatment means to the random variation within the treatment groups. If the treatment has an effect, we expect to see the sample means further apart than could be explained by randomness.

Example 7.5.1   For the Optical Lens Data (p. 26) we have previously computed (Example 3.1.2, p. 39) $MS_e = 1.68$ and $N - I = 24$. Using the ANOVA and $\alpha = 0.01$, one obtains $MS_{treatment} = 7(2.43) = 17.01$ and

$$F = \frac{17.01}{1.68} = 10.125.$$

Since $10.125 > F(0.01; 3, 24) = 4.718$, one would reject H_0. Note that this is the same conclusion we came to in Example 7.2.1 (p. 251) using the ANOM. The information from an ANOVA is usually summarized in an ANOVA table like the one in Table 7.7. The ANOVA table provides the p-value (0.0002) associated with the

Table 7.7. ANOVA Table for Example 7.5.1

```
ANALYSIS OF VARIANCE TABLE FOR DHAZE

SOURCE            DF      SS          MS          F        P
-------------     ----    ----------  ----------  -------  ------
TREATMENT (A)      3      51.0674     17.0225     10.12    0.0002
RESIDUAL          24      40.3521      1.68134
-------------     ----    ----------
TOTAL             27      91.4195
```

F value of 10.12. This can be used to test significance by comparing the p-value with the α level. Since $0.0002 < 0.01$, one would reject H_0.

Unequal Sample Sizes

With unequal sample sizes the equation for the $MS_{treatment}$ (equation (7.5.1)) is modified to obtain

$$MS_{treatment} = \frac{\sum n_i (\bar{y}_i - \bar{y})^2}{I - 1} \tag{7.5.3}$$

which is then compared with MS_e using the F statistic (equation (7.5.2)).

Example 7.5.2   (*Reaction Yield Data, p. 245*)   Consider again the three catalyst types. We have previously computed (see Example 7.3.1, p. 261) $\bar{y}_{\bullet\bullet} = 5.31$ and

$MS_e = 0.607$. Using the ANOVA to test for any differences in the three catalysts, one would compute

$$MS_{treatment} = \frac{3(4 - 5.31)^2 + 4(4.7 - 5.31)^2 + 5(6.58 - 5.31)^2}{2} = \frac{14.70}{2} = 7.35.$$

The test statistic is then

$$F = \frac{7.35}{0.607} = 12.1$$

which for $\alpha = 0.05$ would be compared with $F(0.05; 2, 9) = 4.256$. Since $12.1 > 4.256$, one would reject H_0 and conclude that there are significant differences in the three catalysts. The three types of catalysts do not all achieve the same product yield when used in this particular reaction.

Problems

1. Perform an analysis of variance on the data in Problem 7.2.1 (p. 259) using $\alpha = 0.05$.

2. Perform an analysis of variance on the adhesion data `adhesion.txt` using $\alpha = 0.10$, and explain what your results mean.

3. Perform an analysis of variance on the `gloss.txt` data using a computer. Report the p-value of your test.

4. Perform an analysis of variance on the data from Problem 7.3.3 (p. 265) by hand using $\alpha = 0.01$, and explain what your results mean.

5. Analyze the data in `fish.txt` from Problem 7.1.3 (p. 249) using the ANOVA. Write a brief report of your analysis, include a plot (not an ANOM chart) to enhance the findings of the ANOVA analysis.

6. Perform an analysis of variance on the data in `fluoride.txt`. Compare your results to those obtained in Problem 7.3.1 (p. 265).

7.6 THE EQUAL VARIANCES ASSUMPTION

Both the ANOM and the ANOVA assume that all the samples come from populations with equal variances. Previously, we have used normal probability plots and residual plots to help check this assumption. It is also possible to formally test the hypothesis that the population variances are all equal. In this section, we will discuss two such tests and what can be done when the variances appear to be unequal.

Testing for Equal Variances

The first test is based on the maximum ratio of the sample variances. It is easier to apply, but it is not as powerful as the second test. The second test, the **analysis of means for variances** (ANOMV), is based on the ANOM. It requires more work to apply, but it is better at detecting unequal variances.

The $F_{\max}$ Test

We are interested in testing

$$H_0 : \sigma_1^2 = \cdots = \sigma_I^2 \tag{7.6.1}$$

against the alternative that at least one of the σ_i^2's is different. The $F_{\max}$ test, which is quick and easy to perform, is based on the assumption that the sample sizes are the same for all the treatment groups. The test statistic is

$$F_{\max} = \frac{\max_i s_i^2}{\min_i s_i^2} \tag{7.6.2}$$

and one rejects H_0 if $F_{\max} > F_{\max}(\alpha; I, \nu)$, where the critical values are given in Table B10. The critical values $F_{\max}(\alpha; I, \nu)$ depend on

$$\alpha = \text{the level of significance desired}$$
$$I = \text{the number of variances being compared}$$
$$\nu = \text{the common degrees of freedom for each } s_i^2.$$

Example 7.6.1 (*Tennis Ball Data, p. 253*) To see if it is reasonable to assume that all four tennis ball coverings have the same variability in wear, one would compute

$$F_{\max} = \frac{600.5}{266.8} = 2.25.$$

Each of the four sample variances has $\nu = 5 - 1 = 4$ degrees of freedom, and for $\alpha = 0.\dot{1}$, one finds $F_{\max}(0.1; 4, 4) = 13.88$. Since 2.25 is not greater than 13.88, one would not reject the null hypothesis. There is no evidence that the four types have different variability in wear.

If the treatment groups have different (but not extremely different) sample sizes, a conservative version of the F_{max} test can be performed by computing F_{max} and comparing it with the critical value obtained using the smallest sample size to obtain ν (i.e., $\nu = \min_i\{n_i - 1\}$).

Example 7.6.2 (*Reaction Yield Data, p. 245*)   To check that the three catalysts all have the same variability, one would compute

$$F_{max} = \frac{0.88^2}{0.68^2} = 1.67$$

Since $1.67 \not> F_{max}(0.1; 3, 2) = 42.48$, do not reject the hypothesis of equal variances.

The ANOMV

The ANOM, which is a test of location (i.e., a test on means), can be converted into a test of scale (i.e., a test on variances) and used to test the hypothesis of equal variances. This is accomplished by transforming the observations y_{ij} to

$$t_{ij} = (y_{ij} - \overline{y}_{i\bullet})^2$$

which when averaged become sample variances (but are also means). The ANOMV is a good choice for testing the hypothesis (7.6.1) when the assumption of normality is reasonable.

Letting $\sigma^2 = \sigma_\bullet^2$ and $\tau_i = \sigma_i^2 - \sigma^2$, the hypothesis (7.6.1) can be written as

$$H_0: \ \tau_1 = \cdots = \tau_I = 0$$

and estimates of the τ_i's are

$$\widehat{\tau}_i = \overline{t}_{i\bullet} - \overline{t}_{\bullet\bullet}.$$

When one has samples of equal size n from each of the I populations,

$$\overline{t}_{i\bullet} = \frac{n-1}{n}s_i^2 \quad \text{and} \quad \overline{t}_{\bullet\bullet} = \frac{n-1}{n}MS_e$$

and, therefore,

$$\widehat{\tau}_i = \frac{n-1}{n}(s_i^2 - MS_e).$$

The $\widehat{\tau}_i$'s are the counterparts of the $\widehat{\alpha}_i$'s, and analogous to the ANOM test, one rejects the hypothesis (7.6.1) if for some i the quantity $\widehat{\tau}_i$ (a measure of the discrepancy between the sample variance and the average of the I variances) is "too large" in magnitude. With a little algebra it is possible to show that $\widehat{\tau}_i$ being "too large" in magnitude is equivalent to s_i^2 falling outside the decision lines

$$\begin{array}{l} UDL = U(\alpha; I, \nu)I \ MS_e \\ CL = MS_e \\ LDL = L(\alpha; I, \nu)I \ MS_e. \end{array} \qquad (7.6.3)$$

The critical values $U(\alpha; k, \nu)$ and $L(\alpha; k, \nu)$ are given in Table B11, and depend on

α = the level of significance desired

I = the number of means being compared

ν = the common degrees of freedom for each s_i^2.

Example 7.6.3   The warm-up time in seconds was determined for three different brands of scanners using four scanners of each brand. The results are given in
Table 7.8, together with the sample means and variances for each brand. Averaging

Table 7.8. Warm-Up Times for Three Brands of Scanners

	Brand	
1	2	3
17	20	23
20	20	20
20	38	16
23	26	17
$\overline{y}_{1\bullet} = 20$	$\overline{y}_{2\bullet} = 26$	$\overline{y}_{3\bullet} = 19$
$s_1^2 = 6$	$s_2^2 = 72$	$s_3^2 = 10$

the sample variances to obtain $MS_e = 29.33$ and using the ANOMV and with
$\alpha = 0.1$, one obtains

$$UDL = U(0.1; 3, 3) \, I \, MS_e = (0.7868)(3)(29.33) = 69.24$$

$$CL = MS_e = 29.33$$

$$LDL = L(0.1; 3, 3) \, I \, MS_e = (0.0277)(3)(29.33) = 2.44.$$

From the ANOMV chart given in Figure 7.12 one sees that there is evidence of
unequal variances at the 0.1 level due to the variance of Brand 2 being significantly
large.

Note that the F_{max} test would compare $F_{max} = 72/6 = 12$ with $F_{max}(0.1; 3, 3) =$
16.77, and since $12 \not> 16.77$, it would not find any evidence of unequal variances.

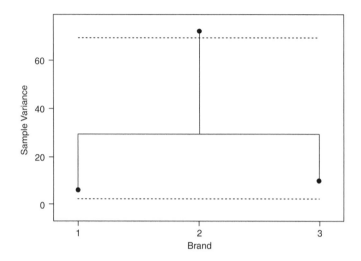

Figure 7.12. ANOMV Chart for the Scanner Data (Example 7.6.3) with $\alpha = 0.1$.

Transformation to Stabilize the Variances

When analysis indicates unequal variances (or non-normality), use of the raw data can give invalid results when testing for equality of means. The problem can sometimes be eliminated by a nonlinear transformation of the data. In general, however, even a nonlinear transformation will have little effect unless the observations vary greatly in magnitude. We will discuss one fairly simple class of variance-stabilizing transformations.

Note: Linear transformations (i.e., $Y \rightarrow aY + b$) have no effect on the variances or the normality of the observations, but can be used for computational convenience (see Problem 7.6.6).

In order to successfully transform unequal variances to equal variances, there must be some relationship between the sample means and the sample variances. As a model for this relationship we will assume that

$$\sigma_i = c\mu_i^p \tag{7.6.4}$$

where c is a constant and p is a parameter to be estimated. This model would hold, for example, if the observations had an exponential distribution; in which case $\sigma_i^2 = \mu_i^2$, and $c = 1$ and $p = 1$.

If model (7.6.4) holds, then applying the transformation

$$g(y) = \begin{cases} y^{1-p} & \text{if } p \neq 1 \\ \ln(y) & \text{if } p = 1 \end{cases} \tag{7.6.5}$$

will stabilize the variances. The only difficulty is that the parameters in equation (7.6.4) are usually unknown. However, when the necessary replicates are available it is possible to estimate p. Taking logs in equation (7.6.4) and replacing the unknown distribution parameters with their sample values gives

$$\ln(s_i) \doteq \ln(c) + p\ln(\overline{y}_{i\bullet}). \tag{7.6.6}$$

Therefore, an estimate of p can be obtained as the slope of the line fitted to a plot of $\ln(s_i)$ versus $\ln(\overline{y}_{i\bullet})$. If the plot of $\ln(s_i)$ versus $\ln(\overline{y}_{i\bullet})$ does not appear to be linear, then one would question the validity of model (7.6.4).

Example 7.6.4   Three brands of gloss white aerosol paints were compared for hiding capability by repeatedly spraying a hiding chart (panels of light to dark gray stripes on a white background) until the pattern was completely obliterated. The amount of paint used was determined by weighing the paint cans both before and after spraying. Four samples of each brand were used and the results are given in Table 7.9 together with the sample mean and variance for each brand. Averaging the three sample variances to obtains $\text{MS}_e = 0.469$, and checking the assumption of equal variances using the ANOMV and $\alpha = 0.1$, one obtains

$$\text{UDL} = U(0.1; 3, 3) \; I \; \text{MS}_e = (0.7868)(3)(0.469) = 1.11$$
$$\text{CL} = \text{MS}_e = 0.469$$
$$\text{LDL} = L(0.1; 3, 3) \; I \; \text{MS}_e = (0.0277)(3)(0.469) = 0.039.$$

Table 7.9. Covering Capabilities (oz./sq. ft.) for Three Brands of Gloss White Aerosol Paints

	Brand	
1	2	3
2.1	4.7	6.4
1.9	3.6	8.5
1.8	3.9	7.9
2.2	3.8	8.8
$\overline{y}_{1\bullet} = 2.0$	$\overline{y}_{2\bullet} = 4.0$	$\overline{y}_{3\bullet} = 7.9$
$s_1^2 = 0.0333$	$s_2^2 = 0.233$	$s_3^2 = 1.14$
$\ln(\overline{y}_{1\bullet}) = 0.693$	$\ln(\overline{y}_{2\bullet}) = 1.39$	$\ln(\overline{y}_{3\bullet}) = 2.07$
$\ln(s_1) = -1.70$	$\ln(s_2) = -0.728$	$\ln(s_3) = 0.0655$

From the ANOMV chart given in Figure 7.13, it appears that the variances are not equal, due to the first variance being significantly small and the third variance being significantly large.

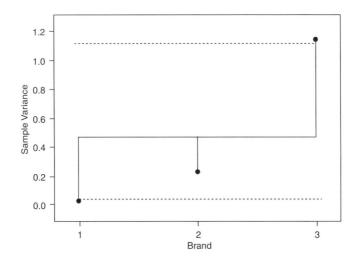

Figure 7.13. ANOMV Chart for the Gloss White Paint Data with $\alpha = 0.1$.

Thus, one would like to find a transformation to stabilize the variances. The $\ln(s_i)$ and $\ln(\overline{y}_{i\bullet})$ are given in Table 7.9 and are plotted in Figure 7.14 together with the least-squares line (see Problem 7.6.5)

$$\widehat{\ln(s_i)} = -2.56 + 1.28 \ln(\overline{y}_{i\bullet}).$$

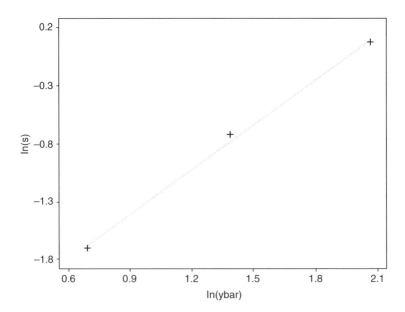

Figure 7.14. Scatter Plot of $\ln(s_i)$ versus $\ln(\overline{y}_{i\bullet})$ Together with the Least-Squares Line.

Therefore, $\widehat{p} = 1.28 \doteq 1.3$ and one would employ the transformation (7.6.5)

$$y_{ij} \rightarrow y_{ij}^{1-1.3} = y_{ij}^{-0.3}.$$

For this limited set of data this is the transformation that will do the best job of making the variances equal, and it does a good job (see Problem 7.6.7). However, it results in very strange units (i.e., $(\text{oz.}/\text{sq. ft.})^{-0.3}$). A transformation that makes more scientific sense would be to use $p = -1$, which would correspond to measuring the paint usage in sq. ft./oz. Using this transform, the transformed (and scaled) values $y'_{ij} = 10/y_{ij}$ are given in Table 7.10 together with their sample means and variances. Analysis of these transformed observations results in the following. Averaging the three variances, one obtains $MS_e = 0.109$, and ANOMV $\alpha = 0.1$ decision lines are

$$\begin{aligned}
\text{UDL} &= U(0.1; 3, 3) \; I \; MS_e = (0.7868)(3)(0.109) = 0.257 \\
\text{CL} &= MS_e = 0.109 \\
\text{LDL} &= L(0.1; 3, 3) \; I \; MS_e = (0.0277)(3)(0.109) = 0.0091.
\end{aligned}$$

Table 7.10. The Transformed Observations $y'_{ij} = 10/y_{ij}$

	Brand	
1	2	3
4.76	2.13	1.56
5.26	2.78	1.18
5.56	2.56	1.27
4.55	2.63	1.14
$\overline{y'}_{1\bullet} = 5.03$	$\overline{y'}_{2\bullet} = 2.52$	$\overline{y'}_{3\bullet} = 1.29$
$s_1^2 = 0.212$	$s_2^2 = 0.078$	$s_3^2 = 0.037$

From the ANOMV chart given in Figure 7.15 there is now no evidence of unequal variances, and a normal probability plot of the transformed data (Figure 7.16) reveals nothing to indicate non-normality.

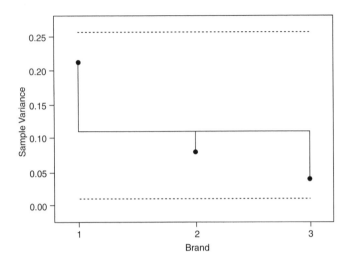

Figure 7.15. ANOMV Chart for the Transformed Data in Example 7.6.4 with $\alpha = 0.1$.

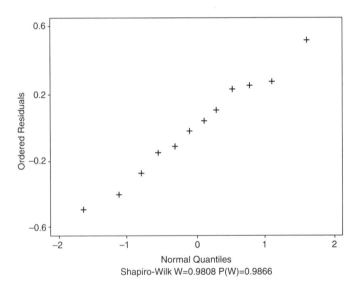

Figure 7.16. A Normal Probability of the Residuals from the Transformed Data.

The $\alpha = 0.001$ ANOM decision lines for the transformed values are

$$\overline{y'}_{\bullet\bullet} \pm h(0.001; 3, 9)\sqrt{\mathrm{MS}_e}\sqrt{\frac{2}{12}}$$

$$2.95 \pm (5.49)\sqrt{0.109}\sqrt{\frac{1}{6}}$$

$$\pm\, 0.740$$

$$(2.21, 3.69)$$

and examination of the ANOM chart in Figure 7.17 reveals that Brand 1 values are significantly large (good) and Brand 3 values are significantly small (bad) at the $\alpha = 0.001$ level. Both

$$\overline{y'}_{1\bullet} - \overline{y'}_{\bullet\bullet} = 5.03 - 2.95 = 2.08 \text{ sq. ft./oz.}$$
$$\overline{y'}_{3\bullet} - \overline{y'}_{\bullet\bullet} = 1.29 - 2.95 = -1.66 \text{ sq. ft/oz.}$$

are of practical significance.

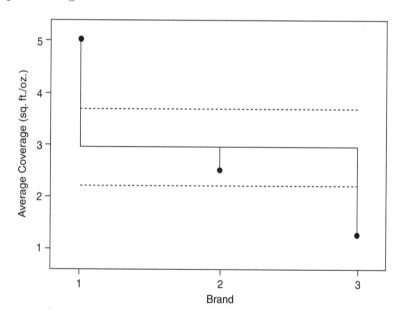

Figure 7.17. ANOM Chart for the Transformed Paint Data with $\alpha = 0.001$.

Problems

1. Test the data in Example 7.2.3 (p. 256) for equal variances using $\alpha = 0.01$.

2. Test the data in Problem 7.2.1 (p. 259) for equal variances using $\alpha = 0.1$.

3. Test the soap packaging data (Example 7.3.2, p. 262) for equal variances using $\alpha = 0.1$.

4. Test the data in Problem 7.3.3 (p. 265) for equal variances using $\alpha = 0.05$.

5. Verify that the least-squares line $(\widehat{\ln(s_i)} = -2.56 + 1.28\ln(\bar{y}_{i\bullet}))$ given in Example 7.6.4 (p. 281) is correct.

6. In Example 7.6.4 (p. 281), the transformed values were multiplied by 10 (a linear transformation) so that they would not all be fractional values. Show that if we had not multiplied by 10 and had just used the transformation $y'_{ij} = 1/y_{ij}$ that the normal probability plot of the residuals, the ANOMV chart, and the ANOM chart would have all looked the same as in the example.

7. Analyze the paint data in Example 7.6.4 (p. 281) using the transformation $y'_{ij} = 10y_{ij}^{-0.3}$. Specifically, do the following.

    (a) Check the transformed values for equal variances using the ANOMV.

    (b) Check the normality assumption using a normal probability plot.

    (c) Test the effect of Brand using the ANOM and $\alpha = 0.001$.

    (d) If there is an effect, check the practical significance of the effect. Note that with the transformed data in such strange units it is best to evaluate practical significance in the original units.

8. For the data in `battery.txt` find a suitable transformation to stabilize the variances.

9. Use equation (7.6.6) to verify that in Example 7.2.3 (p. 256) the log transformation is the correct transformation to stabilize the variances.

10. Consider the data in `saftey.txt`. Is there a difference in the variability of the number of cases reported for each plant location?

7.7 SAMPLE SIZES

Similarly to what was done in Section 5.4, it is possible to obtain sample sizes for both the ANOM and the ANOVA if one specifies both the level of significance and the power. In the multi-sample setting, however, it is a little more complicated since one has to specify the power for a particular configuration of unequal means. Generally, one measures how far a configuration of unequal means is from the equal means configuration by the maximum distance between any two means. That is, one specifies the power in terms of

$$\Delta = \max_{i,j} \frac{|\mu_i - \mu_j|}{\sigma}$$

which is the counterpart to the quantity

$$d = \frac{|\mu_1 - \mu_0|}{\sigma}$$

used in Section 5.4. Since σ is rarely known, one simply specifies a value for Δ, which can be thought of as the maximum difference between any two means (measured in units of σ) before the means are far enough apart that you want to reject H_0. Just as in the case of one- and two-sample t tests, there are tables that give the necessary sample sizes.

Sample Sizes for ANOM

For a given α-level, power, specified detectable difference, and number of treatment groups; the number of observations that are needed in *each* treatment group is given in Table B8.

Example 7.7.1   One is interested in studying the transfer rates for three different brands of hard drives. How many drives of each brand are needed to detect with probability 0.9 a difference of 2σ for any two drives when testing using the ANOM and $\alpha = 0.05$? From Table B8 with $\alpha = 0.05$, $\Delta = 2$, power $= 0.9$, and $k = 3$; one obtains $n = 8$. Thus, one would need $n = 8$ hard drives of each brand.

Example 7.7.2   Suppose that in the previous example one was only able to obtain $n = 4$ drives of each brand. If one was planning to use the ANOM, would it be worthwhile conducting the experiment? Using Table B8 in reverse; one finds that with $\alpha = 0.05$, $\Delta = 2$, $k = 3$, and $n = 4$; the power is only 0.5. Therefore, there seems to be little point in conducting the experiment.

Sample Sizes for ANOVA

For a given α-level, power, specified detectable difference, and number of treatment groups; the number of observations that is needed in *each* treatment group is given in Table B9.

Example 7.7.3   How many batteries would be needed in order to study the cold cranking power (the number of seconds of rated amperage delivered at $0°F$ without falling below 7.2 volts) of five brands if one planned to test using the ANOVA with $\alpha = 0.1$ and wanted to detect a difference of 2σ with a power of 0.9? From Table B9 with Number of Treatments = 5, $\alpha = 0.1$, $\Delta = 2$, and power = 0.9; one obtains $n = 8$. Therefore, $n = 8$ batteries of each brand are required.

Problems

1. For the Optical Lens experiment (Example 2.3.1, p. 26), how many observations with each coating would be necessary to detect a difference in any two coatings of 1.5σ with probability 0.9 when testing using the ANOM and $\alpha = 0.05$?

2. For the Tennis Ball experiment (Example 7.2.2, p. 253), how many observations on each type are needed to detect a difference in any two types of 1.25σ with power 0.8 when testing using the ANOM and $\alpha = 0.1$?

3. An experiment is to be carried out to compare five treatments. How many observations are required on each treatment in order to be able to detect a difference of 2σ between any two treatments with power 0.95 when using $\alpha = 0.01$ and testing with

   (a) the ANOM?

   (b) the ANOVA?

4. For the Reaction Yield experiment (Example 7.0.1, p. 245), what sample size is needed to detect with power $= 0.95$ differences of 1.5σ using the ANOM and $\alpha = 0.05$?

5. For the Optical Lens data (Example 2.3.1, p. 26), using $\alpha = 0.05$, what is the power to detect a difference of $\Delta = 2$

   (a) When using the ANOM?

   (b) When using the ANOVA?

6. For the Optical Lens experiment (Example 2.3.1, p. 26), how many observations with each coating would be necessary to detect a difference in any two coatings of 1.4σ with probability 0.9 when testing using the ANOVA and $\alpha = 0.05$?

7. For the Tennis Ball experiment (Example 7.2.2, p. 253), how many observations on each type are needed to detect a difference in any two types of 1.6σ with power 0.8 when testing using the ANOVA and $\alpha = 0.1$?

8. The environmental consultants mentioned in Problem 7.1.3 (p. 249) want to catch as few as fish as possible in order to detect a difference of 2σ between any two sites with $\alpha = 0.05$ and power$= 0.80$.

   (a) Using the ANOM, how many fish do they need to catch at each site?

   (b) How many are needed if they only want to detect a 3σ difference?

7.8 CHAPTER PROBLEMS

1. Three different alcohols can be used in a particular chemical process. The resulting yields (in %) from several batches using the different alcohols are given below.

Alcohol		
1	2	3
93	95	76
95	97	77
94	87	84

(a) Test whether or not the three populations appear to have equal variances using $\alpha = 0.1$.

(b) In the event that one finds non-constant variances, what would be the next step in the analysis of this data? Be specific.

2. To determine the effect of three phosphor types on the output of computer monitors, each phosphor type was used in three monitors, and the coded results are given below.

Type		
1	2	3
7	4	1
1	2	3
7	3	-1

(a) Test for constant variances at $\alpha = 0.05$.

(b) Find an estimate for the common variance.

(c) Construct a 95% confidence interval for the difference in the first two phosphor types.

(d) Use the ANOM to test for difference in phosphor types at $\alpha = 0.05$.

3. An experiment was done to compare the amount of heat loss for a three types of thermal panes. The inside temperature was kept at a constant 68°F, the outside temperature was kept at a constant 20°F, and heat loss was recorded for three different panes of each type.

Pane Type		
1	2	3
20	14	11
14	12	13
20	13	9

(a) Test for constant variances at $\alpha = 0.05$.

(b) Use the ANOM to test for differences in heat loss due to pane type at $\alpha = 0.05$. What can you conclude from this test?

(c) Construct a normal probability plot of the residuals.

(d) Use the ANOVA to test for differences in heat loss due to pane type at $\alpha = 0.05$. What can you conclude from this test?

4. An experiment was performed to study the effect of light on the root growth of mustard seedlings. Three different amounts of light (measured in hours per day) were used, and seedlings were assigned at random to the three light amounts. After a week the root lengths (in mm) were measured. Summary statistics were computed and the results are given below.

Light Amount	$\bar{x}_{i\bullet}$	s_i	n_i
1	17	2.5	7
2	15	2	7
3	13	1.5	7

(a) Test for unequal variances using the ANOMV and $\alpha = 0.1$.

(b) Use the ANOM to test for differences in root growth using $\alpha = 0.001$. What can you conclude from this test?

(c) Using the ANOM with $\alpha = 0.05$ in this problem, how large a sample size would be needed in order to detect two means that differ by 1.5σ with power at least 0.75?

(d) Use the ANOVA to test for differences in root growth at level $\alpha = 0.001$. What can you conclude from this test?

5. A summary of the results of an experiment comparing five treatments is given below.

Treatment	y	s	n
1	15.5	3.1	6
2	16.3	2.8	6
3	21.4	2.6	6
4	15.8	3.2	5
5	24.3	3.3	4

(a) Using $\alpha = 0.1$ and the ANOMV, test the assumption of equal variances.

(b) Perform an ANOM test on the data using $\alpha = 0.05$, and explain what your results mean.

(c) Using the ANOVA and $\alpha = 0.05$, test for any differences in the treatment effects.

(d) What is the power of the test in part (c) for detecting any two means that differ by at least by 2.5σ?

6. An experiment was conducted to compare three formulations for a lens coating with regard to its adhesive property. Four samples of each formulation were used, and the resulting adhesions are given below.

Formulation		
1	2	3
5	29	133
10	60	119
21	91	49
23	20	21

(a) Construct a normal probability plot of the residuals.

(b) Test the assumption of equal variances using the ANOMV and $\alpha = 0.05$.

(c) Find a transformation to stabilize the variances.

(d) Test for differences in adhesion using ANOM and $\alpha = 0.05$.

(e) Find the power of the test in part (d) for detecting a difference of $\Delta = 2$.

7. Based on the results of the experiment described in Problem 6, a new experiment was conducted using just Formulations 2 and 3. This time all that was recorded was whether the adhesion was acceptable (greater than 15) or not. The results given below are the number of samples in each category.

	Formulation	
	2	3
> 15	20	25
≤ 15	15	5

(a) Test to see if there is any difference in the two formulations using $\alpha = 0.05$.

(b) Check that the sample sizes are large enough to perform the test in part (a).

8. One is interested in determining if there is any difference in the difficulty of final exams given by three different instructors. Five students were selected at random from each of the three sections and their final exam scores are recorded below along with some summary statistics.

	Instructor		
	1	2	3
	47	48	44
	62	78	85
	73	64	98
	84	58	65
	95	60	76
$\bar{y}_{i\bullet}$	72.2	61.6	73.6
s_i^2	349.7	118.8	420.3

(a) Test the assumption of equal variances using the AMONV and $\alpha = 0.1$.

(b) Using $\alpha = 0.1$ and the ANOM, test if there are any differences in final exam score due to instructor.

(c) How large a sample size would be needed in part (b) to be 90% sure of being able to detect a true difference between two of the means of 1.5σ?

(d) Suppose the exam score of 98 from Instructor 3 was recorded in error and the correct value is unavailable. With only the remaining four observations from Instructor 3 the sample mean and variance are 67.5 and 312.33, respectively. Re-do part (b) in this situation using the ANOVA.

9. A pet store was interested in comparing how weight gain was affected by the brand of food. Three brands (Purina, Alpo and Iams) were used. Dogs were given four cups of food per day, and at the end of a week there weight gain was recorded. The results are given below together with some summary statistics.

	Purina	Alpo	Iams
	4.3	1.8	1.2
	5.0	2.4	0.9
	3.7	2.9	0.6
	2.2	1.3	1.7
$\bar{y}$	3.8	2.1	1.1
s^2	1.42	0.4867	?

(a) Test for unequal variances using the ANOMV and $\alpha = 0.05$.

(b) Construct a normal probability plot of the residuals.

(c) Use the ANOM to compare the three brands. What do you conclude using $\alpha = 0.01$?

(d) If one were interested in detecting a difference in weight of 2σ between any two brands with probability 0.8 when using the ANOM and $\alpha = 0.01$, how many observations would be needed?

10. One is interested in studying the effect of depth from the top of the tank on the concentration of a cleaning solution. Three depths were used and the data are summarized below.

i	$\overline{x}_{i\bullet}$	s_i^2	n_i
1	17	2.5	7
2	15	2	8
3	49.5	21.5	7

(a) Test H_0: $\mu_1 = \mu_2$ versus H_a: $\mu_1 \neq \mu_2$ at level $\alpha = 0.05$.

(b) Test for homogeneity of variances using the ANOMV and $\alpha = 0.1$.

(c) If necessary, compute an appropriate transformation to stabilize the variances.

(d) Use the ANOM to test for differences in concentration using $\alpha = 0.05$.

(e) For the test in (d) how large a sample size would be needed in order to detect two means that differ by 1.5σ with power at least 0.75?

11. One is interested in comparing three sections of a statistics course to see if there is any difference in the proportions of students receiving A's. The data are given below.

Section 1	Section 2	Section 3
$n_1 = 16$	$n_2 = 32$	$n_3 = 20$
2 A's	8 A's	5 A's

Test to see if there is any difference using $\alpha = 0.1$.

12. Using the data set `exposure.txt` (see Problem 2.3.4, p. 31), run an ANOM to see if their is a difference in the three test formulations.

13. A filling machine had just undergone maintenance to replace worn parts. To check to see if the machine was running as desired five tubes of product from each of four filling runs were weighed and the data are recorded in the data set `fillweight.txt`.

(a) Is there a difference in the mean weights due to batch? Use both ANOM and ANOVA techniques and compare your results.

(b) Use a runs chart to look at this data over time. What do you notice about each group of five data points?

8

EXPERIMENTS WITH TWO FACTORS

The experimental design discussed in Chapter 7 assumes that each treatment is a different level of a single factor. The underlying model was $Y_{ij} = \mu_i + \epsilon_{ij}$. Often, however, a process to be studied will involve more than one factor, and studying more than one factor at a time leads to much greater efficiency. That is, one can obtain more (or better) information with less work than if one studied the factors separately. The simplest experimental designs for studying two or more factors are complete layouts. Consider the study of the sources of variability in the adhesion of a lens coating (Problem 2.33, p. 31). There are I different pH's being considered and J catalysts. The I pH's are I levels of factor A, and the J catalysts are J levels of factor B. Observations from this experiment could be arranged in an $I \times J$ table consisting of one row for each level of factor A and one column for each level of factor B, where the observations corresponding to the ith level of A and the jth level of B are recorded in cell (i, j). This is called a **two-way layout**, and if every cell in the table contains at least one observation, then it is a **complete layout**. Complete layouts are one example of a class of designs referred to as **factorial designs**. For the adhesion data we have a factorial design with $I = 3$, $J = 2$, and $n = 5$ observations per cell. Designs with the same number of observations (replicates) in each cell are called **balanced** designs. We will assume in this chapter that multi-factor experiments are complete and balanced. The analysis of unbalanced experiments is beyond the scope of this course, and some incomplete designs are discussed in Chapter 9.

8.1 INTERACTION

In addition to the improved efficiency obtained from the use of multi-factor designs, one is also afforded the opportunity to study possible interaction between the factors. As was mentioned in Chapter 2, two factors are said to interact if the behavior of one factor depends on the level of the other factor. Thus, it is not possible to study interaction with one-factor-at-a-time experiments, and the unidentified presence of an interaction can lead to erroneous results. Factorial designs, on the other hand, can estimate interactions, and if none are present they allow the effect of every factor to be evaluated as if the entire experiment were devoted entirely to that factor.

As was noted in Chapter 2, an interaction plot can only indicate the possible presence of an interaction. Whether or not the differences in the slopes of the line segments in an interaction plot are statistically significant depends on the experimental error. In order to indicate the magnitude of the experimental error, confidence intervals are sometimes drawn around the end points of the line segments in an interaction plot. The plot in Figure 8.1 is an example of this for the adhesion data (`adhesion2.txt`). However, since one is actually interested in comparing the slopes of the line segments, confidence intervals on the end points still only provide an indication with regard to the presence of an interaction.

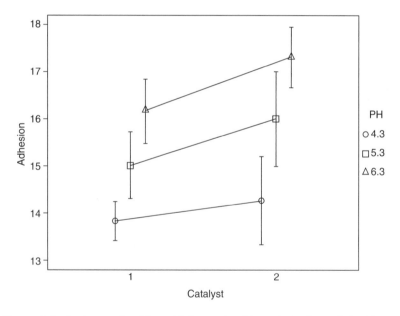

Figure 8.1. An Interaction Plot with Intervals of Length 2σ Around the Means.

8.2 MORE THAN ONE OBSERVATION PER CELL

When there are $n > 1$ observations per cell, statistical tests for interactions as well as for the main effects of both factors can be performed using a two-way ANOVA. This is done using the model

$$Y_{ijk} = \mu + \alpha_i + \beta_j + \alpha\beta_{ij} + \epsilon_{ijk}. \tag{8.2.1}$$

This is similar to model (3.2.1), but it includes an interaction term.

Notation:

$$\alpha\beta_{ij} = \text{ the interaction between the } i^{\text{th}} \text{ level of A and the } j^{\text{th}} \text{ level of B}$$

As usual, we assume the ϵ_{ijk}'s are $N(0, \sigma^2)$ and independent, and one would want to check the reasonableness of these assumptions before proceeding with any analysis based on model (8.2.1). Again as usual, this can be done using a normal probability plot of the residuals and the ANOMV to test for equal variances. For model (8.2.1) the residuals are

$$\begin{aligned}
\widehat{\epsilon}_{ijk} &= y_{ijk} - \widehat{y}_{ijk} \\
&= y_{ijk} - [\widehat{\mu} + \widehat{\alpha}_i + \widehat{\beta}_j + \widehat{\alpha\beta}_{ij}]
\end{aligned}$$

where estimates of the overall mean and the main effects (the α_i's and the β_j's) are given in equations (3.2.2), and estimates of the interaction terms are obtained using the formula

$$\widehat{\alpha\beta}_{ij} = \overline{y}_{ij\bullet} - \overline{y}_{i\bullet\bullet} - \overline{y}_{\bullet j\bullet} + \overline{y}_{\bullet\bullet\bullet}. \tag{8.2.2}$$

Combining all these equations, the residuals for model (8.2.1) reduce to (see Problem 8.2.9)

$$\widehat{\epsilon}_{ijk} = y_{ijk} - \overline{y}_{ij\bullet}. \tag{8.2.3}$$

With model (8.2.1) there are three null hypotheses of interest

$$\begin{aligned}
H_A &: \alpha_i = 0 \text{ for all } i \\
H_B &: \beta_j = 0 \text{ for all } j \\
H_{AB} &: \alpha_i\beta_j = 0 \text{ for all } (i, j).
\end{aligned}$$

These hypotheses correspond to, respectively, Factor A has no effect, Factor B has no effect, and there is no AB interaction. The hypothesis H_{AB} must be tested first since the appropriate subsequent analyses depend on whether or not any interaction is present. In order to do this the $\alpha\beta_{ij}$ terms are combined to form the sum of squares for the AB interaction

$$SS_{AB} = n \sum_{i=1}^{I} \sum_{j=1}^{J} (\widehat{\alpha\beta_{ij}})^2.$$

(8.2.4)

The degrees of freedom for SS_{AB} are the product of the degrees of freedom for Factor A and the degrees of freedom for Factor B, namely, $(I-1)(J-1)$. Therefore, the mean square for the AB interaction is

$$MS_{AB} = \frac{SS_{AB}}{(I-1)(J-1)}$$

(8.2.5)

and the test statistic for testing H_{AB} is

$$F_{AB} = \frac{MS_{AB}}{MS_e}$$

(8.2.6)

where

$$MS_e = \frac{1}{IJ} \sum_{i,j} s_{ij}^2$$

(8.2.7)

is simply the average of the sample variances for all the treatment combinations. The hypothesis H_{AB} is rejected if

$$F_{AB} > F(\alpha; (I-1)(J-1), IJ(n-1)).$$

Example 8.2.1   To analyze the adhesion data (`adhesion2.txt`), one would first check the assumptions of normality and equal variances. A normal probability plot of the residuals from model (8.2.1) is given in Figure 8.2, and there is no indication that normality is a problem.

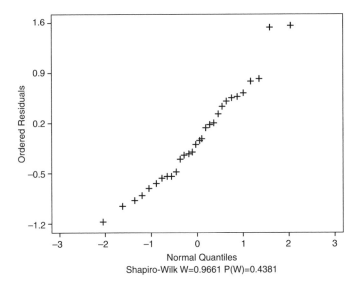

Figure 8.2. A Normal Probability Plot of the Adhesion Data in Example 8.2.1.

To test for equal variances we need a table of the cell variances, which is given in Table 8.1.

Table 8.1. Cell Variances for the Adhesion Data in Example 8.2.1

	Catalyst (B)	
pH (A)	1	2
4.3	0.172	0.896
5.3	0.566	1.023
6.3	0.448	0.415

From the table of variances one finds

$$\text{MS}_e = (0.172 + \cdots + 0.415)/6 = 0.587.$$

Since there are five replicates, the ANOMV decision lines for $\alpha = 0.1$ are

$$\text{UDL} = U(0.1; 6, 4)I\text{MS}_e = (0.4794)(6)(0.587) = 1.69$$
$$\text{CL} = \text{MS}_e = 0.587$$
$$\text{LDL} = L(0.1; 6, 4)I\text{MS}_e = (0.0130)(6)(0.587) = 0.46.$$

It is clear from the ANOMV chart in Figure 8.3 that there is no indication of unequal variances.

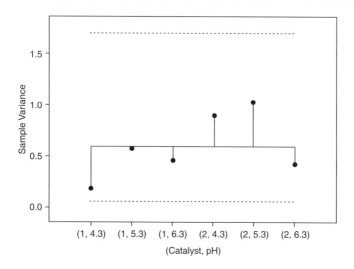

Figure 8.3. An ANOMV Chart for the Variances in Example 8.2.1 with $\alpha = 0.1$.

The next step is to test for interaction between catalyst type and pH level. It is easiest to compute the $\widehat{\alpha\beta}_{ij}$'s by first computing a table of sample means (Table 8.2).

Table 8.2. A Table of the Sample Means for Example 8.2.1

pH (A)	Catalyst (B) 1	2	pH average
4.3	13.824	14.264	14.044
5.3	14.918	16.024	15.471
6.3	16.038	17.330	16.684
Catalyst average	14.927	15.873	15.400

From this table one would compute (using equation (8.2.2))

$$\widehat{\alpha\beta}_{11} = 13.824 - 14.044 - 14.927 + 15.400 = 0.253$$
$$\widehat{\alpha\beta}_{12} = 14.264 - 14.044 - 15.873 + 15.400 = -0.253$$
$$\widehat{\alpha\beta}_{21} = 14.918 - 15.471 - 14.927 + 15.400 = -0.08$$
$$\text{and so forth.}$$

All of the $\widehat{\alpha\beta}_{ij}$'s are given in Table 8.3.

Table 8.3. The $\widehat{\alpha\beta}_{ij}$'s for Example 8.2.1

	Catalyst (B)	
pH (A)	1	2
4.3	0.253	−0.253
5.3	−0.08	0.08
6.3	−0.173	0.173

Using equations (8.2.4)–(8.2.6), one obtains

$$SS_{AB} = (5)[0.253^2 + (-0.08)^2 + \cdots + 0.173^2] = 1.00$$

$$MS_{AB} = \frac{1.00}{2} = 0.50$$

$$F_{AB} = \frac{0.50}{0.587} = 0.85.$$

Since $0.85 \ngtr F(0.1; 2, 24) = 2.538$, there is no significant interaction between pH level and catalyst type.

Example 8.2.2   A 3×2 factorial experiment was designed to study the effect of three alcohols and two bases on the percent yield for a particular chemical process. The process was run four times with each alcohol/base combination, and the results are given in Table 8.4. The 24 ($= 3 \times 2 \times 4$) trials were run in a random order.

Note: Each of the four replicates (i.e, the four trials with each alcohol/base combination) required a complete run of the process. Simply taking four measurements from one run of the process with a particular alcohol/base combination would not suffice since that would not incorporate the batch-to-batch variability. Also, the trials were randomized (run in random order) to average out the effects of any unaccounted for factors.

Table 8.4. Results for the Experiment Described in Example 8.2.2

		Base			
		1		2	
Alcohol	1	91.3	89.9	87.3	89.4
		90.7	91.4	91.5	88.3
	2	89.3	88.1	92.3	91.5
		90.4	91.4	90.6	94.7
	3	89.5	87.6	93.1	90.7
		88.3	90.3	91.5	89.8

From this table, a table of sample means (Table 8.5) and a table of sample variances (Table 8.6) were calculated. A normal probability plot of the residuals from model (8.2.1) is given in Figure 8.4 and shows the assumption of normality is reasonable.

Table 8.5. The Average Yield for Each Alcohol/Base Combination Together with Row Means, Column Means, and the Grand Mean

		Base		
		1	2	
Alcohol	1	90.825	89.125	89.975
	2	89.800	92.275	91.037
	3	88.925	91.275	90.100
		89.850	90.892	90.371

Table 8.6. Sample Variances for Each Cell in Example 8.2.2.

		Base	
		1	2
Alcohol	1	0.476	3.243
	2	2.020	3.096
	3	1.456	1.963

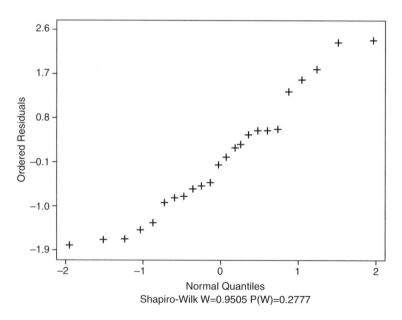

Figure 8.4. A Normal Probability Plot of the Residuals in Example 8.2.2.

From Table 8.6 one can compute

$$MS_e = \frac{0.476 + \cdots + 1.963}{6} = 2.042$$

and the ANOMV decision lines

$$UDL = U(0.1; 6, 3)IMS_e = (0.5310)(6)(2.042) = 6.51$$
$$CL = MS_e = 2.042$$
$$LDL = L(0.1; 6, 3)IMS_e = (0.0066)(6)(2.042) = 0.081.$$

From the ANOMV chart given in Figure 8.5, one sees that there is no evidence of unequal variances at the 0.1 level.

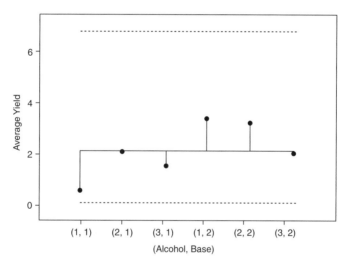

Figure 8.5. An ANOMV Chart for the Variances in Table 8.6 with $\alpha = 0.1$.

Using equation (8.2.2) and the table of means (Table 8.5), one can compute Table 8.7, and from Table 8.7

$$SS_{AB} = 4(2)(1.371^2 + 0.717^2 + 0.654^2) = 22.57$$
$$MS_{AB} = \frac{22.57}{2} = 11.29.$$

Finally,

$$F_{AB} = \frac{11.29}{2.042} = 5.53$$

and since $5.53 > F(0.05; 2, 18) = 3.555$, there is a significant AB interaction. This confirms what one would have expected by looking at an interaction plot (Figure 8.6).

Table 8.7. The $\widehat{\alpha\beta}_{ij}$'s for Example 8.2.2

		Base	
		1	2
	1	1.371	−1.371
Alcohol	2	−0.717	0.717
	3	−0.654	0.654

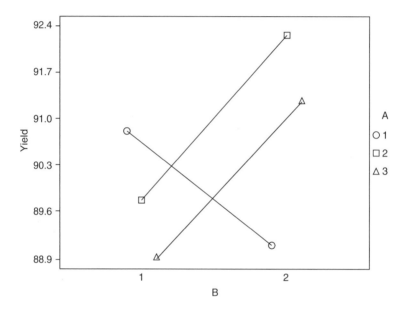

Figure 8.6. An Interaction Plot for Example 8.2.2.

Further Analysis When There is no AB Interaction

When there is no significant AB interaction, one can study the effects of the two factors separately. This is best done using the ANOM. Let f represent the number of levels for the factor of interest. That is, $f = I$ for factor A and $f = J$ for factor B. The ANOM decision limits in this case are

$$\bar{y}_{\bullet\bullet\bullet} \pm h(\alpha; f, IJ(n-1))\sqrt{\mathrm{MS}_e}\sqrt{\frac{f-1}{N}} \,. \tag{8.2.8}$$

One could also test for significant main effects using the ANOVA, but as in the one-way layout, this will not provide any information as to which levels are different when a significant effect is found. To test the significance of factor A, one would compute

$$SS_A = nJ \sum_{i=1}^{I} (\hat{\alpha}_i)^2$$

$$MS_A = \frac{SS_A}{I-1}$$

and

$$F_A = \frac{MS_A}{MS_e}.$$

The effect of factor A would be declared significant at level α if

$$F_A > F(\alpha; I-1, IJ(n-1)).$$

Computationally, MS_A is most easily obtained by noting that

$$\boxed{MS_A = nJ(\text{variance of the } \overline{y}_{i\bullet\bullet}\text{'s}).} \qquad (8.2.9)$$

Similarly, for factor B one has

$$SS_B = nI \sum_{j=1}^{J} (\hat{\beta}_j)^2$$

$$MS_B = \frac{SS_B}{J-1}$$

$$F_B = \frac{MS_B}{MS_e}$$

and the effect of factor B would be declared significant at level α if

$$F_B > F(\alpha; J-1, IJ(n-1)).$$

Again, MS_B is most easily obtained by noting that

$$\boxed{MS_B = nI(\text{variance of the } \overline{y}_{\bullet j\bullet}\text{'s}).} \qquad (8.2.10)$$

Example 8.2.3   Since in Example 8.2.1 we found no significant interaction between pH level and catalyst type, we can study the two factors separately. Using the ANOM for factor A (pH), one obtains the decision lines

$$15.40 \pm h(0.001; 3, 24)\sqrt{MS_e}\sqrt{\frac{2}{30}}$$

$$\pm 4.16\sqrt{0.587}\sqrt{\frac{2}{30}}$$

$$\pm 0.82$$

$$(14.58, 16.22)$$

and the ANOM chart is shown in Figure 8.7. There is an effect on adhesion due to

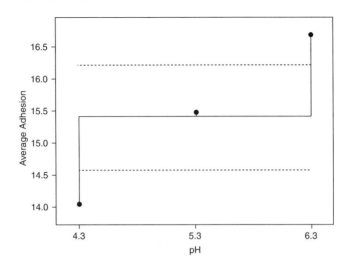

Figure 8.7. An ANOM Chart for pH (Example 8.2.1) with $\alpha = 0.001$.

pH level since the average adhesions for pH $= 4.3$ and for pH $= 6.3$ fall outside the decision lines. Because there are a different number of levels for pH and catalyst, new ANOM decision lines need to be calculated to check for a catalyst effect. For factor B (catalyst) one obtains the decision lines

$$15.40 \pm h(0.01; 2, 24)\sqrt{\mathrm{MS}_e}\sqrt{\frac{1}{30}}$$
$$\pm 2.80\sqrt{0.587}\sqrt{\frac{1}{30}}$$
$$\pm 0.39$$
$$(15.01, 15.79)$$

and the resulting ANOM chart is shown in Figure 8.8.

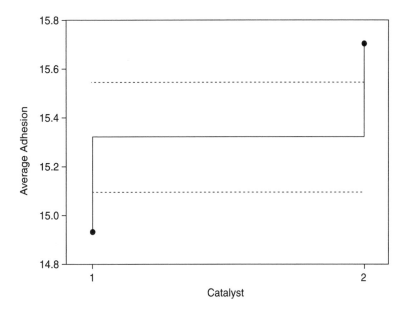

Figure 8.8. An ANOM Chart for Catalyst (Example 8.2.1) with $\alpha = 0.01$.

Since the average adhesion for Catalyst 1 falls below the lower decision limit, and the average for Catalyst 2 falls above the upper decision limit, Catalyst has an ($\alpha = 0.05$) effect.

Using the ANOVA, one obtains

$$\mathrm{MS}_A = 5(2)(\text{variance of the } \overline{y}_{i\bullet\bullet}\text{'s}) = 5(2)(1.75) = 17.5$$
$$\mathrm{MS}_B = 5(3)(\text{variance of the } \overline{y}_{\bullet j\bullet}\text{'s}) = 5(3)(0.4475) = 6.71$$
$$F_A = \frac{17.5}{0.587} = 29.8$$
$$F_B = \frac{6.71}{0.587} = 11.4.$$

F_A and F_B could be compared with $F(0.01; 2, 24) = 5.614$. All of the ANOVA results for this data set are compiled in Table 8.8.

Note: When a factor has only two levels, little is gained by drawing the ANOM chart. If the factor has a significant effect (easily determined from an ANOVA table), then the larger mean will plot above the UDL and the smaller mean will plot below the LDL.

Table 8.8. An ANOVA Table for Adhesion (Example 8.2.1)

```
ANALYSIS OF VARIANCE TABLE FOR ADHESION

SOURCE          DF      SS          MS          F       P
-----------     ----    ----------  ----------  ------- ------
PH (A)          2       34.9243     17.4622     29.77   0.0000
CAT (B)         1       6.71187     6.71187     11.44   0.0025
A*B             2       1.00338     0.50169      0.86   0.4377
RESIDUAL        24      14.0779     0.58658

-----------     ----    ----------
TOTAL           29      56.7175
```

Computation of SS_{AB}

The various effect sums of squares (degrees of freedom) in an ANOVA table sum to the total sum of squares (total degrees of freedom). When using model (8.2.1), this means that

$$SS_A + SS_B + SS_{AB} + SS_e = SS_{total} \qquad (8.2.11)$$

and

$$(I-1) + (J-1) + (I-1)(J-1) + IJ(n-1) = N - 1.$$

(See, e.g., the ANOVA table in the above example.) Rearranging equation (8.2.11), one obtains

$$\boxed{SS_{AB} = SS_{total} - SS_A - SS_B - SS_e} \qquad (8.2.12)$$

which can sometimes be used to compute SS_{AB}. However, care must be taken when using this formula because, like the "computing formula" for the sample variance, equation (8.2.12) is numerically unstable.

Example 8.2.4   Consider the following (fictitious) data from a 2×3 factorial experiment (Table 8.9).

Table 8.9. Fictitious Data From a 2×3 Factorial Experiment

		Factor B		
		1	2	3
Factor A	1	1.1	2.2	3.0
		2.0	3.1	4.2
		3.2	4.0	5.2
	2	4.2	4.8	6.1
		5.2	5.9	7.3
		6.3	7.0	7.9

Table 8.10. Sample Means for Example 8.2.4

		Factor B			
		1	2	3	
Factor A	1	2.10	3.10	4.13	3.11
	2	5.23	5.90	7.10	6.08
		3.67	4.50	5.62	4.595

Table 8.11. Sample Variances for Example 8.2.4

		Factor B		
		1	2	3
Factor A	1	1.11	0.81	1.21
	2	1.10	1.21	0.84

From Table 8.9, one would compute

$$SS_{total} = (N - 1)(\text{variance of the } y_{ijk}\text{'s}) = 17(3.750) = 63.750$$

and the tables of sample means (Table 8.10) and variances (Table 8.11). From Tables 8.10 and Table 8.11 one would compute mean squares (equations (8.2.9), (8.2.10), and (8.2.7)), and then sums of squares by multiplying the means squares by their degrees of freedom:

$$MS_A = 3(3)(4.41) = 39.694$$
$$MS_B = 3(2)(0.958) = 5.746$$
$$MS_e = \frac{1.11 + \cdots + 0.84}{6} = 1.047$$
$$SS_A = 1(39.694) = 39.694$$
$$SS_B = 2(5.746) = 11.492$$
$$SS_e = 12(1.047) = 12.564.$$

Finally, from equation (8.2.12)

$$SS_{AB} = 63.750 - 39.694 - 11.492 - 12.564 = 0.000.$$

Unfortunately, comparing this with the computer generated value of 0.08333 (see the ANOVA in Table 8.12), one finds that SS_{AB} computed using equation (8.2.12) is wrong in the first significant figure.

Table 8.12. An ANOVA Table for the Fictitious Data in Example 8.2.4

ANALYSIS OF VARIANCE TABLE FOR Y

SOURCE	DF	SS	MS	F	P
A (A)	1	39.6050	39.6050	37.80	0.0000
B (B)	2	11.4878	5.74389	5.48	0.0204
A*B	2	0.08333	0.04167	0.04	0.9611
RESIDUAL	12	12.5733	1.04778		
TOTAL	17	63.7494			

Table 8.13. $\widehat{\alpha\beta}_{ij}$'s for Example 8.2.4

		\multicolumn{3}{c}{Factor B}
		1	2	3
Factor A	1	−0.085	0.085	−0.005
	2	0.075	−0.085	−0.005

On the other hand, computing SS_{AB} using equation (8.2.4) results in Table 8.13 and

$$SS_{AB} = 3[(-0.085)^2 + \cdots + (-0.005)^2] = 3(0.02735) = 0.082.$$

Equation (8.2.12) is less of a problem when SS_{AB} is the same order of magnitude as the other sums of squares.

Example 8.2.5   Using equation (8.2.12) to compute SS_{AB} for the chemical yield data (Example 8.2.2), one would obtain

$$MS_A = 4(2)(0.3369) = 2.695$$
$$MS_B = 4(3)(0.5429) = 6.515$$
$$MS_e = 2.042$$
$$SS_A = 2(2.695) = 5.390$$
$$SS_B = 1(6.515) = 6.515$$

$$SS_e = 18(2.042) = 36.756$$
$$SS_{total} = 23(3.097) = 71.231$$

and

$$SS_{AB} = 71.231 - 5.390 - 6.515 - 36.756 = 22.57$$

which agrees with the value obtained using equation (8.2.4).

Further Analysis When There is an AB Interaction

If the ANOVA test shows that there is significant interaction, one cannot really talk about the "main effect" of one factor or the other since the effect of one factor then depends on the particular level of the other factor. Instead, one must somehow reduce the data to one or more one-way layouts where the problem of an interaction doesn't exist. One possibility is to split the data into subsets corresponding to the levels of one of the factors (generally the factor with fewer levels). If we split the data according to the levels of factor B, then we can study the effects of factor A separately for each level of factor B. At each level of factor B one has a one-way layout with the treatments being the levels of factor A. More specifically, for each level of factor B one would fit the model

$$Y_{ik} = \mu + \alpha_i + \epsilon_{ik}. \qquad (8.2.13)$$

Note that this would require J separate analyses.

Example 8.2.6 In Example 8.2.2 (p. 301), we found there was a significant interaction between the Alcohols and the Bases. Since there are only two Bases (but three Alcohols), it would make sense to consider the effect of the Alcohols separately for each Base using model (8.2.13). This is best done using the ANOM. Using only the data for Base 2 from Example 8.2.2, one obtains

$$MS_e = \frac{3.243 + 3.069 + 1.963}{3} = 2.767$$

and $\alpha = 0.1$ decision lines

$$90.892 \pm h(0.1; 3, 9)\sqrt{2.767}\sqrt{\frac{2}{12}}$$
$$\pm 2.34(0.6791)$$
$$\pm 1.59$$
$$(89.30, 92.48).$$

From the ANOM chart given in Figure 8.9 one sees that the Alcohols are significantly different due to Alcohol 1 producing significantly low average yields.

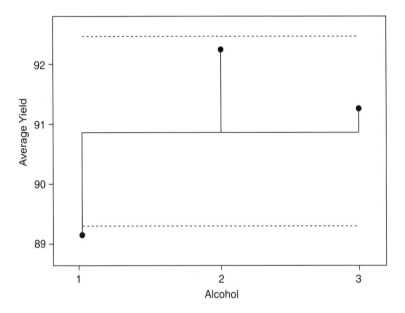

Figure 8.9. An ANOM Chart Using Data for Base 2 Only with $\alpha = 0.1$.

A similar analysis using only the data for Base 1 results in the decision lines

$$89.85 \pm h(0.1; 3, 9)\sqrt{1.317}\sqrt{\frac{2}{12}}$$
$$\pm\, 2.34(0.4685)$$
$$\pm\, 1.10$$
$$(88.75, 90.95).$$

From the ANOM chart in Figure 8.10 one sees that the Alcohols are not significantly different.

Note: As expected, the results for the two analyses are different. If the behavior of the Alcohols was the same for both levels of Base, we would not have found a significant interaction.

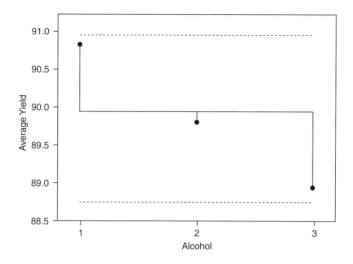

Figure 8.10. An ANOM Chart Using Data for Base 1 Only with $\alpha = 0.1$.

A second possibility is to consider each of the $I \times J$ treatment combinations as one level of a single factor. That is, one could model the results of the experiment as

$$\boxed{Y_{ijk} = \mu + \alpha_{ij} + \epsilon_{ijk}} \qquad (8.2.14)$$

where α_{ij} is the main effect of the $(i, j)^{\text{th}}$ treatment combination. Which technique is more appropriate depends on the question(s) one is interested in answering. For the chemical process data in Example 8.2.2, model (8.2.14) is probably more appropriate since the goal of studying the process would most likely be to find the alcohol/base combination that results in the maximum yield.

Example 8.2.7   Analyzing the chemical process data in Example 8.2.2 (p. 301) using model (8.2.14), one would use $\text{MS}_e = 2.042$ from the analysis in Example 8.2.2 and compute ANOM decision lines (there are $3 \times 2 = 6$ treatment combinations)

$$90.371 \pm h(0.05; 6, 18)\sqrt{2.042}\sqrt{\frac{5}{24}}$$
$$\pm 2.91(0.652)$$
$$\pm 1.90$$
$$(88.47, 92.27).$$

From the ANOM chart in Figure 8.11, one sees that the treatment combinations are significantly different at the $\alpha = 0.05$ level due to the Alcohol 2 and Base 2 combination producing significantly high average yields.

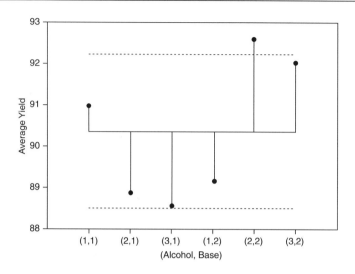

Figure 8.11. An ANOM Chart for Example 8.2.7 with $\alpha = 0.05$.

It is important to realize that if one obtains F_{AB} from an ANOVA table and there is a significant interaction, the statistics F_A and F_B are no longer meaningful.

Example 8.2.8   We know from Example 8.2.6 that Alcohol has a significant $\alpha = 0.1$ effect when used with Base 2. However, the ANOVA table for the complete data set (Example 8.2.2), which is given in Table 8.14, indicates (p-value $= 0.2915$) that there is no Alcohol effect. Also, Table 8.14 indicates that Base is significant at the 0.1 level. However, studying the Base effects separately for each Alcohol, one finds that with Alcohol 1, Base is not significant at the 0.1 level; with Alcohol 2, Base is significant at the 0.1 level; and with Alcohol 3, Base is significant at the 0.05 level. (See Tables 8.15–8.17.)

Table 8.14. An ANOVA Table for the Complete Data Set in Example 8.2.2

ANALYSIS OF VARIANCE TABLE FOR YIELD

SOURCE	DF	SS	MS	F	P
A (A)	2	5.39583	2.69792	1.32	0.2915
B (B)	1	6.51042	6.51042	3.19	0.0910
A*B	2	22.5658	11.2829	5.53	0.0135
RESIDUAL	18	36.7575	2.04208		
TOTAL	23	71.2296			

Table 8.15. An ANOVA Table with Data For Alcohol 1 Only

ANALYSIS OF VARIANCE TABLE FOR YIELD

SOURCE	DF	SS	MS	F	P
B (A)	1	5.78000	5.78000	3.11	0.1283
RESIDUAL	6	11.1550	1.85917		
TOTAL	7	16.9350			

Table 8.16. An ANOVA Table with Data For Alcohol 2 Only

ANALYSIS OF VARIANCE TABLE FOR YIELD

SOURCE	DF	SS	MS	F	P
B (A)	1	12.2513	12.2513	4.79	0.0712
RESIDUAL	6	15.3475	2.55792		
TOTAL	7	27.5987			

Table 8.17. An ANOVA Table with Data For Alcohol 3 Only

ANALYSIS OF VARIANCE TABLE FOR YIELD

SOURCE	DF	SS	MS	F	P
B (A)	1	11.0450	11.0450	6.46	0.0440
RESIDUAL	6	10.2550	1.70917		
TOTAL	7	21.3000			

One would not want to apply model (8.2.14) before testing for (and finding) significant interaction since that might lead incorrectly to the conclusion that there were no significant effects.

Example 8.2.9   An experiment was done to study the repair time (in minutes) for three different brands of computers. Since the repair time might depend on how the computers were configured, the type of computer (business machine, expensive home machine, inexpensive home machine) was considered as a second factor. The data are given Table 8.18 and in `computer.txt`. The ANOVA table (Table 8.19) shows no interaction and indicates that Type has a significant effect. An ANOM

Table 8.18. Repair Time in Minutes for Three Brands of Computer Drives

		Type		
		1	2	3
	1	52.80	52.80	56.80
		58.00	44.20	59.60
		47.60	49.00	53.60
		55.20	55.80	55.00
Brand	2	52.00	53.40	47.60
		44.00	62.00	57.20
		43.40	49.60	55.60
		50.20	48.80	56.20
	3	41.60	44.80	49.30
		51.80	46.00	44.60
		46.00	48.00	56.00
		53.20	41.80	46.80

Table 8.19. ANOVA Table for the Data in Example 8.2.9

ANALYSIS OF VARIANCE TABLE FOR Y

SOURCE	DF	SS	MS	F	P
BRAND (A)	2	99.4117	49.7058	2.40	0.1094
TYPE (B)	2	219.345	109.673	5.31	0.0114
A*B	4	113.083	28.2708	1.37	0.2713
RESIDUAL	27	558.148	20.6721		
TOTAL	35	989.988			

chart (see Problem 8.2.8) would confirm the effect of Type and show that the effect was due to Type 3 requiring a significantly short average repair time. However, analyzing these data using model (8.2.14) and ANOM, one obtains

$$50.84 \pm h(0.1; 9, 27 \to 24)\sqrt{20.67}\sqrt{\frac{8}{36}}$$

$$\pm 2.69(2.143)$$

$$\pm 5.77$$

$$(45.07, 56.61).$$

The ANOM chart given in Figure 8.12 indicates none of the Brand/Type combinations had a significant effect.

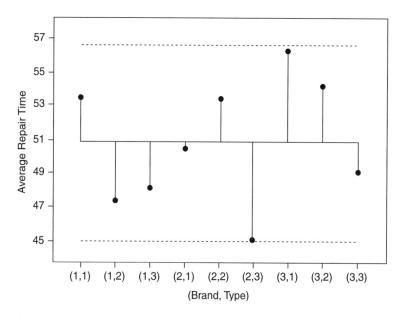

Figure 8.12. An ANOM Chart for Brand/Type Combinations in Example 8.2.9 with $\alpha = 0.1$.

Problems

1. The data below are a summary of the results of a two-factor experiment with five replicates.

Means

		Factor B	
		1	2
	1	12	19
Factor A	2	20	27
	3	6	13
	4	11	17

Variances

		Factor B	
		1	2
	1	25	16
Factor A	2	25	9
	3	9	12
	4	16	9

Use the data to do the following by hand.

(a) Test for equal variances.

(b) Draw an interaction plot and comment on it.

(c) Calculate MS_{AB} and test for interaction.

(d) Using $\alpha = 0.01$, test to see whether Factor A has an important effect. Your analysis will depend on the result of part (c).

(e) Using $\alpha = 0.01$, test to see whether Factor B has an important effect. Your analysis will depend on the result of part (c).

2. The data below are the results of a two-factor experiment.

		Factor B	
Factor A	1	2	3
	6	20	24
1	4	14	25
	5	15	26
	8	18	22
2	4	17	19
	6	16	25

Use the data to do the following by hand.

(a) Calculate the cell means, row means, column means, grand mean, and the cell variances.

(b) Make an interaction plot for the data and comment on it.

(c) Estimate all the parameters in model (8.2.1).

(d) Calculate the residuals.

(e) Make a normal probability plot of the residuals and comment on it.

(f) Test for equal variances.

(g) Calculate MS_{AB} and test to see if there is a significant interaction at the $\alpha = 0.05$ level.

(h) Using $\alpha = 0.05$, test to see if either factor is important. Explain what your results mean.

3. One is interested in studying the effects of depth from the top of the tank and position in the tank on the concentration of a cleaning solution. Three depths, two positions, and six replicates were used; and the data are summarized below.

Means

		Depth			
		1	2	3	
Position	1	4.5	16	16.5	12.33
	2	20	11	12	14.33
		12.25	13.5	14.25	13.33

Variances

		Depth		
		1	2	3
Position	1	1.5	30	25
	2	44	11	13.5

(a) Test the assumption of equal variances using $\alpha = 0.05$.

(b) Using what would be the counterpart to equation (7.6.6) for a two-way layout, compute an appropriate transformation to stabilize the variances.

4. Example 2.3.2 (p. 28) discusses comparing various chemicals for two components in an automotive paint. Each treatment combination had two replicates, and the data are given in Example 2.3.2 and in `lw.txt`. Some summary statistics are given below.

Means

		Component 1				
		1	2	3	4	
	1	9.55	5.60	9.50	7.25	7.98
Component 2	2	7.40	4.80	6.05	3.50	5.44
	3	11.35	7.75	12.15	10.95	10.55
		9.43	6.05	9.23	7.23	7.99

Variances

		Component 1			
		1	2	3	4
	1	1.445	0.180	2.000	1.125
Component 2	2	1.620	0.720	1.805	0.020
	3	0.045	0.245	1.125	1.445

(a) Make an interaction plot and comment on it.

(b) Check the assumptions of normality and equal variances.

(c) Calculate MS_{AB} and test for significant interaction using $\alpha = 0.01$.

(d) Test to see which factors are important using $\alpha = 0.01$. Explain what your results mean.

(e) Write a paragraph summarizing the results of the experiment.

5. Using the thin film data `thinfilm.txt`, do the following.

(a) Calculate the residuals for model (8.2.1), and make a normal probability plot of them.

(b) Test for equal variances.

(c) Make an interaction plot and comment on it.

(d) Calculate MS_{AB} and test for significant interaction using $\alpha = 0.01$.

(e) Test to see which factors are important using $\alpha = 0.01$. Explain what your results mean.

(f) Write a paragraph summarizing the results of the experiment.

6. The data set `deink.txt` contains the results from an experiment on de-inking of newspaper. The factors studied were the amount of alkali in the solution and the hardness of the water used. The response is a measure of brightness of the resulting pulp (that would then be used in new newsprint).

(a) Using a computer, make an interaction plot for the data and then comment on it.

(b) Using a computer, perform a two-factor analysis of variance.

(c) Make any necessary follow-up comparisons.

(d) Write a short report of the results.

7. In Example 3.2.4 (p. 45), the drying ability of several brands of gas and electric lab ovens was measured. Using these data (`oven.txt`), do the following.

(a) Make a normal probability plot of the residuals and comment on it.

(b) Explain what interaction between Brand and Type stands for, in the context of the problem. Test for this interaction and report your results.

(c) For each brand, make a 95% confidence interval for the difference between gas and electric ovens.

(d) Suppose a lab manager wishes to purchase an oven of Brand C, because of its excellent repair record. Should the manager buy a gas oven, or an electric one? How much difference in drying ability can the manager expect, by choosing one type or the other?

(e) Suppose a lab manager must buy an electric oven because gas lines are not available in the lab. What brand would you recommend, and why?

8. For the data in `computer.txt` (see Example 8.2.9, p. 315), do the following.

   (a) Check the assumption of normality with a normal probability plot.

   (b) Check the assumption of equal variances using the ANOMV.

   (c) Construct an ANOM chart for Type using $\alpha = 0.05$. What do you conclude?

9. Verify equation (8.2.3).

8.3 ONLY ONE OBSERVATION PER CELL

The experiments we have looked at so far contain **replication**. That is, several observations (experimental units) were assigned to each treatment combination. Replication is always a good statistical practice. Without replication, we cannot be sure that the good results we got from a particular set of conditions are really due to the conditions rather than just a random fluctuation.

Often, however, unreplicated experiments are performed. Usually this is done to conserve resources, either when the total number of treatment combinations is large, or when the experiment is a preliminary one.

One of the complications of an unreplicated experiment is that we have no way to calculate the MS_e, which we have been using to estimate σ^2, the random variability among the experimental units. In an unreplicated experiment, we must find some other way to estimate σ^2. This is often done by assuming there is no interaction between the two factors, which results in the model

$$Y_{ij} = \mu + \alpha_i + \beta_j + \epsilon_{ij} \ .$$
(8.3.1)

If the assumption of no interaction is true, then the MS_{AB} is actually an estimate of σ^2 and can be used as the MS_e in calculations. Naturally, because the MS_e is obtained in a different fashion, it will have different degrees of freedom associated with it than when it is computed by pooling sample variances (equation (8.2.7)). The MS_e is computed exactly as the MS_{AB} was computed using equations (8.2.4) and (8.2.5). This time, however, one is computing the residuals

$$\widehat{\epsilon}_{ij} = y_{ij} - \overline{y}_{i\bullet} - \overline{y}_{\bullet j} + \overline{y}_{\bullet\bullet}$$
(8.3.2)

which are squared and summed to obtain

$$SS_e = \sum_{i,j} (\widehat{\epsilon}_{ij})^2$$
(8.3.3)

and

$$MS_e = \frac{SS_e}{(I-1)(J-1)} \ .$$
(8.3.4)

Example 8.3.1 (*Particle Size Data, p. 73*) Let us reconsider the particle size data in Example 3.5.5 from a different point of view. Recall that an experiment was performed to understand how two processing parameters (a flow rate and a vacuum) affected the final particle size of a product. The data from the experiment are given

Table 8.20. Particle Sizes for Different Vacuum/Flowrate Combinations

| | Flow Rate | | | |
Vacuum	85	90	95	Average
20	6.3	6.1	5.8	6.0667
22	5.9	5.6	5.3	5.6
24	6.1	5.8	5.5	5.8
Average	6.1	5.8333	5.533	5.822

in Table 8.20, and for this particular product smaller particle sizes are favored (so smaller y values are desired). The engineer did not anticipate any interaction between the two factors, and the interaction plot in Figure 8.13 shows no evidence of interaction.

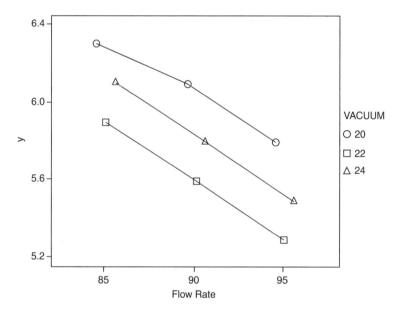

Figure 8.13. An Interaction Plot for the Data in Example 8.3.1.

From equation (8.3.2) one obtains

$$\widehat{\epsilon}_{11} = 6.3 - 6.0667 - 6.1 + 5.822 = -0.045$$

and all of the residuals are given in Table 8.21. Using these residuals, one obtains

$$SS_e = (-0.045)^2 + \cdots + (-0.011)^2 = 0.00444$$

and

$$MS_e = \frac{0.00444}{(2)(2)} = 0.00111.$$

Table 8.21. The Residuals for Example 8.3.1

	Flow Rate		
Vacuum	85	90	95
20	−0.045	0.022	0.022
22	0.022	−0.011	−0.011
24	0.022	−0.011	−0.011

Since interaction is not an issue (we have assumed it does not exist), we can analyze the effects of flow rate and vacuum using the ANOM. Decision limits for flow rate (using $\alpha = 0.001$) are

$$\bar{y}_{\bullet\bullet} \pm h(0.001; 3, 4)\sqrt{0.00111}\sqrt{\frac{2}{9}}$$
$$5.822 \pm 10.6(0.0157)$$
$$\pm 0.166$$
$$(5.66, 5.99).$$

Since flow rate and vacuum have the same number of levels, their decision limits would be the same.

The two ANOM charts are given in Figures 8.14 and 8.15, from which one sees that flow rate is significant at the $\alpha = 0.001$ level due to the 85 rate being significantly bad and the 95 rate being significantly good. The vacuum is significant at the $\alpha = 0.001$ level due to the 20 setting being significantly bad and the 22 setting being significantly good. To minimize the particle size, one would use a flow rate of 95 and a vacuum setting of 22.

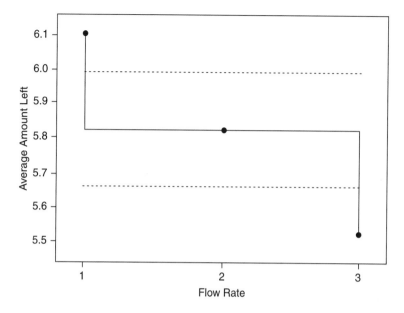

Figure 8.14. An ANOM Chart for Flow Rate (Example 8.3.1) with $\alpha = 0.001$.

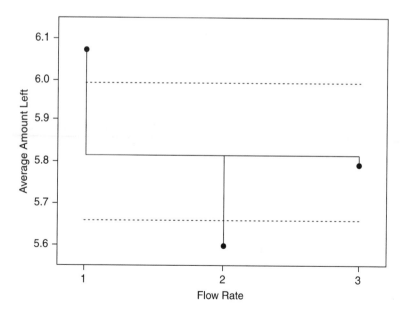

Figure 8.15. An ANOM Chart for Vacuum (Example 8.3.1) with $\alpha = 0.001$.

Alternatively, one could perform an ANOVA by computing

$$\boxed{\text{MS}_A = J(\text{variance of the } \bar{y}_{i\bullet}\text{'s})} \tag{8.3.5}$$

$$\boxed{\text{MS}_B = I(\text{variance of the } \bar{y}_{\bullet j}\text{'s})} \tag{8.3.6}$$

and then

$$F_A = \frac{\text{MS}_A}{\text{MS}_e}$$
$$F_B = \frac{\text{MS}_B}{\text{MS}_e} .$$

This, however, will not provide information on which levels cause any significant differences that are found.

Example 8.3.2   For the particle size data one would compute

$$\text{MS}_A = 3(0.0548) = 0.1644$$
$$\text{MS}_B = 3(0.08046) = 0.2414$$
$$F_A = \frac{0.1644}{0.00111} = 148$$
$$F_B = \frac{0.2414}{0.00111} = 217$$

and the complete ANOVA is shown in Table 8.22.

Table 8.22. An ANOVA Table for the Data in Example 8.3.1

ANALYSIS OF VARIANCE TABLE FOR Y

SOURCE	DF	SS	MS	F	P
VACUUM (A)	2	0.32889	0.16444	148.00	0.0002
FLOWRATE (B)	2	0.48222	0.24111	217.00	0.0001
A*B	4	0.00444	0.00111		
TOTAL	8	0.81556			

Problems

1. The output in milliamperes of a newly designed electronic component was measured at three temperatures and three humidity levels. One component was studied at each temperature/humidity combination and the results are given in the table below.

		Temperature (°C)		
		10	20	30
Relative	20	169	173.5	185
Humidity (%)	30	189	201	203.5
	40	200	202	208

(a) Make an interaction plot for these data. What do you conclude?

(b) Find the residuals when using model (8.3.1).

(c) If appropriate, construct $\alpha = 0.05$ ANOM charts to study the effects of the two factors.

(d) Perform an ANOVA on the data using $\alpha = 0.05$.

2. Silicon crystals were grown using the Czochralski process and treatment conditions consisting of combinations of type of impurity used (factor A) and rate of cooling (factor B). A two-way layout was used with one observation in each cell. The (coded) electrical properties of the resulting crystals are given below.

		Cooling		
		1	2	3
Impurity	1	2	12	4
Type	2	28	32	30

(a) Make an interaction plot for this data. What do you conclude?

(b) Find the residuals when using model (8.3.1).

(c) If appropriate, construct $\alpha = 0.1$ ANOM charts to study the effects of the two factors.

(d) Perform an ANOVA on the data using $\alpha = 0.1$.

8.4 BLOCKING TO REDUCE VARIABILITY

In order to decrease the variability associated with the factors of interest in an experiment, the experimental trials can sometimes be split into sub-groups in which the experimental conditions are more homogeneous than they are over the entire experiment. These sub-groups are called blocks, and a factor used to accomplish this sub-division is called a blocking factor. The simplest example of blocking is a paired experiment, such as Example 6.1.1, where one was interested in comparing the durability of two UV coatings for eyeglasses, and the blocks were the test subjects. Failure to account for changes in the experimental conditions (the blocks) would contaminate the experiment.

Two different consequences are possible depending on exactly how the experiment was conducted, both bad. If one UV coating was used on glasses worn by test subjects who were meticulous about cleaning and care of their lenses, and the other type was used on glasses worn by subjects who took much less care of their lenses, one would not be able to distinguish between differences due to the UV coatings and differences due to their environment. This is a particularly bad design since one would at the very least want to randomize over any factor (in this case the subjects) that might affect the outcome. Alternatively, if one randomized over the subjects, that is, the type of coating was assigned at random to the subjects (this would be referred to as a **completely randomized design**), then the additional variability due to the differences in the way the subjects use their glasses would make it more difficult (or maybe impossible) to find differences in the UV coatings when they really existed.

The appropriate design in this case is called a **randomized block design**. This is a design where the different experimental conditions are accounted for by assigning each level of the factor of interest to each block, and additional randomization is done within the blocks. In the lens example, this would amount to having each subject with one lens of each type, and the randomization would consist of assigning the two UV coating types to the left and right eye at random.

Other examples of blocking factors are batches of raw material, technicians, and days on which experimental trials are performed. In each instance, the experimenter is not specifically interested in the effect of the blocking factor, but in separating the variability associated with it from the variability associated with the factor(s) of interest. It is possible to have more than one blocking factor in an experiment, such as the situation where experimental trials are performed by different technicians over several days. The analysis associated with a blocking factor is the same as for any other factor. The only difference is that because of their nature blocking factors are assumed to not interact with the other factors.

Example 8.4.1 In Example 6.1.1 (p. 218) we analyzed the data by pairing. It is also possible to analyze the data using the ANOVA with type of UV coating as the factor of interest and the test subject as the blocking factor. The data, reproduced in the correct format for this analysis, together with row means, column means, and the grand mean are given in Table 8.23. From this table one can compute Table 8.24 of the residuals (equation (8.3.2)) and (equations (8.3.4) and (8.3.5))

$$MS_e = [0.09^2 + \cdots + 0.01^2]/9 = 0.0298$$
$$MS_A = 10(0.0242) = 0.242$$
$$F_A = \frac{0.242}{0.0298} = 8.12.$$

Table 8.23. Durability Measurements for Different UV Coatings

		1	2	3	4	5	6	7	8	9	10	
						Test Subject						
UV Coating	A	8.9	9.4	11.2	11.4	13	6.4	13.4	5.6	4.8	15.8	9.99
	B	8.5	9.3	10.8	11.6	12.9	6.5	13.1	5.1	4.3	15.6	9.77
		8.7	9.35	11	11.5	12.95	6.45	13.25	5.35	4.55	15.7	9.88

Table 8.24. The Residuals from the ANOVA in Example 8.4.1

		1	2	3	4	5	6	7	8	9	10
						Test Subject					
UV Coating	A	0.09	−0.06	0.09	−0.21	−0.06	−0.16	0.04	0.14	0.14	−0.01
	B	−0.09	0.06	−0.09	0.21	0.06	0.16	−0.04	−0.14	−0.14	0.01

Note that $F_A = 8.12$ is the square of $t = 2.85$ computed in Example 6.1.1. Further, for a two-sided test at level 0.01 the appropriate critical values are $F(0.01; 1, 9) = 10.561$ and $t(0.005; 9) = 3.25$ for the two test statistics, respectively, and $10.561 = (3.25)^2$. Thus, for paired data the ANOVA is equivalent to performing a t test on the differences.

The advantage of the ANOVA approach is that it is easily extended to the situation where one has more than two levels for the factor of interest and pairing is no longer feasible. Actually, in that situation it would be preferable to use the ANOM since it would indicate which levels account for any significant differences found.

Example 8.4.2 (*De-Inking Formulation Data*) The R&D group of a chemical manufacturer is working on three formulas to be used to de-ink newspapers (for recycling). To aid in the choice of formulation, the chemists conduct a small experiment in which five newspaper specimens (from five different newspapers) are divided into thirds, and one group is treated with each de-inking formulation (assigned at random). Each group is then graded as to the quality of its color. The data are recorded in Table 8.25 as well as in `deink2.txt`. This is a randomized block design where the five newspapers are the blocks, and the de-inking formulations are the treatments (the levels of the factor of interest). The residuals (equation (8.3.2)) are shown in Table 8.26 and (equations (8.3.3) and (8.3.4))

$$SS_e = [(-2.47)^2 + \cdots + (3.27)^2] = 246.13$$
$$MS_e = \frac{246.13}{8} = 30.77.$$

Table 8.25. Color Quality for Different De-Inking Processes

De-Inking Formulation	Newspaper					Average
	1	2	3	4	5	
1	64	68	66	72	67	67.4
2	71	74	79	81	62	73.4
3	85	87	81	76	81	82.0
Average	73.3	76.3	75.3	76.3	70.0	74.3

Table 8.26. The Residuals for Example 8.4.2

De-Inking Formulation	Newspaper				
	1	2	3	4	5
1	-2.47	-1.47	-2.47	2.53	3.87
2	-1.47	-1.47	4.53	5.53	-7.13
3	3.93	2.93	-2.07	-8.07	3.27

Using $\alpha = 0.05$, the ANOM decision lines are

$$74.27 \pm h(0.05; 3, 8)\sqrt{30.77}\sqrt{\frac{2}{15}}$$
$$\pm 2.86(2.026)$$
$$\pm 5.79$$
$$(68.48, 80.06)$$

and the ANOM chart is given in Figure 8.16. From the ANOM chart one sees that the formulations are significantly different ($\alpha = 0.05$) due to Formulation 1 producing significantly less brightness and Formulation 3 producing significantly more brightness.

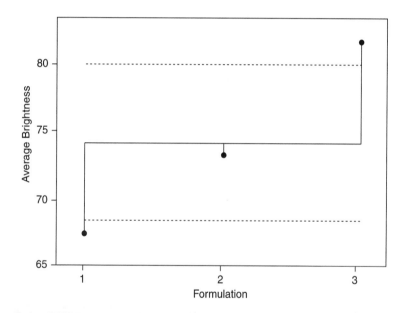

Figure 8.16. An ANOM Chart for Example 8.4.2 with $\alpha = 0.05$.

Problems

1. An engineer wants to compare UV absorbance of five different lens coating materials (labeled A through E). Once coated, a lens must be cured in an oven. Due to processing restrictions, no one oven has space to run the entire experiment. The engineer decides to choose three ovens at random, and cure five lenses, one of each coating type, in each oven. For each coating type the lenses are assigned to the ovens at random. The data are given below.

Oven	Coating				
	A	B	C	D	E
1	1.68	1.83	1.71	1.52	1.84
2	1.64	1.93	1.82	1.61	1.82
3	1.76	1.73	1.74	1.77	1.97

(a) Describe, as completely as possible, what type of experiment this is. What are the factors in the experiment?

(b) The engineer has used randomization in two places in the experiment. Explain why this was done, and what problems might arise if it were not done.

(c) Perform an ANOM on the data using $\alpha = 0.05$, and explain what your results mean.

2. The data set `sarea.txt` contains the surface area (m^2/g) of three batches of silica measured by four different lab technicians.

(a) What type of experiment is this?

(b) Analyze the data, and explain what the results mean.

3. The data set `pigment.txt` contains the results of an experiment to determine if changing the pigment used in a white paint would improve the yellowing of the paint. Yellowing is measured by Δb, the change in color of the paint over time. Three pigments were to be studied: the current pigment (labeled C) and two new pigments (A and B). A lab batch of paint only contains enough material for three tests so the experimenters used three lab batches of paint, and subdivided each paint batch into three parts so each pigment could be tested with each batch.

(a) What is the block and what is the treatment in this experiment?

(b) Perform an ANOM on the data and report the results.

4. One test of pizza cheese, involves determining the "stretch" of the cheese when the pizza is removed from the oven. The stretch is determined by using a fork to pick up the cheese and stretch it until it breaks. The length of

cheese is then measured. There are three technicians who run this test. For a fixed amount of cheese, the bake time and temperature of the pizza have an effect on the stretch. Design an experiment to quantify the effect of time and temperature on stretch. Keep in mind that the technicians could influence the results.

8.5 CHAPTER PROBLEMS

1. An experiment was conducted to study the effect of oats and bran on weight gain per week (measurements in grams). Two kinds of animals were used in the study: mice and horses. The data were collected in a factorial design and are given below.

	Bran	Oats
Mice	0.5	1
	1	2
	1.5	3
Horses	9534	2724
	11350	4540
	13166	6356

   (a) Check if the observations in the four cells have constant variance.

   (b) Comment on possible reasons for any non-constant variances.

   (c) Find a reasonable transformation to stabilize the variances.

   (d) Perform the transformation on the data and re-do part (a) to see if the transformation was successful.

2. Tomato plants were grown in a greenhouse under treatment conditions consisting of combinations of two fertilizers (Factor A) and three soil types (Factor B). Two plants were grown under each of the six treatment combinations using a completely randomized design. The (coded) yields of tomatoes are recorded below.

	Factor B		
Factor A	1	6	10
	1	4	8
	8	4	1
	10	6	1

   (a) Construct the ANOVA table.

   (b) Can the table in part (a) be used to test the effect of Factor A? If so, what do you conclude; and if not, describe how you would test the effect of factor A.

3. An experiment was conducted to study the effects of type of ignition system (factor A) and timing level (factor B) on the percentage of smoke emitted by a particular type of engine. Two types of ignition systems and three timing levels were studied using two replicates and a completely randomized design.

	Factor B		
Factor A	2	2	2
	2	4	6
	1	1	0
	3	3	2

(a) Test for an interaction.

(b) Use the ANOM to study the effect of Factor B.

4. Two factors were studied to determine their effect on a response. Four readings were made under each set of conditions. The experiment was completely randomized and the coded results are given below.

	B	
	−1	2
	2	4
	3	3
A	1	3
	6	8
	7	10
	9	8
	4	5

(a) Test the assumption of equal variances using $\alpha = 0.1$.

(b) Analyze the data using $\alpha = 0.05$.

5. The output of computer monitors was determined for two phosphor types (Factor A) and at two temperatures (Factor B). Three readings were made under each set of conditions. The experiment was completely randomized and the coded results are given below.

	B	
	−1	1
	2	2
A	3	4
	6	4
	7	8
	9	10

(a) Test the assumption of equal variances using $\alpha = 0.1$.

(b) Analyze the data using $\alpha = 0.01$.

6. A study was conducted to investigate the stability over time of the pH of a drug. Vials (2 ml) of the drug were stored at either 30 or 40°C and at specified time points (1, 3, 6, and 9 months) specimens were tested and the pH was recorded. At each temperature/time combination six replicate observations

were obtained. For your convenience tables of sample means and variances are given below.

Sample Means

		Time				
		1	3	6	9	
Temperature	30° C	3.8017	3.7150	3.7083	3.6133	3.7096
	40° C	3.5317	3.6850	3.5617	3.5633	3.5854
		3.6667	3.7000	3.6350	3.5883	3.6475

Sample Variances

		Time			
		1	3	6	9
Temperature	30°C	0.01826	0.01844	0.00974	0.02366
	40°C	0.19750	0.01018	0.02122	0.04545

(a) Parts (b) through (f) of this problem are various analyses that one would carry out.

     i. Place them in an order (from first to last) that it would be reasonable to conduct them, and then work them in that order.

     ii. If possible write down a different, but equally acceptable, order.

(b)   i. Using $\alpha = 0.1$ construct an ANOM chart to study the effect of Time. What do you conclude from the chart?

     ii. Explain why it is not necessary to construct an ANOM chart to study the effect of temperature.

(c) The column means (i.e, the average pH at each time) are decreasing with increasing time.

     i. Using least-squares, fit a line to these values.

     ii. For the model in part (i) compute an estimate of σ^2.

(d)   i. Construct the ANOVA table.

     ii. Test for interaction between time and temperature using $\alpha = 0.05$.

(e) Given below is a normal probability plot of the residuals obtained using the model

$$Y_{ijk} = \mu + \alpha_i + \beta_j + \alpha\beta_{ij} + \epsilon_{ijk}$$

except that the point corresponding to the smallest residual is missing. The smallest residual comes from the 40°C/1 month cell, which contains the pH values

$$2.63, \quad 3.71, \quad 3.65, \quad 3.8, \quad 3.7, \quad 3.7.$$

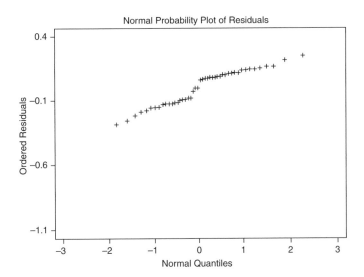

Normal Probability Plot of Residuals

    i. At what coordinates should the missing residual be plotted?

    ii. Test for equal variances at the $\alpha = 0.1$ level.

(f)   i. What does part (e) suggest about the data point corresponding to the smallest residual?

    ii. Changing the value of 2.63 to 3.63 results in both the assumptions of normality and equal variances appearing reasonable. Check this.

    iii. Before proceeding with the other parts to this problem, make the necessary changes in the tables of sample means and sample variances. Indicate precisely which values you have changed.

7. The data set **stretch.txt** contains data from an experiment used to study the stretch in hot pizza cheese. Long stretch is a desired property for commercial pizza. Measuring stretch is not a precise science. One uses an implement to lift the cheese and stretch it, measuring the length of the cheese just prior to it breaking. Because of the noise in this measurement, five measurements were taken on each pizza studied. Factors considered were bake temperature and amount of cheese on the pizza.

    (a) Identify the type of experimental design used.

    (b) Analyze the data.

    (c) Summarize your findings in a brief report which includes graphics and an explanation of how the multiple measurements on each pizza were treated and why.

9

MULTI-FACTOR EXPERIMENTS

An experiment need not be limited to only one or two factors. If there are many factors that potentially affect the response, it is more efficient to design an experiment to examine as many of them as possible rather than running individual experiments for each factor. A well-designed experiment can detect significant factors even if there are only a few experimental units in each treatment group.

9.1 ANOVA FOR MULTI-FACTOR EXPERIMENTS

An experiment with many factors has many possible sources of interaction: not only the two-way interaction we have already discussed, but also three-way and even higher-order interactions. To detect high-order interactions with any assurance, the experiment must be quite large. Often, one assumes that high-order interactions will not be important, and this assumption reduces the necessary size of the experiment. When this is the case, the variances associated with the high-order interactions are not considered separately. Instead, these variances are lumped in with the random variability. This type of data is generally analyzed on a computer.

Example 9.1.1 (*Battery Separator Data*)   In this experiment, $Y =$ electrical resistance of a battery separator (extruded from a rubber, oil, and silica mix). The goal of the experiment was to study the effects of three process factors on the electrical resistance of the final product. The three factors in the experiment are A = type of silica added to the mix, B = temperature of a water bath, and C = amount of time the material spends in the water bath. Each factor occurs at two levels, designated "High" and "Low". The data are given in Table 9.1 and also in `separate.txt`.

Table 9.1. Battery Separator Data

		C = Low	C = High	
A	B = Low	B = High	B = Low	B = High
	40.9	46.3	48.6	69.0
Low	42.2	47.0	49.5	66.3
	41.3	48.2	46.6	66.1
	36.5	53.3	59.6	75.2
High	34.8	55.4	56.4	72.5
	35.7 ·	56.3	58.8	73.2

The two Silica/Time interaction plots given in Figures 9.1 and 9.2 are not similar, and this is an indication of a possible three-way interaction.

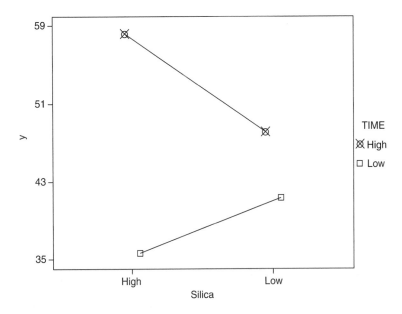

Figure 9.1. Interaction Plot with Temperature at the Low Level.

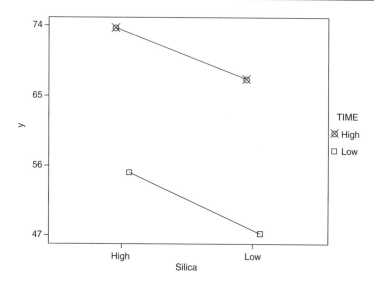

Figure 9.2. Interaction Plot with Temperature at the High Level.

A model for these data that includes the three-way interaction is

$$Y_{ijkl} = \mu + \alpha_i + \beta_j + \gamma_k + \alpha\beta_{ij} + \alpha\gamma_{ik} + \beta\gamma_{jk} + \alpha\beta\gamma_{ijk} + \epsilon_{ijkl}$$

and the analysis of variance for this model is given in Table 9.2. We can see from Table 9.2 that all of the interactions are important.

Table 9.2. An ANOVA Table for the Battery Separator Data

ANALYSIS OF VARIANCE TABLE FOR Y

SOURCE	DF	SS	MS	F	P
SILICA (A)	1	129.270	129.270	73.75	0.0000
TEMP (B)	1	1318.68	1318.68	752.28	0.0000
TIME (C)	1	1732.30	1732.30	988.24	0.0000
A*B	1	38.2538	38.2538	21.82	0.0003
A*C	1	78.8438	78.8438	44.98	0.0000
B*C	1	31.9704	31.9704	18.24	0.0006
A*B*C	1	110.510	110.510	63.04	0.0000
RESIDUAL	16	28.0467	1.75292		
TOTAL	23	3467.88			

Further analysis requires splitting the data into subsets where there are no interactions. There are several possible ways to do this, but after trying a few possibilities, it appears that the least amount of splitting is required if one first splits the data into two parts using the two levels of Temperature. Analyzing the half of the data with Temperature at the low level, one obtains Table 9.3.

Table 9.3. An ANOVA Using the Data with Temperature at the Low Level Only

```
ANALYSIS OF VARIANCE TABLE FOR Y (Temp at the low level)

SOURCE          DF      SS          MS          F       P

SILICA (A)      1       13.4408     13.4408       8.75  0.0182
TIME (B)        1       646.801     646.801     421.14  0.0000
A*B             1       188.021     188.021     122.42  0.0000
RESIDUAL        8       12.2867     1.53583

TOTAL          11       860.549
```

With this half of the data one still finds a significant interaction, so further splitting is necessary. Splitting the data based on the two levels of Time, one obtains the ANOVA in Table 9.4. It is now clear that for low Temperature and low Time, Silica has a very significant effect. Since there are only two levels of Silica, a scatter plot of the means for the two Silica levels (Figure 9.3) provides all the information on how the levels behave and the practical significance of the difference. We see from the plot that (when Temperature and Time are both at the low level) the low level of silica leads to low resistance and the high level of silica leads to high resistance. We see from Figures 9.3 and 9.4 that when Temperature is at the low level, changing Time from the low level to the high level reverses the effect of Silica. In this case, the low level of Silica leads to high resistance and the high level of Silica leads to low resistance.

Table 9.4. An ANOVA Using the Data with Temperature and Time at their Low Levels Only

```
ANALYSIS OF VARIANCE TABLE FOR Y (Temp low and Time low)

SOURCE          DF      SS          MS          F       P

SILICA (A)      1       50.4600     50.4600      86.50  0.0007
RESIDUAL        4       2.33333     0.58333

TOTAL           5       52.7933
```

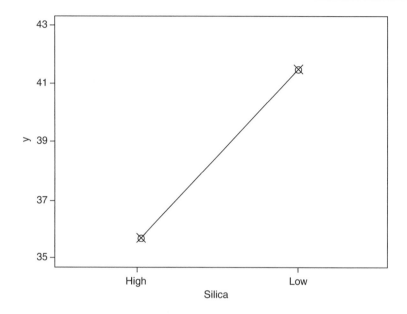

Figure 9.3. A Scatter Plot of the Means for the Two Silica Levels.

A similar analysis with Time at the high level results in Table 9.5 and Figure 9.4.

Table 9.5. An ANOVA Using the Data with Temperature at the Low Level and Time at the High Level

```
ANALYSIS OF VARIANCE TABLE FOR Y (Temp low and Time high)

SOURCE          DF      SS          MS           F        P
-------------   ----    ----------  -----------  -------  ------
SILICA (A)       1      151.002     151.002      60.68    0.0015
RESIDUAL         4      9.95333     2.48833
-------------   ----    ----------
TOTAL            5      160.955
```

Analyzing the other half of the data with Temperature at the high level, one obtains Table 9.6 and Figures 9.5 and 9.6. This time there is no interaction, and we can study the (significant) main effects of both Silica and Time when Temperature is at the high level. From the plots shown above we see that when Temperature is at the high level, high levels of Silica and Time lead to high resistance, and low levels of Silica and Time lead to low resistance.

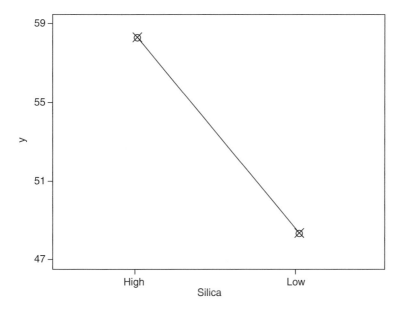

Figure 9.4. A Scatter Plot with Low Level of Temperature and High Level of Time.

Table 9.6. An ANOVA Using Just the Data with Temperature at the High Level

ANALYSIS OF VARIANCE TABLE FOR Y

SOURCE	DF	SS	MS	F	P
SILICA (A)	1	154.083	154.083	78.21	0.0000
TIME (B)	1	1117.47	1117.47	567.24	0.0000
A*B	1	1.33333	1.33333	0.68	0.4345
RESIDUAL	8	15.7600	1.97000		
TOTAL	11	1288.65			

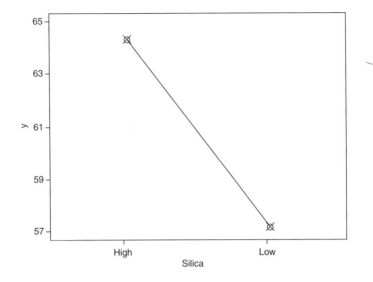

Figure 9.5. The Main Effect of Silica When Temperature is at the High Level.

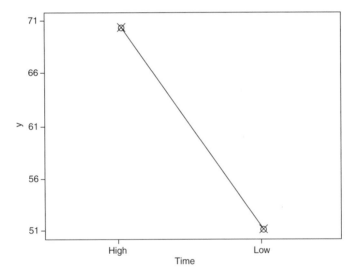

Figure 9.6. The Main Effect of Time When Temperature is at the High Level.

The results could be summarized as follows. A $2 \times 2 \times 2$ factorial experiment with three replicates was run to study the effects of Silica, Temperature, and Time on the electrical resistance of a battery separator. Since all the interactions were significant, the data were split into subsets for further analysis. With Temperature at the high level, there was still a significant Silica/Time interaction, so this half of the data was further split using the levels of Time. For low Temperature and low Time, Silica had a significant effect with low levels resulting in low resistance and high levels resulting in high resistance. For low Temperature and high Time, the effect of Silica was reversed. With Temperature at the high level there was no Silica/Time interaction, and low levels of Silica and Time led to significantly low resistance, while high level led to significantly high resistance.

Blocking Revisited

A multi-factor experiment can have blocks as well as treatments. Just as before, the purpose of blocking is to account for any effects that cannot (or should not) be held constant in the experiment. This reduces the random variability associated with the experimental units, and increases the power of the hypothesis tests. It is usual to assume that blocks and treatments do not interact.

Example 9.1.2 (*Powder Application Data*)   A producer of washing machines uses an industrial powder coating (i.e., paint) on the machines. Powder coatings are applied electrostatically as a solid and then baked. An experiment was performed to find what (if any) were the effects of applicator type (gun or bell) and particle size of the paint on the defective rate in the painting process. The measured response was $Y = \%$ defective paint jobs per hour. Factor A is the type of applicator (at two levels), Factor B is the particle size of the paint (also at two levels). Three operators were used to run the paint process. (The operators are blocks.) Each operator worked for 1 hour with each combination of applicator type and particle size. The resulting data are given in Table 9.7.

Table 9.7. Defective Paint Jobs Per Hour

	Operator 1		Operator 2		Operator 3	
	Applicator		Applicator		Applicator	
Size	A	B	A	B	A	B
1	9	5	7	2	8	4
2	9	5	8	3	7	4

The model for these data is

$$Y_{ijk} = \mu + \alpha_i + \beta_j + \gamma_k + \alpha\beta_{ij} + \epsilon_{ijk}$$

where

$$\mu = \text{the overall mean}$$
$$\alpha_i = \text{the main effect of the } i^{\text{th}} \text{ applicator type } (i = 1, 2)$$
$$\beta_j = \text{the main effect of the } j^{\text{th}} \text{ particle size } (j = 1, 2)$$

$$\gamma_k = \text{the main effect of the } k^{\text{th}} \text{ operator } (k = 1, 2, 3)$$
$$\alpha\beta_{ij} = \text{the applicator type by particle size interaction}$$
$$\epsilon_{ijk} = \text{the random error.}$$

The ANOVA for this model is given in Table 9.8.

Table 9.8. ANOVA for the Defective Paint Jobs Per Hour in Example 9.1.2

ANALYSIS OF VARIANCE TABLE FOR Y

SOURCE	DF	SS	MS	F	P
APPLICATOR (A)	1	52.0833	52.0833	125.00	0.0000
SIZE (B)	1	0.08333	0.08333	0.20	0.6704
OPERATOR (C)	2	8.16667	4.08333	9.80	0.0129
A*B	1	0.08333	0.08333	0.20	0.6704
ERROR	6	2.50000	0.41667		
TOTAL	11	62.9167			

It is clear that the particle size and the applicator type do not interact, the particle size has no effect, and the applicator type has a large effect. Operator (the blocking variable) is also significant, however, the p-value associated with Operator is much smaller than that associated with Applicator. Hence, to reduce the defective rate the first order of business would be to use the better of the two applicator types. Of secondary importance would be to look at how the operators are running the applicators differently from one another.

Since there are only two applicators, Figure 9.7 shows not only that Applicator A results in a higher defective rate, but also allows for assessing the practical significance of the difference (in this case an average of $8 - 3.8 = 4.2\%$). Since there are three levels of Operator, the ANOM is a good way to study the effects. Decision lines (using $\alpha = 0.05$) are

$$\bar{y}_{\bullet\bullet\bullet} \pm h(0.05; 3, 6)\sqrt{0.4167}\sqrt{\frac{2}{12}}$$
$$5.92 \pm 3.07(0.2635)$$
$$\pm 0.81$$
$$(5.11, 6.73).$$

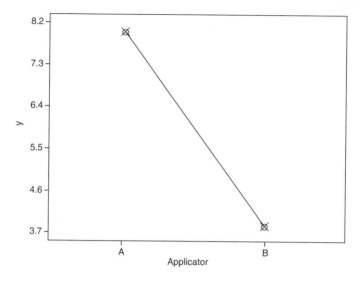

Figure 9.7. A Plot of the Applicator Effect.

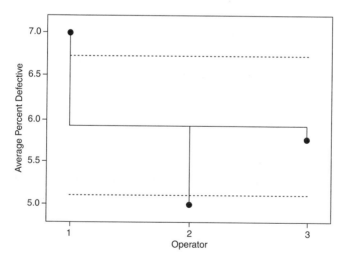

Figure 9.8. An ANOM Chart for the Operators in Example 9.1.2 with $\alpha = 0.05$.

The ANOM chart is given in Figure 9.8. From the ANOM chart one sees that Operator 1 produces significantly more defective paint jobs than average and that Operator 2 produces significantly fewer defective paint jobs than average.

Problems

1. The data set `stab.txt` contains the results of a study to investigate the stability over time of the active ingredient in a chemical product. The product has a nominal 20 mg/ml of the active ingredient. Five-milliliter samples of the chemical were stored at either 23°C (i.e., room temperature) or at 35°C (an extreme condition) in either 0% humidity or 85% relative humidity (again, normal, and extreme conditions). At specified time points (1, 4, and 8 weeks) samples were tested and the actual level of active ingredient was recorded.

   (a) What are the three factors in this experiment?

   (b) Make interaction plots for the three possible two-way interactions. What do they indicate?

   (c) Use a computer to analyze the data, starting with a three-factor analysis of variance.

   (d) Write a short report summarizing the experiment and its results.

2. The data set `defoam.txt` contains the height of a solution containing a "defoamer" in a 50-ml graduated cylinder after being heated to a particular temperature. The goal of a defoamer is to control the amount of foam in a solution (defoamers are often used, for example, in fermentation processes to avoid overflowing kettles). The defoamers studied here had three different pH's, three different concentrations, and were heated to three different temperatures. All three factor levels are given simply as L, M, and H.

   (a) Make interaction plots of temperature versus pH for the three concentrations. What do the plots indicate?

   (b) Analyze the data, starting with the model suggested by the results of part (a). Report the results.

3. The data set `chemreact.txt` is the result of an experiment designed to improve the yield of a chemical reaction. The factors under consideration were: temperature (120/140°C), time (10/30 min.), and catalyst (1/2).

   (a) Make separate interaction plots for the two catalysts. What do they indicate?

   (b) Analyze the data, starting with a model that can be used to confirm the results of part (a).

   (c) Write a report of the results in language the chemist would be able to understand.

4. The data set `purity.txt` is data from a study to ascertain the effects (if any) of two factors (Factor 1 and Factor 2) on the purity of a product from a filtration process. In order to complete all the experiments more quickly, two different technicians were used.

   (a) Analyze the data using $\alpha = 0.05$ and considering Technician to be a blocking factor.

(b) Re-analyze the data using $\alpha = 0.05$ but without considering Technician to be a factor.

(c) Which analysis gives better information? Explain why.

9.2 2^K FACTORIAL DESIGNS

We have looked at some fairly general multi-factor experiments, both with and without blocking. Next we will consider some special cases of these situations that are useful if one is doing *exploratory* work on a phenomenon. That is, we will assume we are in the situation where there are many possibly important factors, and our primary goal is to find out which, if any, of these factors are important. The type of design we choose for this type of experiment will be dependent on the number of experimental trials we can afford to conduct. Such studies are preliminary in the sense that once important factors have been identified, more work may be necessary to better estimate the effects of those factors.

2^k factorial designs are a special class of multi-factor designs in which each of the k factors occur at exactly two levels resulting in 2^k trials in the experiment. The two levels used in such a design are often designated as "high" and "low". By only looking at two levels (rather than three or more) we can include more factors in an experiment of a given size (i.e., for a given number of trials). This is desirable when one is in "exploratory mode". For example, suppose a scientist is interested in studying the effects of four factors, but due to time constraints can do no more than 18 trials. With a 2^4 design only 16 trials are needed, so all four factors could be studied. However, if three levels were used for two of the factors and two levels were used for the remaining two factors, then $3 \times 3 \times 2 \times 2 = 36$ trials would be needed. Furthermore, we will see that with 2^k designs the factor coefficients are easily interpreted and the data are easily graphed.

The disadvantage of this setup is more subtle. Suppose that, for a particular factor, the levels you have chosen as "high" and "low" have the same response, while a level such as "medium" has a different response. Because you have not observed this level, you may erroneously conclude that the factor has no effect. This scenario is illustrated in Figure 9.9, and although it is possible to choose factor levels in this manner, it is unlikely. It is more likely that the levels chosen (based on knowledge of the phenomenon under study) would fall such that some effect is identified, as illustrated by the factor levels labeled Low1 and Low2. In each of these two cases an effect is apparent, however, the experimenter would not know of the quadratic effect of the factor without further study.

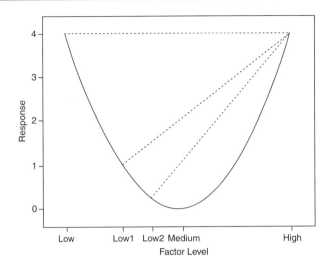

Figure 9.9. An Example Where the Response is the Same at the "High" and "Low" Levels.

If each (or even some) of the treatment combinations are replicated, the analysis can be performed in the usual way. However, often such experiments are unreplicated, out of a desire not to "waste" experimental units looking at the same treatment combination repeatedly. When there is no replication, we cannot estimate σ^2 (i.e., obtain the MS_e) in the usual way. Instead, one generally assumes that the highest-order interaction is not important (i.e., does not really exist), and then uses its sum of squares and any other sums of squares of the same magnitude to obtain SS_e. Recall that this is what we did in Section 8.3 when we had two factors and only one observation per cell.

Example 9.2.1 (*Windshield Molding Data*)   A study was undertaken to reduce the occurrence of dents in a windshield molding manufacturing process. The dents are caused by pieces of metal or plastic carried into the dies during the stamping and forming operations. The three factors to be investigated are:

$$A = \text{poly-film thickness}$$
$$B = \text{oil mixture ration for surface lubrication}$$
$$C = \text{operator glove type.}$$

For each treatment combination 1000 moldings were fabricated, and Y = number of defect-free moldings produced was measured.

The data are given in Table 9.9, where -1 denotes the low level and $+1$ denotes the high level of each factor. An initial ANOVA to obtain all the sums of squares is shown in Table 9.10.

Table 9.9. Windshield Molding Data

A	B	C	Y
−1	−1	−1	917
1	−1	−1	600
−1	1	−1	953
1	1	−1	750
−1	−1	1	735
1	−1	1	567
−1	1	1	977
1	1	1	647

Table 9.10. An Initial ANOVA to Obtain all the Sums of Squares

```
ANALYSIS OF VARIANCE TABLE FOR Y
```

SOURCE	DF	SS	MS	F	P
A (A)	1	129540	129540	13.60	0.1685
B (B)	1	32258.0	32258.0	3.39	0.3168
C (C)	1	10804.5	10804.5	1.13	0.4799
A*B	1	288.000	288.000	0.03	0.8904
A*C	1	60.5000	60.5000	0.01	0.9494
B*C	1	2312.00	2312.00	0.24	0.7085
A*B*C	1	9522.00	9522.00		
TOTAL	7	184785			

Since all of the two-factor interactions have sums of squares even smaller than SS_{ABC}, they should also be considered as error, resulting in

$$MS_e = \frac{288 + 60.5 + 2312 + 9522}{4} = 3045.6 \qquad \text{with 4 df.}$$

A second ANOVA using this new MS_e is given in Table 9.11.

Table 9.11. A Second ANOVA Using the New MS_e

ANALYSIS OF VARIANCE TABLE FOR Y

SOURCE	DF	SS	MS	F	P
A (A)	1	129540	129540	42.53	0.0029
B (B)	1	32258.0	32258.0	10.59	0.0312
C (C)	1	10804.5	10804.5	3.55	0.1327
A*B*C	4	12182.5	3045.62		
TOTAL	7	184786			

Note: The Source labeled as A*B*C *in the above table is actually the error sum of squares, which is obtained by pooling the sums of squares for all the terms not included in the model. This labeling is a quirk of the particular computer program used to produce the table.*

It is clear from looking at the p-values that Factor A has the largest effect (p-value = 0.0029), Factor B has a more moderate effect (p-value = 0.0312), and Factor C has no effect (p-value = 0.1327).

While an ANOVA table indicates which factors and interactions have statistically significant effects, it does not give the experimenter any practical information such as the magnitude or direction of the effect. There are a number of ways to obtain this information. Often simple data plots of the significant effects, such as were used in Example 9.1.2, provide sufficient information for the experimenter. However, at other times the actual effects need to be calculated. Recall that the effect of a factor is the measurement of how the response changes as one moves from the low level of the factor to the high level of the factor. Thus, one way to calculate this effect is to take the average of all responses when the factor is at the high level and subtract the average of all responses when it is at the low level.

For factor A in Example 9.2.1 one would calculate the main effect as

$$\text{Effect of A} = \frac{600 + 750 + 567 + 647}{4} - \frac{917 + 953 + 735 + 977}{4} = -254.5.$$

Thus, as one increases the poly-film thickness from its low level to its high level, the average number of defect-free moldings decreases by roughly 254.

The low and high levels are obvious for main effects, but what about interaction effects? To calculate the effect of an interaction in this manner one must first determine which responses correspond to the "high" level and which correspond to the "low" level of the interaction. This is easily done by multiplying the columns of +1's and −1's that indicate the high and low levels for factors A and B. This new column is an indicator of which runs correspond to the "high" and "low" levels for the interaction. For Example 9.2.1 the columns of $+1$'s and -1's for the main effects of factor A and factor B and for the AB interaction are given in Table 9.12.

Table 9.12. Computation of the Interaction Effect in Example 9.2.1

A	B	AB	Y
-1	-1	1	917
1	-1	-1	600
-1	1	-1	953
1	1	1	750
-1	-1	1	735
1	-1	-1	567
-1	1	-1	977
1	1	1	647

Thus, the AB interaction effect would be calculated as

$$\text{Effect of AB} = \frac{917 + 750 + 735 + 647}{4} - \frac{600 + 953 + 567 + 977}{4} = -12.$$

Recall that the initial ANOVA table for this example indicated that the AB interaction was not significant.

Yates' Algorithm

A more convenient way to calculate effects by hand is to use **Yates' algorithm**. Yates' algorithm utilizes a notation system that is not dependent on writing out long strings of $+1$'s and -1's as were used above.

Notation: In experiments with a large number of factors writing out long strings of $+1$'s and -1's is tedious, so a simpler notation has been developed. Since we only need to indicate which factors are at their "Low" level and which factors are at their "High" level for each treatment combination, we can create labels as follows. Designate each factor with a lower case letter (e.g., a for Factor A), and indicate which level it is at with an exponent of either 0 (the low level) or 1 (the high level). Thus, Factor A at the high level, Factor B at the low level, and Factor C at the low level would be written as $a^1 b^0 c^0 = a$. Similarly, ab would denote a treatment combination with Factors A and B both at their high levels, and all other factors at their low levels. An observation where all factors are at their low levels is denoted by (1).

Example 9.2.2 (*Windshield Molding Data, p. 351*) Using this notation, we could represent the responses as shown in Table 9.13.

Table 9.13. Yates' Notation for the Windshield Molding Data

Treatment Combination	(1)	a	b	ab	c	ac	bc	abc
Y	917	600	953	750	735	567	977	647

Some statistical computer packages do not calculate the treatment and interaction effects automatically, although they do generate analysis of variance tables. Hand calculation of the estimates of the treatment effects is most easily done using Yates' method. Once the treatment effects are calculated, one can easily compute the corresponding sums of squares and the resulting ANOVA table. When computing by hand, this is far easier than using the general ANOVA formulas.

Yates' algorithm proceeds as follows. First, all the treatment combinations are written in a column in standard order (sometimes referred to as Yates' order). Standard orders are obtained by starting with the ordered set

$$(1), a.$$

Each element of the set is then multiplied by the next factor designator (b in this case), and the new elements are added to the original set in the order they were obtained. This results in

$$(1), a, b, ab.$$

Repeating the process for a third factor gives

$$(1), a, b, ab, c, ac, bc, abc$$

and so forth. With the treatment combinations column in standard order one adds and subtracts their respective values in pairs, as indicated in Example 9.2.3 below, to produce column (1). The algorithm is then repeated on column (1) to produce column (2). For a 2^k design the process continues until column (k) is obtained. The first number in column (k) is the sum of all the observations, and the rest are 2^{k-1} times the estimates of the treatment effects (see Example 9.2.3).

Notation: A, B, and AB (for example) are being used to denote the effects of factor A, factor B, and the AB interaction, respectively.

Example 9.2.3 (*Yates' Algorithm*) This is Yates' method applied to a 2^2 design with $n = 1$ observation per cell. The design layout is given in terms of $+1$'s and -1's in Table 9.14 and then Yates' notation and algorithm are given in Table 9.15.

Table 9.14. The Design Layout for Example 9.2.3 Using +1's and −1's

A	B	Y
−1	−1	210
+1	−1	240
−1	+1	180
+1	+1	200

Table 9.15. Yates' Notation and Algorithm for Example 9.2.3

Treatment Combinations	Response	(1)	(2)	
(1)	210	450	830	$= \sum_{i,j} Y_{ij}$
a	240	380	50	$= 2A = 2^{k-1}A$
b	180	30	−70	$= 2B$
ab	200	20	−10	$= 2AB$

———→ designates adding the two values

− − − → designates subtracting (bottom−top) the two values

From the table for Yates' algorithm it is easy to calculate the effect of A as 25, the effect of B as −35, and the interaction effect as −5.

Example 9.2.4 (*Cement Thickening Time*) An experiment was conducted to study the effects of temperature, pressure, and stirring time on the thickening time of a certain type of cement. The coded results are given in Table 9.16.

Table 9.16. Thickening Times for a Certain Type of Cement

	Stirring Time			
	1		2	
	Pressure		Pressure	
Temperature	1	2	1	2
1	5	17	−3	13
2	−11	−1	−15	−5

Using Yates' method on these data with the factors Temperature, Pressure, and Stirring Time represented by A, B, and C, respectively; one obtains Table 9.17.

Table 9.17. Yates' Algorithm for the Cement Thickening Time Data

Treatment Combinations	Response	(1)	(2)	(3)	SS
(1)	5	−6	10	0	−
a	−11	16	−10	−64	512
b	17	−18	−34	48	288
ab	−1	8	−30	−8	8
c	−3	−16	22	−20	50
ac	−15	−18	26	4	2
bc	13	−12	−2	4	2
abc	−5	−18	−6	−4	2

The sums of squares were obtained using the equation

$$SS_{effect} = \frac{(column(k))^2}{n2^k}.$$ (9.2.1)

For example,

$$SS_A = \frac{(-64)^2}{2^3} = 512.$$

The complete ANOVA is given in Table 9.18. From Table 9.18, one can see that both Temperature and Pressure have a significant effect on the thickening time of the cement. From Yates' algorithm table the effect of A is $-64/4 = -16$ so increasing the temperature decreases the thickening time. Similarly, the effect of B is $48/4 = 12$ so increasing the pressure increases the thickening time.

Table 9.18. An ANOVA Table for Example 9.2.4

Source	SS	df	MS	F	Prob > F
A (Temperature)	512	1	512	256	0.040
B (Pressure)	288	1	288	144	0.053
C (Stirring Time)	50	1	50	25	0.126
AB	8	1	8	4	0.295
BC	2	1	2	1	0.500
AC	2	1	2	1	0.500
Error	2	1	2		
Total	864	7			

Problems

1. The data set `mw.txt` contains data on the Mw of a paint as a function of five process factors. (Mw influences certain quality characteristics of the paint.) For proprietary reasons the factors are labeled A–E, with no explanation of what they might represent. Each factor occurs at two levels, designated as "high" and "low", again with no indication of what that stands for. Perform an ANOVA on these data, fitting main effects, two-factor, and three-factor interactions. Report the results.

2. Re-analyze the `mw.txt` data, this time fitting only main effects and two-factor interactions. Compare the results to the previous analysis.

3. The data set `tpaste.txt` contains the results of an experiment to study the turbidity of a toothpaste formulation. The three factors studied in the experiment were: NaCl level, reaction temperature, and the addition rate of a particular component in the formulation. Perform an ANOVA and report the results.

4. Re-analyze the `chemreact.txt` data (see Problem 9.1.3, p. 348), this time fitting only the main effects. Does this change your conclusions? Explain.

5. The data set `bath.txt` contains the results of an experiment to investigate the effects of a portion of a production process where the product passes through a water bath. The two factors studied were the time in the bath and the temperature of the bath water. The response of interest was the electrical resistance of the final product. Analyze the data and report the results. The data are given below.

Time	Temperature	Electrical Resistance
−1	−1	36
1	1	47
−1	1	43
1	−1	41

6. A process engineer wants to study a new filling machine. The controls on the machine include the material flow rate, viscosity of the material, and fill head opening size. Design an experiment (using −1 for the low levels and +1 for the high levels) to study the effects of these settings on the amount of material dispensed in 10 seconds.

9.3 FRACTIONAL FACTORIAL DESIGNS

An unreplicated 2^k factorial design is one method for reducing the size (and hence the cost) of an experiment, while still making it possible to examine a number of factors. We can reduce the number of observations needed even further by making use of the fact that many of the higher-order interactions are probably not important, and not observing every possible combination of factor levels. However, we cannot choose the combinations that we *do* observe haphazardly. We must choose them very carefully so that all of the main effects (and other effects of interest) can be estimated. Such a design is called a **fractional factorial** design, because only a fraction of the total number of treatment combinations will be used. One of the most widely used fraction factorials is the *half-fraction*, where exactly half of the possible treatment combinations are used.

Generating Fractions of 2^k Factorial Designs

By far the most useful designs in the physical sciences and engineering discussed thus far are the 2^k factorial designs. Fractions of 2^k factorial designs are possibly even more useful since they require fewer experimental trials. For example, consider the case of four factors, each at two levels. A complete 2^4 factorial design would require 16 runs, while using a one-half fraction of a 2^4 design (denoted by 2^{4-1}), the four factors could be studied in only eight runs.

A 2^{4-1} fractional factorial design can be constructed as follows. Starting with a 2^3 design, one assumes the ABC interaction does not exist and uses it to measure the effect of a fourth factor (D). The effect of D is said to be confounded (confused) with the ABC interaction, and what is being measured is actually the effect of D plus the ABC interaction effect. The resulting design is shown in Table 9.19. The plus and minus signs represent the high and low levels of the factors, respectively; and the A, B, and C columns are the complete 2^3 factorial design. The $D = ABC$ column is obtained by multiplying the signs for the A, B, and C columns together $((+)(+) = +, (+)(-) = -$, etc.). The four columns of signs specify the design by indicating what levels the factors should be set at for each of the eight runs. For example, the third row $(- + - +)$ indicates a run with A and C at the low level and B and D at the high level. The far right-hand column duplicates this information (for convenience) using the notation for treatment combinations.

Having confounded D with ABC, it is necessary to find out what other effects have been confounded. This is referred to as the **confounding pattern**, or **alias structure**, for the design, and is obtained as follows. Starting with $D = ABC$, which is called the **generator** for the design, both sides of the equation are multiplied by D to obtain $D^2 = ABCD$. For fractions of 2^k designs any factor raised to an even power is replaced by an identity operator I $(AI = IA = A)$, and any factor raised to an odd power is replaced by that factor without an exponent. (One is actually

Table 9.19. A 2^{4-1} Fractional Factorial Design

	A	B	C	$D = ABC$	
(1)	$-$	$-$	$-$	$-$	(1)
a	$+$	$-$	$-$	$+$	$a(d)$
b	$-$	$+$	$-$	$+$	$b(d)$
ab	$+$	$+$	$-$	$-$	ab
c	$-$	$-$	$+$	$+$	$c(d)$
ac	$+$	$-$	$+$	$-$	ac
bc	$-$	$+$	$+$	$-$	bc
abc	$+$	$+$	$+$	$+$	$abc(d)$

Table 9.20. Confounding Pattern for the 2^{4-1} Design in Table 9.19

$$
\begin{aligned}
I &= ABCD \\
A &= BCD \\
B &= ACD \\
C &= ABD \\
D &= ABC \\
AB &= CD \\
AC &= BD \\
BC &= AD
\end{aligned}
$$

considering the exponents modulo two, which means each exponent is replaced with the remainder after dividing the exponent by two.) Thus, $D^2 = ABCD$ becomes $I = ABCD$, and this is called the **defining relation** or **defining contrast** for the design. The confounding pattern is obtained by further multiplication using the same rule for exponents. For example, $AI = A^2BCD$ becomes $A = BCD$, indicating that A and BCD are confounded. The complete confounding pattern for the 2^{4-1} design shown above is given in Table 9.20. It is also possible to construct other fractions of a 2^k design. For example, a 2^{5-2} design (a one-quarter replicate of a 2^5 design) can be constructed using the generators $D = AB$ and $E = AC$. In general, a 2^{k-p} design can be constructed using p generators.

Design Resolution

The various effects (e.g., ABD and ACE) in the defining relation are referred to as **words**, and the length of the shortest word in the defining relation is the **design resolution**. Knowing the design resolution provides information on what effects are confounded with one another. Designs of resolution III confound main effects with two-factor interaction, designs of resolution IV confound two-factor interactions with other two-factor interactions, and designs of resolution V confound two-factor interactions with three-factor interactions. Thus, higher-resolution designs confound higher-order interaction effects.

Example 9.3.1   The 2^{4-1} design in Table 9.19 has defining relation $I = ABCD$, which has only one word that is of length 4. Therefore, that design is a resolution IV design. Note from the confounding pattern in Table 9.20 that this design (as expected) confounds two-factor interactions with other two-factor interactions.

Yates' Algorithm for Fractional Factorial Designs

Analysis of 2^{k-p} designs is easily accomplished (without a computer) using Yates' algorithm. Instead of writing the treatment combinations down in standard order, however, the order is obtained from the standard order for a complete 2^f factorial design, where $f = k-p$. One uses the standard order for the complete 2^f design, and then adds the designators for the p additional factors to obtain the set of treatment combinations that was actually run. An example of this is given in Table 9.19 where the leftmost column contains the treatment combinations for a 2^3 design and the right-most column contains the actual runs in the correct order for Yates' algorithm. The analysis is then done as if one had a complete 2^f design. Each effect estimate and sum of squares resulting from Yates' algorithm measures the same effect it would have in the complete 2^f design plus whatever effects are confounded with it. For example, using the design in Table 9.19, the sum of squares in row $a(d)$ measures the effect of A plus the effect of the BCD interaction (see the confounding pattern in Table 9.20).

Example 9.3.2   An experiment was run to assess the effect of four mixing factors on the percent of abrasive material in a gel product. The design used was the 2^{4-1} design given in Table 9.19. Table 9.21 contains the measured values, the analysis of the results using Yates' algorithm (including the sums of squares), and the effects being measured.

Testing for significant effects in a 2^{k-p} design (with $n = 1$) is not quite so straightforward as for a complete 2^k design. The highest-order (k-way) interaction may be confounded with an effect of interest, such as in Table 9.21 where ABC is confounded with D. Therefore, instead of using the k-way interaction as a measure of error, one uses the smallest sum of squares together with the other sums of squares that are of approximately the same order of magnitude. A graphical procedure can be employed to help determine which sums of squares should be used. If an effect does not really exist, and hence its estimator is actually measuring error, then that estimator should behave as an $N(0, \sigma^2)$ random variable. Thus, when the (ordered) estimates of the effects are plotted in a normal probability plot, the estimates of non-significant effects will tend to fall along a straight line, while the estimates of

Table 9.21. Yates' Method Applied to the 2^{4-1} Design in Example 9.3.2, the Resulting Sums of Squares, and the Effect Being Measured

Treatment Combinations	Response	(1)	(2)	(3)	SS	Effect Measured
(1)	3.6	13.6	24.8	45.3	–	$I + ABCD$
$a(d)$	10.0	11.2	20.5	−0.5	0.03125	$A + BCD$
$b(d)$	8.0	10.8	1.6	−3.5	1.53125	$B + ACD$
ab	3.2	9.7	−2.1	−4.5	2.53125	$AB + CD$
$c(d)$	7.6	6.4	−2.4	−4.3	2.31125	$C + ABD$
ac	3.2	−4.8	−1.1	−3.7	1.71125	$AC + BD$
bc	3.7	−4.4	−11.2	1.3	0.21125	$BC + AD$
$abc(d)$	6.0	2.3	6.7	17.9	40.05125	$ABC + D$

significant effects will not. The sums of squares associated with those estimates that tend to fall along a straight line should be used to form SS_e.

Actually, since column (f) from Yates' algorithm is the effects multiplied by a constant (i.e., 2^{f-1}), it is easier to simply plot these values than the effects themselves. The effect of the constant just changes the scale of one axis, and does not affect which points seem to fall along a straight line.

Example 9.3.3 A normal probability plot of the estimates of the effects×4 in Example 9.3.2 (column (3) in Table 9.21) is given in Figure 9.10.

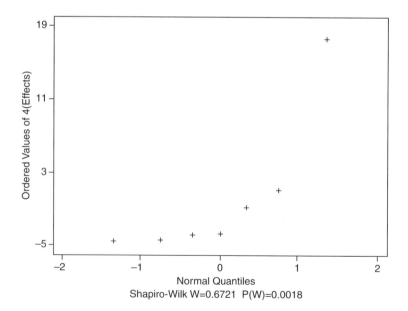

Figure 9.10. A Normal Probability Plot of the Estimates of the Effects×4 in Example 9.3.2.

The plot suggests that $D + ABC$ is the only significant effect and that the sums of squares associated with all the other effects can be used to form SS_e. This results in the ANOVA shown in Table 9.22.

Table 9.22. An ANOVA for Example 9.3.2

Source	SS	df	MS	F	Prob $> F$
$D + ABC$	40.051	1	40.051	28.86	0.00171
Error	8.328	6	1.388		
Total	48.379	7			

Once again, it must be emphasized that the purpose of a fractional experiment is to investigate as many factors as possible while keeping the size of the experiment small. Only a limited amount of information is available from such an experiment. In the example above it is possible that some three-way interaction that we have not considered is the most important effect of all. The purpose of a fractional factorial experiment is to locate some factors that are important, and eliminate other factors as less important. Later, more efficient experiments can be designed using this knowledge. Thus, by proceeding in a step-wise fashion, we avoid "wasting" experimental runs on areas that will turn out not to be of interest.

Problems

1. The data set **bright.txt** contains the results of a half-fraction experiment to study the effects of several variables on the removal of ink from newsprint. The variables of interest are: Alkali Type (A/B), %Alkali (25/75), Agitation Speed (Slow/Fast), Time (30 min./45 min.), and Water Hardness (150/250). The response of interest is the resulting brightness of the pulp (higher values are desired). Analyze the data and write a summary of the results.

2. The data set **moisture.txt** is the result of an experiment to increase the moisture content of a silica product. This experiment was run in a pilot plant at a research location. The experimental factors were Temperature (°F), Speed, Percent Solids, and pH. The data are given below.

Temp	Speed	%Solids	pH	%Moisture
200	1	10	6.5	6.3
300	1	10	7	4.4
200	2	10	7	5.3
300	2	10	6.5	3.1
200	1	15	7	5.9
300	1	15	6.5	4.1
200	2	15	6.5	4.4
300	2	15	7	3.1

   (a) What type of design was used?

   (b) Analyze the data. What factors seem to be important?

   (c) A moisture level of at least 6.5% is desired. Has this been achieved in this experiment? What settings should the plant use or what further research would you recommend?

3. The data set **reflect.txt** contains the results of an experiment to test various components in an anti-reflective coating system (for eye glass lenses). The response variable is the amount of reflectance (lower values are desired). The factors in the experiment are: Binder Type (A/B); Basecoat Type (A/B); Etch Time (L/H), and the Ratio of binder to basecoat (1=60/40; 2=85/15). Analyze the data. Suggest where further research should be concentrated.

9.4 CHAPTER PROBLEMS

1. An experimenter is interested in constructing a design to study four factors in eight runs.

   (a) Construct a 2^{4-1} fractional factorial design using the generator $D = AC$.

   (b) Write down the confounding pattern for the design in part (a).

   (c) Which is better, the design in part (a) or the design obtained using the generator $D = ABC$? Why?

   Suppose the design in part (a) was run and the responses in the correct order for Yates' algorithm were $\{1, 16, 23, 14, 22, 35, 40, 83\}$.

   (d) Find the treatment effects.

   (e) Find the sums of squares for the effects.

   (f) Use a normal probability plot to help you decide which sums of squares should be used for error.

   (g) Test for any significant effects at the 5% level.

2. The adhesive force of gummed material was determined at two humidity (Factor A) and two temperature (Factor B) conditions. Two readings were made under each set of conditions. The experiment was completely randomized and the coded results are given below.

	B	
	−1	1
A	1	3
	1	1
	1	5

   (a) Find the residuals for the model $E[Y] = \alpha_i + \beta_j + \alpha\beta_{ij}$.

   (b) Analyze the data using Yates' algorithm and $\alpha = 0.05$.

3. One is interested in studying the effect of tool type (Factor A), tool condition (Factor B), and feed rate (Factor C) on the horsepower necessary to remove one cubic inch of metal per minute. One observation was taken for each treatment combination in a completely randomized order. The coded results are given below.

		Feed Rate			
		1		2	
		Condition		Condition	
		New	Old	New	Old
Type	1	45	71	80	125
	2	−5	45	53	95

(a) Write down the responses in Yates' (standard) order.

(b) If column (2) from Yates' algorithm is $\{156, 353, -376, -357, 76, 87, 24, -33\}$, find all the sums of squares.

(c) Construct an ANOVA table and test for significant ($\alpha = 0.05$) effects.

4. One is interested in constructing a design to study six two-level factors in eight runs.

   (a) Construct a 2^{6-3} fractional factorial design using the generators $D = AC$, $E = AB$, and $F = BC$.

   (b) Determine what effects are confounded with the main effect of factor A.

5. One is interested in studying four factors in eight runs using a 2^{4-1} fractional factorial design with the generator $D = ABC$. Suppose this design was run and the responses in the correct order for Yates' algorithm were $\{45, -5, 71, 45, 80, 53, 125, 95\}$.

   (a) Find the treatment effects and the sums of squares for the effects.

   (b) Use a normal probability plot to help you decide which sums of squares should be used for error.

   (c) Are there any significant effects?

6. One is interested in constructing a design to study three factors (each at two levels) in only four runs.

   (a) Write down such a design (i.e., for each of the four runs specify the levels of the three factors).

   (b) Write down the confounding pattern for the design in part (a).

   Suppose the design in part (a) was run and the responses (in the correct order for Yates' algorithm) were $\{5, 7, 9, 13\}$.

   (c) Compute the sums of squares for the three factors.

   (d) Construct a normal probability plot to determine which sums of squares to use for error, and test for any main effects using $\alpha = 0.05$.

7. A study was conducted of the effects of three factors on the growth of a bacteria. A 2^3 factorial experiment was designed and run. The coded results are given below.

		Factor C			
		1		2	
		Factor B		Factor B	
		1	2	1	2
		2	0	0	1
	1	2	4	0	1
		3	1	-2	1
Factor A		1	3	2	5
		0	-1	-1	4
	2	1	0	2	4
		1	3	1	3
		2	2	2	5

(a) Write down the model for this design.

(b) Test the assumption of equal variances using $\alpha = 0.1$.

(c) Construct a normal probability plot to check the normality assumption.

(d) Compute the effect for each treatment combination.

(e) Compute the sum of squares for each treatment combination.

(f) Compute MS_e.

(g) Construct the complete ANOVA table.

(h) What does the ANOVA table tell you about interactions at $\alpha = 0.05$? Why?

(i) What does the ANOVA table tell you about the effect of Factor B at $\alpha = 0.05$? Why?

(j) Construct the appropriate ANOVA table(s) to study the effect of Factor B at $\alpha = 0.05$.

(k) How many observations in each cell would have been required in order to be able to detect a difference in any two μ_{ijk} of 2σ with probability 0.9 when testing using the ANOVA at level 0.05?

(l) Construct a 90% prediction interval for the coded bacteria growth when Factors B and C are both at the high level.

8. An experiment using a 2^3 factorial design was used to study the effects of temperature, concentration, and catalyst on the yield of a chemical process. The results are given below.

	Concentration			
	1		2	
	Catalyst		Catalyst	
Temperature	1	2	1	2
1	120	108	104	90
2	144	136	166	160

(a) Labeling the factors as A – Temperature, B – Catalyst, and C – Concentration, use Yates' algorithm to find all the sums of squares.

(b) Construct a normal probability plot of the effects to determine which effects are really measuring error.

(c) Test the significance of the AC interaction at level 0.05.

9. A study was done to analyze the service times (in minutes) for computer disk drive repairs made by a small company. The two factors in the experiment were the brand of disk drive and the technician performing the repair. Each technician repaired five drives of each brand, and the summary statistics are given below.

Means

	Technician		
	1	2	3
Brand 1	59.8	48.4	60.2
Brand 2	47.8	61.2	60.8
Brand 3	58.4	56.2	49.6

Variances

	Technician		
	1	2	3
Brand 1	61.7	45.8	53.7
Brand 2	55.7	53.7	39.7
Brand 3	72.8	64.7	20.3

(a) If one were going to compare nine brand/technician combinations using the ANOM with $\alpha = 0.05$ and wanted to have a power of 0.8 if any two means were 2σ apart, how large a sample would be needed for each brand/technician combination?

(b) Test the assumption of equal variances.

(c) Test for interaction.

(d) If you were the person responsible for assigning the technicians to repair jobs, which brand would you want Technician 3 to repair? Justify your answer by showing that Technician 3 is significantly better ($\alpha = 0.05$) at repairing that brand.

10

INFERENCE FOR REGRESSION MODELS

10.1 INFERENCE FOR A REGRESSION LINE

In Section 3.3, we discussed the method of **least squares** for estimating a regression line for bivariate data. This estimated regression line is really an estimate of the relationship between x and Y. The y-intercept is an estimate of the value of Y when $x = 0$, and the slope is an estimate of how much Y changes when x is increased. To assess how well the regression line describes the data, residuals and the coefficient of determination, R^2, were considered. In this chapter, we consider inference for the estimated model coefficients and for values predicted from the model. We will also introduce a test for lack of fit of the model.

Recall that the model for a regression line is

$$Y = \beta_0 + \beta_1 x + \epsilon$$

and that once the parameters have been estimated we have the fitted model

$$\widehat{y} = \widehat{\beta}_0 + \widehat{\beta}_1 x.$$

Since $\widehat{\beta}_0$ and $\widehat{\beta}_1$ (and hence $\widehat{y}$) are calculated from a sample, they contain uncertainty. That is, if we were to calculate $\widehat{\beta}_0$ and $\widehat{\beta}_1$ from a second set of data the results would be slightly different.

Example 10.1.1 (*Repair Test Temperature Data, pp. 12, 53, 77*)   Previously, we calculated the least-squares line relating time to the desired test temperature for a repair test panel removed from a freezer. The fitted regression line was

$$\widehat{y} = -37.49 - 1.864x$$

where $Y = $ time (to $-10°C$) and $x = $ temperature. We estimate that as the starting temperature increases by $1°C$, the panel takes 1.86 minutes less to warm to the desired temperature. (We also estimate that a panel with a starting temperature of $0°C$ takes -37.49 minutes to reach $-10°C$, but this is nonsense.)

The relationship in Example 10.1.1 was determined from a sample of 13 test panels. If we were to run the same experiment with 13 different panels, how might the regression line change? A confidence interval for β_1 quantifies the uncertainty in using $\widehat{\beta}_1$ to estimate β_1. A hypothesis test on β_1 assesses whether or not a particular value is a plausible one. Confidence intervals and hypothesis test procedures are also available for β_0, but are often less interesting.

Inference for β_1

Although it may not be obvious from the formula, $\widehat{\beta}_1$ is actually a linear combination of the Y_i's. It follows that $\widehat{\beta}_1$ will be approximately normally distributed, and it can be shown that

$$E(\widehat{\beta}_1) = \beta_1 \tag{10.1.1}$$

$$\mathrm{Var}(\widehat{\beta}_1) = \frac{\sigma^2}{S_{xx}}. \tag{10.1.2}$$

Confidence Intervals for β_1

Using equations (10.1.1) and (10.1.2), a $(1 - \alpha)100\%$ confidence interval for β_1 is

$$\boxed{\widehat{\beta}_1 \pm t(\alpha/2; n-2)\sqrt{\frac{MS_e}{S_{xx}}}.} \tag{10.1.3}$$

Example 10.1.2 (*Repair Test Temperature Data pp. 12, 53, 77*) Consider again the bivariate sample of time and temperatures of the test panels. We have already seen there appears to be a negative linear relationship between these two variables, and that the slope of the regression line is $\widehat{\beta}_1 = -1.864$. In the process of computing $\widehat{\beta}_1$ (p. 53) we also obtained $S_{xx} = 51.91$, and in the process of computing r (p. 77) we found $SS_e = 6.52$. Thus, $MS_e = 6.52/11 = 0.59$, and a 95% confidence interval for β_1 is (using formula (10.1.3))

$$-1.86 \pm t(0.025; 11)\sqrt{\frac{0.59}{51.91}}$$

$$\pm 2.201(0.12)$$

$$\pm 0.26$$

$$(-2.12, -1.6).$$

Therefore, for each degree warmer a panel is when it is removed from the freezer it takes somewhere between 1.6 and 2.12 seconds less to warm to the desired $-10°C$.

Hypothesis Tests on β_1

Hypothesis testing is often used in regression problems to decide whether or not the independent variable x is actually related to (or affects) the dependent variable Y. If changing x does not affect Y, then it must be the case that $\beta_1 = 0$ in the linear model. It follows that we can test whether or not x and Y are related by testing

$$H_0\colon \ \beta_1 = 0 \qquad \text{versus} \qquad H_a\colon \ \beta_1 \neq 0.$$

In other situations one might be interested in testing whether Y increases linearly as x increases ($H_a\colon \ \beta_1 > 0$) or whether Y decreases linearly as x increases ($H_a\colon \ \beta_1 < 0$).

While testing to see if $\beta_1 = 0$ is probably the most common test on β_1, there are also instances when one would want to test if β_1 was equal to some specific value β_{10}. The statistic for testing $H_0\colon \ \beta_1 = \beta_{10}$ is (again using equations (10.1.1) and (10.1.2))

$$t = \frac{\widehat{\beta}_1 - \beta_{10}}{\sqrt{\dfrac{\mathrm{MS}_e}{S_{xx}}}} \tag{10.1.4}$$

and would be compared to a critical value from the t distribution. The different possibilities are given in Table 10.1.

Table 10.1. Rejection Regions for Hypothesis Tests of $H_0\colon \ \beta_1 = \beta_{10}$

Alternate Hypothesis	Rejection Region
$H_a\colon \ \beta_1 < \beta_{10}$	Reject H_0 if $t < -t(\alpha; n-2)$
$H_a\colon \ \beta_1 > \beta_{10}$	Reject H_0 if $t > t(\alpha; n-2)$
$H_a\colon \ \beta_1 \neq \beta_{10}$	Reject H_0 if $|t| > t(\alpha/2; n-2)$

Example 10.1.3 (*Film Build Data*) A study was performed to quantify how the film build of an automotive paint affects appearance. Typically, higher film builds (i.e., thicker paint) result in better appearance. The two measured variables were x = film build and Y = gloss. The data are given in Table 10.2 and in **fbuild.txt**. The scatter plot of the data in Figure 10.1 confirms the expected linear relationship.

Table 10.2. Gloss Measurements for Different Film Builds

$x = $ film build (mm)	$y = $ gloss
0.7	85
2.3	87
1.5	86
3.1	88
2.4	87
2.9	88
3.4	89
3.8	90
3.7	89

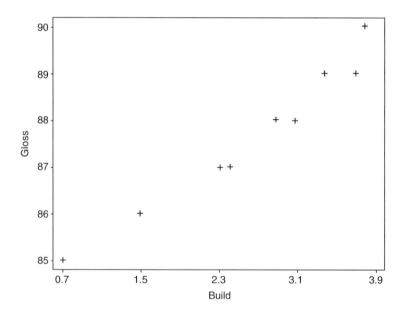

Figure 10.1. A Scatter Plot of the Film Build Data in Example 10.1.3.

Some summary statistics are $\bar{x} = 2.64$, $\bar{y} = 87.67$, $S_{xx} = 8.56$, and $S_{xy} = 12.84$.
So

$$\widehat{\beta}_1 = \frac{12.84}{8.56} = 1.50$$

$$\widehat{\beta}_0 = 87.67 - 1.50(2.64) = 83.7$$

and the fitted regression line is

$$\widehat{y} = 83.7 + 1.50x.$$

This means that the estimated rate of gloss improvement is 1.50 units/mm increase in paint thickness. The residuals are

$$\widehat{\epsilon}_1 = 85 - [83.7 + 1.50(0.7)] = 0.25$$

$$\vdots$$

$$\widehat{\epsilon}_9 = 89 - [83.7 + 1.50(3.7)] = -0.25$$

and a normal probability plot of the residuals is shown in Figure 10.2. It looks quite linear, so there is no reason to think the residuals are not normally distributed.

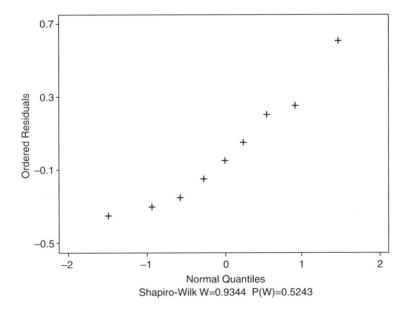

Figure 10.2. A Normal Probability Plot of the Residuals in Example 10.1.3.

We can now proceed to do inference on the slope. The estimated error variance is

$$MS_e = \frac{(0.25)^2 + \cdots + (-0.25)^2}{7} = 0.11$$

and testing to confirm that higher film build results in higher gloss, one would test $H_0:$ $\beta_1 = 0$ versus $H_a:$ $\beta_1 > 0$. Using equation (10.1.4) and Table 10.1 (with $\alpha = 0.01$), one would compute

$$t = \frac{1.50}{\sqrt{\frac{0.11}{8.56}}} = 13.23$$

which is greater than $t(0.01; 7) = 2.998$. Therefore, we can reject the null hypothesis and conclude that higher film build results in higher gloss.

Alternatively, as with tests on means, one could compute a p-value rather than comparing with a critical value. The p-values for the three possible alternatives are given in Table 10.3, where $X \sim t(n-2)$.

Table 10.3. p-values for Hypothesis Tests of $H_0:$ $\beta_1 = \beta_{10}$, where $X \sim t(n-2)$

Alternative Hypothesis	p -Value
$H_a:$ $\beta_1 < \beta_{10}$	$P[X < t]$
$H_a:$ $\beta_1 > \beta_{10}$	$P[X > t]$
$H_a:$ $\beta_1 \neq \beta_{10}$	$2P[X > |t|]$

Example 10.1.4   In Example 3.3.3 (p. 55), we looked at the set of data reproduced in Table 10.4 and modeled it using the curve $Y = \beta_0 x^{\beta_1}$. In order to estimate the parameters of the model using techniques for estimating the parameters of a line, we transformed it (by taking logs). Any inference on the parameter estimates for the transformed model depends on having errors of the usual form. In other words, we are actually assuming the transformed model (including error) is

$$Y^* = \beta_0^* + \beta_1 x^* + \epsilon^*$$

where $Y^* = \ln Y$, $x^* = \ln x$, $\beta_1^* = \ln \beta_1$, and $\epsilon^* = \ln \epsilon$. This amounts to assuming the original model is

$$Y = \beta_0 x^{\beta_1} \epsilon$$

where ϵ has a log-normal distribution.

Table 10.4. Data From Example 3.3.3

x	9	2	5	8	4	1	3	7	6
y	253.5	7.1	74.1	165.2	32.0	1.9	18.1	136.1	100.0

In the original model, β_1 represents the growth rate of Y in relation to x. For example, if $\beta_1 = 2$, then Y grows at the rate of x^2. We previously found $S_{xx} = 4.138$, and the fitted model was

$$\widehat{y}^* = 0.51 + 2.251x^*.$$

Therefore, one question of interest is whether $\beta_1 > 2$. For this model we can compute $MS_e = 0.0160$, and to perform the test of interest,

$$H_0: \beta_1 = 2 \qquad \text{versus} \qquad H_a: \beta_1 > 2$$

one would compute (using equation (10.1.4))

$$t = \frac{\widehat{\beta}_1 - \beta_{10}}{\sqrt{\dfrac{MS_e}{S_{xx}}}} = \frac{2.251 - 2}{\sqrt{\dfrac{0.016}{4.138}}} = 4.04.$$

Using Table 10.3 and a computer, one finds that

$$P[t(7) > 4.04] = 0.00247$$

and since $0.00247 < 0.01$, we would reject H_0 at the 0.01 level and conclude that Y grows at a faster rate than x^2.

Inference for $E(Y\,|\,x)$

In a bivariate population each value of x defines a subset of the whole population. In some situations one will need to estimate the mean response for a sub-population.

Notation: The sub-population of responses for a specific value of x is denoted $\{Y|\,x\}$ or $\{Y|\,x = x_0\}$. These are read as "Y given x" and "Y given $x = x_0$".

Example 10.1.5   If $Y = $ compressive strength of concrete and $x = $ curing time, then $\{Y|\,x = 20\}$ denotes the compressive strength of the sub-population that was cured for 20 days.

Example 10.1.6   If $Y = $ yield of a chemical process and $x = $ temperature of the process, then $\{Y|\,x = 150\}$ denotes the yield of the sub-population that was processed at 150 degrees.

The expected value of a sub-population is unknown, but it can be estimated using

$$\widehat{E}(Y|\,x = x_0) = \widehat{\beta}_0 + \widehat{\beta}_1 x_0.$$

It is also possible to show that

$$E[\widehat{E}(Y|\,x = x_0)] = E(\widehat{\beta}_0 + \widehat{\beta}_1 x_0) = \beta_0 + \beta_1 x_0 \qquad (10.1.5)$$

$$\text{Var}[\widehat{E}(Y|\,x = x_0)] = MS_e\left[\frac{1}{n} + \frac{(x_0 - \overline{x})^2}{S_{xx}}\right]. \qquad (10.1.6)$$

Confidence Intervals

Using equations (10.1.5) and (10.1.6), a $(1 - \alpha)$-level confidence interval for $E(Y \mid x = x_0)$ is

$$\widehat{E}(Y \mid x = x_0) \pm t(\alpha/2; n - 2)\sqrt{\mathrm{MS}_e \left[\frac{1}{n} + \frac{(x_0 - \overline{x})^2}{S_{xx}}\right]}. \qquad (10.1.7)$$

Example 10.1.7 (*Film Build Data, p. 371*)   Consider the sub-population of test panels that have a paint thickness of 2.6 mm. Our estimate of the mean of this sub-population is

$$\widehat{E}(Y \mid x = 2.6) = 83.7 + 1.50(2.6) = 87.6.$$

A 90% confidence interval for the mean of this sub-population is (using formula (10.1.7))

$$87.6 \pm 1.895\sqrt{0.11 \left[\frac{1}{9} + \frac{(2.6 - 2.64)^2}{8.56}\right]}$$

$$\pm 0.210$$

$$(87.39, 87.81).$$

Note: $E(Y \mid x = 0) = \beta_0$. *So formula (10.1.7) with $x_0 = 0$ will give a confidence interval for β_0.*

Hypothesis Tests

Hypothesis tests for $H_0 : E(Y \mid x_0) = \mu_0$ are conducted using the test statistic (based on equations (10.1.5) and (10.1.6))

$$t = \frac{\widehat{E}(Y \mid x_0) - \mu_0}{\sqrt{\mathrm{MS}_e \left[\frac{1}{n} + \frac{(x_0 - \overline{x})^2}{S_{xx}}\right]}} \qquad (10.1.8)$$

and the appropriate critical values and p-values are given in Tables 10.5 and 10.6, respectively.

Table 10.5. Rejection Regions for Hypothesis Tests of H_0: $E(Y \mid x_0) = \mu_0$

Alternate Hypothesis	Rejection Region
H_a: $E(Y \mid x_0) < \mu_0$	Reject H_0 if $t < -t(\alpha; n - 2)$
H_a: $E(Y \mid x_0) > \mu_0$	Reject H_0 if $t > t(\alpha; n - 2)$
H_a: $E(Y \mid x_0) \neq \mu_0$	Reject H_0 if $|t| > t(\alpha/2; n - 2)$

Table 10.6. p-values for Hypothesis Tests of H_0: $E(Y|\, x_0) = \mu_0$, where $X \sim t(n-2)$

Alternative Hypothesis	p-Value
H_a: $E(Y|\, x_0) < \mu_0$	$P[X < t]$
H_a: $E(Y|\, x_0) > \mu_0$	$P[X > t]$
H_a: $E(Y|\, x_0) \neq \mu_0$	$2P[X > |t|]$

Example 10.1.8   For the yellowing data in `yellow.txt` it was hoped that the material was stable. If that is true, then the initial color measurement being zero would result in the color measurement after 1 month also being zero. This corresponds to H_0: $\beta_0 = 0$. We can test this hypothesis using the statistic (10.1.8) with $x_0 = 0$ and $\mu_0 = 0$, namely,

$$t = \frac{\widehat{E}(Y|\, x_0) - \mu_0}{\sqrt{\text{MS}_e \left[\frac{1}{n} + \frac{(x_0 - \bar{x})^2}{S_{xx}} \right]}}$$

$$= \frac{\widehat{\beta}_0 - 0}{\sqrt{\text{MS}_e \left[\frac{1}{n} + \frac{(\bar{x})^2}{S_{xx}} \right]}}.$$

For this data set $n = 23$, $\widehat{\beta}_0 = 0.083$, $\text{MS}_e = 0.00945$, $\bar{x} = 0.6817$, and $S_{xx} = 0.2486$ (see Problem 10.1.10). Therefore,

$$t = \frac{0.083}{\sqrt{0.00945 \left[\frac{1}{23} + \frac{(0.6817)^2}{0.2486} \right]}} = 0.617$$

and (Table 10.6)

$$p\text{-value} = 2P[t(21) > 0.617] = 2(0.27193) = 0.5439.$$

There is no evidence that β_0 is different from zero.

Alternatively, most computer programs would provide the information necessary to test H_0: $\beta_0 = 0$. See, for example, the output in Table 10.7, which also has the t- and p-values for testing H_0: $\beta_1 = 0$ (see Problem 10.1.11).

Table 10.7. Linear Regression for Example 10.1.8

UNWEIGHTED LEAST SQUARES LINEAR REGRESSION OF B2

PREDICTOR VARIABLES	COEFFICIENT	STD ERROR	STUDENT'S T	P
CONSTANT	0.08301	0.13455	0.62	0.5439
B1	0.91460	0.19511	4.69	0.0001

R-SQUARED	0.5113	RESID. MEAN SQUARE (MSE)	0.00945
ADJUSTED R-SQUARED	0.4881	STANDARD DEVIATION	0.09719

SOURCE	DF	SS	MS	F	P
REGRESSION	1	0.20756	0.20756	21.97	0.0001
RESIDUAL	21	0.19836	0.00945		
TOTAL	22	0.40592			

Inference for Y

Once it has been confirmed that a line is an appropriate regression model, it is often of interest to use the model to predict future observations. For a specific level of x (call it x_0) the predicted value of Y is

$$\widehat{y} = \widehat{E}(Y \mid x_0) = \widehat{\beta}_0 + \widehat{\beta}_1 x_0.$$

As with prediction intervals for values from a single population (see equation (5.3.2)), we need to account for both the variability of the new observation and the variability of our estimator $\widehat{Y}$ (equation (10.1.6)). Thus, a $(1 - \alpha)$-level prediction interval for Y is

$$\widehat{y} \pm t(\alpha/2; n - 2)\sqrt{\mathrm{MS}_e \left[1 + \frac{1}{n} + \frac{(x_0 - \overline{x})^2}{\mathrm{S}_{xx}}\right]}. \tag{10.1.9}$$

Example 10.1.9 (*Film Build Data, p. 371*)   For this set of data $\overline{x} = 2.64$, $\mathrm{S}_{xx} = 8.56$, $\mathrm{MS}_e = 0.11$, and the least-squares regression line is

$$\widehat{y} = 83.7 + 1.50x.$$

The gloss at 3 mm would be predicted to be

$$\hat{y} = 83.7 + 1.50(3) = 88.2$$

and a 90% prediction interval is (using formula (10.1.9) and $t(0.05; 7) = 1.895$)

$$88.2 \pm 1.895 \sqrt{0.11 \left[1 + \frac{1}{9} + \frac{(3 - 2.64)^2}{8.56} \right]}$$

$$\pm 0.667$$

$$(87.53, 88.87).$$

The gloss at 4 mm would be predicted to be

$$\hat{y} = 83.7 + 1.50(4) = 89.7$$

and a 90% prediction interval is

$$89.7 \pm 1.895 \sqrt{0.11 \left[1 + \frac{1}{9} + \frac{(4 - 2.64)^2}{8.56} \right]}$$

$$\pm 0.724$$

$$(88.98, 90.42).$$

The second prediction interval is slightly wider because 4 mm is further from $\bar{x}$ than 3 mm.

Planning an Experiment

One can influence the width of these confidence and prediction intervals and the power of these hypothesis tests by how the x's are chosen. The width of the intervals and the power of the tests depend on S_{xx}. (They also depend on MS_e, but that is not under our control.) The larger S_{xx} is, the more precisely we can estimate β_1 (see formula (10.1.3)), which in turn leads to more precise predictions for Y and more precise estimates of $E(Y \mid x)$. Because S_{xx} measures the spread of the x's, it is a good idea when planning a study to make the range of x's as large as possible. In addition, placing more x values near the extremes of the range will increase S_{xx} still further.

Inference on Y and $E(Y \mid x)$ can be made still more precise by choosing the x values so that $\bar{x}$ is close to the specific value x_0 of interest (see formulas (10.1.9) and (10.1.7)).

Also, if more than one observation is taken at some of the x values, it is possible to test whether a line is the appropriate model.

Testing for Lack of Fit

The MS_e is an estimate of σ^2 based on the assumption that the linear model is correct. If one has more than one observation at some x values, then the sample variances computed at each x where there are multiple observations can be pooled to obtain an estimate of σ^2 that does not depend on the linear model being correct. By comparing the two estimates one can test the appropriateness of the linear model. More specifically, when there are multiple observations at some x values, the SS_e actually consists of two pieces. We will need the following notations.

Notation:

$$n_i = \text{the number of observations at } x_i$$
$$I = \text{the number of different } x_i \text{ values}$$
$$n = \sum_{i=1}^{I} n_i = \text{the total number of observations}$$
$$y_{ij} = \text{the response of the } j^{\text{th}} \text{ replicate at } x_i$$
$$\bar{y}_i = \text{the average response at } x_i$$

The residual associated with y_{ij} is $y_{ij} - \widehat{y}_i$, and SS_e can be written as

$$
\begin{aligned}
\mathrm{SS}_e &= \sum_{i=1}^{I} \sum_{j=1}^{n_i} (y_{ij} - \widehat{y}_i)^2 \\
&= \sum_{i=1}^{I} \sum_{j=1}^{n_i} [(y_{ij} - \bar{y}_i) + (\bar{y}_i - \widehat{y}_i)]^2 \\
&= \sum_{i=1}^{I} \sum_{j=1}^{n_i} (y_{ij} - \bar{y}_i)^2 + \sum_{i=1}^{I} \sum_{j=1}^{n_i} (\bar{y}_i - \widehat{y}_i)^2 \\
&= \sum_{i=1}^{I} \sum_{j=1}^{n_i} (y_{ij} - \bar{y}_i)^2 + \sum_{i=1}^{I} n_i (\bar{y}_i - \widehat{y}_i)^2
\end{aligned}
$$

where the third equality follows from the second because the missing cross-product term sums to zero (see Problem 10.1.12).

Thus, we have partitioned SS_e into two pieces

$$\boxed{\mathrm{SS}_e = \mathrm{SS}_{\mathrm{pe}} + \mathrm{SS}_{\mathrm{LF}}} \tag{10.1.10}$$

where

$$\boxed{\mathrm{SS}_{\mathrm{pe}} = \sum_{i=1}^{I} \sum_{j=1}^{n_i} (y_{ij} - \bar{y}_i)^2} \tag{10.1.11}$$

and

$$\boxed{\mathrm{SS}_{\mathrm{LF}} = \sum_{i=1}^{I} n_i (\bar{y}_i - \widehat{y}_i)^2.} \tag{10.1.12}$$

The sums of squares SS_{pe} and SS_{LF} are referred to as the **sum of squares for pure error** and the **sum of squares for lack of fit**, respectively. Their corresponding mean squares are

$$MS_{pe} = \frac{SS_{pe}}{n - I} \qquad (10.1.13)$$

and

$$MS_{LF} = \frac{SS_{LF}}{I - 2}. \qquad (10.1.14)$$

One can test the hypothesis $H_0\colon E(Y) = \beta_0 + \beta_1 x$ by computing

$$F_{LF} = \frac{MS_{LF}}{MS_{pe}} \qquad (10.1.15)$$

and comparing it with $F(\alpha; n - I, I - 2)$.

Example 10.1.10 In the data set `phmeas.txt` there are two observations with `phold` $= 5.63$ (call this x_1). Therefore, one can test $H_0\colon E(Y) = \beta_0 + \beta_1 x$ using the statistic F_{LF} (equation (10.1.15)). For this data set $n = 11$, $I = 10$, $\overline{y}_1 = 5.725$, and the fitted line is $\widehat{y} = 1.945 + 0.6689x$ (see Problem 10.1.3). Using equations (10.1.11) and (10.1.13),

$$MS_{pe} = \frac{(5.71 - 5.725)^2 + (5.74 - 5.725)^2}{1} = 0.00045$$

and using equations (10.1.12) and (10.1.14) with Table 10.8,

$$MS_{LF} = \frac{0.0002 + \cdots + 0.0000}{8} = 0.0023.$$

Finally (equation (10.1.15)),

$$F_{LF} = \frac{0.0023}{0.00045} = 5.11$$

and since $5.11 < F(0.1; 8, 1) = 59.439$, there is no evidence of lack of fit.

Note: As an alternative, in Example 10.1.10 one could have computed SS_{LF} using equation (10.1.10) and $SS_e = 0.019$ (see Problem 10.1.3). This would result in $SS_{LF} = 0.01855$ and $MS_{LF} = 0.0023$. However, as with equation (8.2.11) and the "computing formula" for the sample variance, equation (10.1.10) can be numerically unstable.

Table 10.8. Values for Computing Lack of Fit

i	n_i	x_i	$\overline{y}_i$	$\widehat{y}_i$	$(\overline{y}_i - \widehat{y}_i)^2$
1	2	5.63	5.725	5.711	0.0002
2	1	6.35	6.16	6.210	0.0011
3	1	6.01	5.95	5.958	0.0002
4	1	6.15	6.05	6.062	0.0001
5	1	6.00	6.01	5.951	0.0027
6	1	6.11	6.06	6.032	0.0008
7	1	5.93	5.83	5.899	0.0067
8	1	5.84	5.81	5.832	0.0017
9	1	6.11	6.10	6.032	0.0046
10	1	6.07	6.01	6.003	0.0000

Problems

1. Use the data below to do the following.

x	6	4	1	3	2	7
y	12	11	4	7	4	16

   (a) By hand, calculate the equation for the regression line.

   (b) Calculate the six residuals.

   (c) Make a normal probability plot of the residuals and explain what it indicates.

   (d) Estimate σ^2 and explain what it indicates.

   (e) Test the hypothesis

$$H_0: \beta_1 = 0 \qquad \text{versus} \qquad H_a: \beta_1 \neq 0$$

   using $\alpha = 0.05$. Explain what your results mean.

   (f) Construct a 95% confidence interval for the population trend (slope) and explain what it indicates.

2. Using the data from Problem 3.3.1 (p. 64), do the following.

   (a) Report the regression line calculated previously. Use the residuals (also already found) to make a normal probability plot and explain what it indicates.

   (b) Use the SS_e already calculated to estimate σ^2 and explain what it indicates.

   (c) Test to see if x and Y are related using $\alpha = 0.10$, and explain what your results mean.

   (d) Construct a 90% confidence interval for the slope of the regression line and explain what it indicates.

3. Using the data from `phmeas.txt`, do the following.

   (a) Find the least-squares equation.

   (b) Estimate σ^2.

   (c) Test whether or not the pH measurement taken with the old instrument is related to the pH measurement taken with the new instrument using $\alpha = 0.01$. Also report the p-value of the test. Explain what your results indicate.

   (d) Construct a 99% confidence interval for the slope of the regression line and explain what it means.

4. Using the data from `adhesion2.txt` do the following.

   (a) Make normal probability plots of the residuals for both sets of data (catalysts 1 and 2) and comment on them.

(b) For each catalyst test for lack of fit in the simple linear model.

(c) For each catalyst test whether adhesion is related to pH. Explain what your results mean.

5. Using the data `lwsw.txt`, do the following.

    (a) Using short wave (SW) as the independent variable, test the simple linear model for lack of fit.

    (b) Test whether or not long wave (LW) and short wave (SW) are related using $\alpha = 0.05$, and explain what your results mean.

    (c) Construct a 95% confidence interval for the slope of the regression line and explain what it means.

6. Use the data below to do the following.

x	3	2	7	1	8	5	6
y	4	5	-3	8	-7	1	-2

    (a) Fit a straight line to the data and calculate the residuals.

    (b) Make a normal probability plot of the residuals. Do the residuals appear to be normally distributed?

    (c) Does x affect Y? Test, using $\alpha = 0.10$, and explain what your results mean.

    (d) Predict Y when $x = 4$. Construct a 90% prediction interval for Y and explain what it indicates.

    (e) Predict Y when $x = 8$. Construct a 90% prediction interval for this prediction also.

    (f) Are predictions more precise when $x = 8$ or when $x = 4$? Explain. What is the underlying reason for this?

    (g) Estimate the sub-population mean for the sub-population that has $x = 2.5$. Construct a 90% confidence interval for the mean and explain what it indicates.

    (h) Estimate the sub-population mean for the sub-population that has $x = 8$. Construct a 90% confidence interval for this mean also.

7. Using the data from `phmeas.txt`, do the following.

    (a) Predict the pH measurement with the new instrument based on a reading of 6 with the old instrument. Construct a 95% prediction interval and explain what it indicates.

    (b) Predict the pH measurement with the new instrument based on a reading of 5.6 with the old instrument. Construct a 95% prediction interval and explain what it indicates.

    (c) For what level of pH can we predict most precisely, and why?

(d) Construct a 95% confidence interval for the average pH reading with the new instrument for the sub-population that has a pH reading of 5.8 on the old instrument. Explain what this interval indicates.

(e) Construct a 99% confidence interval for the average pH reading with the new instrument for the sub-population that has a pH reading of 6.3 on the old instrument. Explain what this interval indicates.

8. Using the vitamin data `vitamin.txt`, do the following.

(a) An elderly patient is entering treatment with a starting calcium level of 17.2. Predict her final calcium level under each treatment regimen. Assuming that calcium level increase is the purpose of the treatment, which regimen would you recommend, and why?

(b) Construct 95% prediction intervals for each prediction in (a). Under which regimen is prediction most precise? Explain.

(c) Another patient has a starting calcium level of 29. Predict his final calcium level under each regimen and construct a 95% prediction interval for each.

9. Use the data from `adhesion2.txt`, to do the following.

(a) For each catalyst type estimate the average adhesion for the sub-population of formulations with a pH level of 5.2. Construct 99% confidence intervals for each estimate. Explain what these intervals indicate.

(b) For each catalyst type estimate the average adhesion for the sub-population of formulations with a pH level of 6.2. Construct 99% confidence intervals for each estimate. Explain what these intervals indicate.

10. Verify that in Example 10.1.8: $\widehat{\beta}_0 = 0.083$, $MS_e = 0.00945$, $\bar{x} = 0.6817$, and $S_{xx} = 0.2486$.

11. Verify that the t- and p-values given in Table 10.7 (p. 378) for testing $H_0 : \beta_1 = 0$ are correct.

12. Verify that $SS_e = SS_{pe} + SS_{LF}$.

10.2 INFERENCE FOR OTHER REGRESSION MODELS

In Section 3.4, we discussed fitting polynomial models such as

$$Y = \beta_0 + \beta_1 x + \beta_2 x^2 + \epsilon \quad \text{(a parabola)}$$

and

$$Y = \beta_0 + \beta_1 x + \beta_2 x^2 + \beta_3 x^3 + \epsilon \quad \text{(a cubic curve)}$$

as well as surface models such as

$$Y = \beta_0 + \beta_1 x_1 + \beta_2 x_2 + \epsilon \quad \text{(a plane)}$$

$$Y = \beta_0 + \beta_1 x_1 + \beta_2 x_2 + \beta_3 x_1 x_2 + \epsilon \quad \text{(a twisted plane)}$$

and

$$Y = \beta_0 + \beta_1 x_1 + \beta_2 x_2 + \beta_3 x_1 x_2 + \beta_4 x_1^2 + \beta_5 x_2^2 + \epsilon \quad \text{(a paraboloid)}.$$

Because fitting such models entails a large amount of algebra, it is usually done on a computer. Some computer packages require that the user create a spreadsheet column for each term in the model and specify in the regression model which columns are to be used as independent variables. The computer then calculates parameter estimates, standard errors of parameter estimates, and test statistics for testing whether or not parameters are equal to zero. Thus, most inference for such models can be done by simply reading the computer output.

Inference for the β_i's

The output of essentially all linear regression software includes estimates of the β_i's, their standard errors (i.e., the estimated standard deviations of the estimators), and test statistics for testing whether $\beta_i = 0$.

Example 10.2.1   (*Particle Size Data, p. 62*)   In Example 3.3.6, we considered an experiment on the relationship between a vacuum setting and particle size for a granular product. We fit the model $Y = \beta_0 + \beta_1 x + \beta_2 x^2 + \epsilon$ and the computer output for this model is given in Table 10.9.

Table 10.9. Linear Regression for the Particle Size Data

UNWEIGHTED LEAST SQUARES LINEAR REGRESSION OF Y

PREDICTOR VARIABLES	COEFFICIENT	STD ERROR	STUDENT'S T	P
CONSTANT	-74.2500	5.24071	-14.17	0.0000
X	7.42107	0.48221	15.39	0.0000
X_SQUARED	-0.17054	0.01094	-15.58	0.0000

R-SQUARED	0.9731	RESID. MEAN SQUARE (MSE)	0.05365
ADJUSTED R-SQUARED	0.9654	STANDARD DEVIATION	0.23163

SOURCE	DF	SS	MS	F	P
REGRESSION	2	13.5734	6.78671	126.49	0.0000
RESIDUAL	7	0.37557	0.05365		
TOTAL	9	13.9490			

The fitted model is

$$\widehat{y} = -74.25 + 7.42x - 0.17x^2.$$

One question of interest is whether or not the x^2 term is needed (i.e., whether a parabolic model is needed rather than a straight-line model). This question can be phrased as testing

$$H_0: \ \beta_2 = 0 \qquad \text{versus} \qquad H_a: \ \beta_2 \neq 0.$$

The test statistic (from Table 10.9) is $t = -15.58$, which has a p-value smaller than 0.0001. Thus, we would reject the null hypothesis and conclude that the parabolic model is needed. (In this case, the scatter plot on p. 62 clearly shows a parabolic curve, so the test is not as necessary as it would be with some sets of data.)

A confidence interval for any parameter can also be easily found. A 95% confidence interval for β_2 is

$$-0.17054 \pm t(0.025; n-2)(0.01094)$$
$$\pm 2.306(0.01094)$$
$$\pm 0.02523$$
$$(-0.196, -0.145).$$

Inference for Y and $E(Y \mid x)$

Most computer programs also provide prediction intervals for a new value Y and confidence intervals for $E(Y \mid x)$.

Example 10.2.2 (*Particle Size Data, p. 62*) One example of such output for this data set is given in Table 10.10 from which one would find a 95% prediction interval for Y when $x = 21$ is $(5.7809, 6.9917)$, and a 95% confidence interval for $E(Y \mid x = 21)$ is $(6.1283, 6.6442)$.

Table 10.10. Predicted and Fitted Y values for the Particle Size Data

```
PREDICTED/FITTED VALUES OF Y


LOWER PREDICTED BOUND      5.7808      LOWER FITTED BOUND      6.1283
PREDICTED VALUE            6.3863      FITTED VALUE            6.3863
UPPER PREDICTED BOUND      6.9917      UPPER FITTED BOUND      6.6442
SE (PREDICTED VALUE)       0.2560      SE (FITTED VALUE)       0.1091


UNUSUALNESS (LEVERAGE)     0.2219
PERCENT COVERAGE             95.0
CORRESPONDING T              2.36


PREDICTOR VALUES: X = 21.000, X_SQUARED = 441.00
```

Testing for Lack of Fit

The procedure described in Section 10.1 for testing lack of fit with the simple linear model can also be applied to other regression models. The only differences are that (i) with k independent variables one must have multiple observations where the $(x_1, \ldots, x_k)$ values are all the same; and (ii) in general, the df associated with SS_{LF} are $I - p$, where p is the number of parameters (i.e, β_i's) in the model. Thus, one would:

1. Compute MS_{pe} by pooling the appropriate sample variances.
2. Compute SS_{pe} by multiplying MS_{pe} by $n - I$.
3. Compute SS_{LF} using either equation (10.1.10) or equation (10.1.12).
4. Compare the test statistic (10.1.15) with $F(\alpha; n - I, I - p)$.

Example 10.2.3 (*Particle Size Data, p. 62*) For this data set there are $n_i = 2$ observations at each of $I = 5$ levels of $(x_1, x_2) = (x, x^2)$. The five sample variances are

$$0.02 \quad 0.125 \quad 0.045 \quad 0.02 \quad 0.045$$

which result in

$$\mathrm{MS}_{\mathrm{pe}} = \frac{0.02 + \cdots + 0.045}{5} = 0.051$$

and

$$\mathrm{SS}_{\mathrm{pe}} = 0.051(5) = 0.255.$$

From Table 10.9 we have $\mathrm{SS}_e = 0.37557$, and therefore,

$$\mathrm{SS}_{\mathrm{LF}} = 0.37557 - 0.255 = 0.12057$$
$$\mathrm{MS}_{\mathrm{LF}} = 0.12057/2 = 0.0603$$
$$F_{\mathrm{LF}} = \frac{0.0603}{0.051} = 1.18.$$

Since $1.18 < F(0.1; 2, 5) = 3.780$, there is no evidence of lack of fit.

Problems

1. A study was performed to predict the gloss (an appearance measure) of an automotive paint job based on application factors. The measured variables were:

   DISTANCE = distance from the applicator to the car
   CHARGE = electrical charge used on the applicator
   FLOWRATE = flow rate of the material through the applicator
   GLOSS = appearance measure.

   Experience suggests that useful predictor variables are x_1 = DISTANCE, x_2 = CHARGE, and x_3 = FLOWRATE. The data are stored in applicat.txt.
   (a) Fit the model $Y = \beta_0 + \beta_1 x_1 + \beta_2 x_2 + \beta_3 x_3 + \epsilon$ to the data. Report the fitted equation.
   (b) Make a normal probability plot of your residuals and comment on it.
   (c) Test to see if $\beta_3 = 0$. Report the p-value of your test and explain your results.
   (d) Construct a 95% confidence interval for β_1 and explain what it indicates.
   (e) Construct a 90% confidence interval for β_2, and explain what it indicates.
   (f) Predict Gloss for a car that has Distance = 12, Charge = 40, and Flow Rate = 100. Construct a 99% prediction interval for this prediction.

2. In Problem 3.4.2 (p. 69), you analyzed the data set sales.txt by fitting the model $Y = \beta_0 + \beta_1 x_1 + \beta_2 x_2 + \epsilon$.
   (a) For each independent variable perform a hypothesis test to see if the variable makes a significant contribution to the model. Write a sentence or two summarizing your results.
   (b) Predict the sales level when the power index is 110 and the amount of capital is 85. Construct a 95% prediction interval for this point.
   (c) Predict the sales level when the power index is 115 and the amount of capital is 81. Construct a 95% prediction interval for this point.

3. For the experimental data in curl.txt do the following.
   (a) Fit the model $Y = \beta_0 + \beta_1 x_1 + \beta_2 x_2 + \epsilon$.
   (b) Test the importance of each independent variable and report the p-values.
   (c) Test for lack of fit.

10.3 CHAPTER PROBLEMS

1. An experiment was done to determine the amount of heat loss for a particular type of thermal pane as a function of the outside temperature. The inside temperature was kept at a constant 68°F, and heat loss was recorded for three different outside temperatures.

$$\text{Outside temperature:} \quad 80 \quad 50 \quad 20$$
$$\text{Heat loss:} \quad 5 \quad 10 \quad 12$$

   (a) Find a 95% confidence interval for the rate at which heat loss increases as the outside temperature decreases.

   (b) Find a 95% confidence interval for the expected heat loss when the outside temperature is 0°F.

2. An experiment was performed to study the effect of light on the root growth of mustard seedlings. Four different amounts of light (measured in hours per day) were used, and after a week the root length (in mm) were measured. The results are given below.

$$\text{Root Length:} \quad 6 \quad 30 \quad 15 \quad 55$$
$$\text{Amount of Light:} \quad 0 \quad 2 \quad 6 \quad 12$$

   (a) Fit the simple linear model.

   (b) Compute the residuals.

   (c) A normal probability plot of the residuals would have the point corresponding to the largest residual at what (x, y) coordinates?

   (d) Test the hypothesis that more light increases growth using $\alpha = 0.05$.

   (e) Find a 95% prediction interval for the amount of root growth in a week of a seedling subjected to 15 hours of light per day.

3. The data given below are the number of hours (x) a student spent studying and his/her grade on a particular exam (y).

$$x: \quad 0 \quad 2.3 \quad 4 \quad 6.7 \quad 7$$
$$y: \quad 6 \quad 45 \quad 70 \quad 95 \quad 89$$

   Test whether more studying increases a student's grade using $\alpha = 0.1$.

4. A producer of electronic components knows that the cost per unit changes as the volume of production increases. Let $Y = $ the unit cost and $x = $ the number produced (in millions). The manufacturer believes that the relationship between x and Y is of the form $E[Y] = \beta_0 + \beta_1 x$. Assume that the Y's are independent and normally distributed with common variance. The results of

3 days' productions are given below.

$$x: \quad 1 \quad 2 \quad 3$$
$$y: \quad 3.1 \quad 1.8 \quad 1.1$$

(a) Find the least squares regression line.

(b) Find an estimate for σ^2.

(c) Test H_0: $\beta_1 = 0$ against H_a: $\beta_1 \neq 0$ at the 0.05 significance level.

(d) Find a 95% prediction interval for a new observation Y when $x = 2$.

5. An experiment was performed to study the effect of light on certain bacteria. Several colonies of the bacteria were exposed to different amounts of light, and the percent increase in the colony size was recorded. The results are given below.

$$\text{Amount of Light:} \quad 2 \quad 2 \quad 4 \quad 4 \quad 6 \quad 6$$
$$\text{Percent Increase:} \quad 11 \quad 12 \quad 33 \quad 37 \quad 50 \quad 46$$

(a) Use the ANOM and $\alpha = 0.05$ to test for significant differences.

(b) Compute the fitted regression line.

(c) Compute the MS_e for the fitted line.

(d) Test to see if an increase in light results in an increase in growth at $\alpha = 0.01$.

6. A study was conducted to investigate the stability over time of the active ingredient in a drug. The drug has a nominal 30 mg/ml of active ingredient. 2 ml vials of the drug were stored at either $30°C$ or $40°C$ and at specified time points (1, 3, 6, and 9 months) specimens were tested and the actual level of active ingredient was recorded. At each temperature/time combination six replicate observations were obtained. For your convenience tables of sample means and variances are given below.

Sample Means

		Time				
		1	3	6	9	
Temperature	30°C	30.085	30.035	29.903	29.782	29.951
	40°C	30.213	30.003	29.827	29.747	29.948
		30.149	30.019	29.865	29.764	29.949

Sample Variances

		Time			
		1	3	6	9
Temperature	30°C	0.00238	0.02460	0.01446	0.00246
	40°C	0.08426	0.03466	0.01126	0.00538

(a) Parts (b) through (f) of this problem are various analyses that one would carry out.

    i. Place them in an order (from first to last) that it would be reasonable to conduct them.

    ii. If possible write down a different, but equally acceptable, order.

(b)   i. Using $\alpha = 0.01$ construct an ANOM chart to study the effect of Time. What do you conclude from the chart?

    ii. Give two reasons why it is not necessary to construct an ANOM chart to study the effect of Temperature.

(c) The column means (i.e., the average amount of active ingredient at each time) are decreasing with increasing time.

    i. Using least squares, fit a line to these values.

    ii. Using this line, construct a 90% confidence interval for the rate at which the active ingredient is decreasing. The residuals for this line are $0.02, -0.014, -0.024, 0.019$.

(d)   i. Construct the ANOVA table.

    ii. Test for interaction between Time and Temperature using $\alpha = 0.05$.

(e) Given below is a normal probability plot of the residuals obtained using the model

$$Y_{ijk} = \mu + \alpha_i + \beta_j + \alpha\beta_{ij} + \epsilon_{ijk}$$

except that the point corresponding to the largest residual of 0.59 is missing.

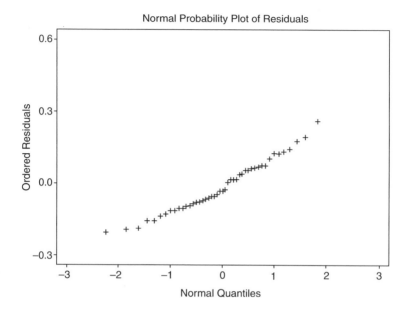

Normal Probability Plot of Residuals

i. At what coordinates should the missing residual be plotted?

ii. What does this plot suggest about the original data point corresponding to the residual 0.59?

(f) i. For the model in part (e) test for equal variances at the $\alpha = 0.05$ level.

ii. For the model in part (c) what sample statistics (be specific) would be needed to test for equal variances, and if one could compute $F_{\max}$, what critical value would it be compared with for $\alpha = 0.1$?

11

RESPONSE SURFACE METHODS

Often when studying continuous factors, after finding their effects to be significant, one is interested in finding conditions (levels of the factors) that lead to a particular response, usually a maximum or a minimum. The responses of an experiment when considered as a function of the possible levels of the factors are called a **response surface**, and designs used to study a response surface are called **response surface designs**. Response surfaces are usually much too complex to be able to easily model the entire surface with a single function, so a simple model (one that is linear in the factor levels) is used to find the local slope (the slope in a specified area) of the surface and point the direction to the maximum (minimum) response. Then, when the desired area (say of the maximum) is reached, a more complex model with quadratic terms is used to provide a more accurate representation of the response surface in the area of its maximum value.

11.1 FIRST-ORDER DESIGNS

A first-order design is one that is constructed in order to fit a first-degree polynomial model (i.e., a model that is linear in the factor levels). For k factors the first-degree polynomial model is

$$Y = \beta_0 + \beta_1 x_1 + \cdots + \beta_k x_k + \epsilon. \tag{11.1.1}$$

For the purpose of conducting significance tests, we will assume (as usual) that $\epsilon \sim N(0, \sigma^2)$. A model of this kind is used to estimate the local slope of a response surface. In addition to allowing for the fitting of model (11.1.1), a desirable first-order design should also provide an estimate of experimental error (an estimate of σ^2) and a means of assessing the adequacy of the model.

A good starting point for a first-order design is a 2^k factorial or a 2^{k-p} fractional factorial design. Model (11.1.1) is easily fitted in this case since when the levels of the x_i are coded as -1 and 1, estimates of the β_i (call them $\widehat{\beta}_i$) are for $i \geq 1$ the estimates of the corresponding factor effects divided by two (i.e., $\widehat{\beta}_1 = A/2, \widehat{\beta}_2 = B/2$, etc.), and $\widehat{\beta}_0$ is the overall average. For a 2^k design the $\widehat{\beta}_i$ can all be obtained by dividing the appropriate values in column (k) from Yates' algorithm by $n2^k$.

Example 11.1.1 (*Cement Thickening Time, p. 356*) In Example 9.2.4, a 2^3 factorial design was used to study the effects of Temperature, Pressure, and Stirring Time on the thickening time of a certain type of cement. The values in column (3) from Yates' algorithm for the three main effects were -64, 48, and -20; for A, B, and C, respectively. The overall average was zero because of how the values were coded. Thus,

$$\widehat{\beta}_0 = 0$$
$$\widehat{\beta}_1 = -64/8 = -8$$
$$\widehat{\beta}_2 = 48/8 = 6$$
$$\widehat{\beta}_3 = -20/8 = -2.5$$

and the fitted first-order model is

$$\widehat{y} = -8x_1 + 6x_2 - 2.5x_3.$$

A significant AB interaction effect would indicate that the cross-product term $\beta_{12}x_1x_2$ ($\widehat{\beta}_{12} = AB/2$) is needed in the model, and the first-order model does not adequately describe that region of the response surface. Similarly, any other significant two-factor interaction would indicate that the first-order model was not adequate.

Example 11.1.2 (*Cement Thickening Time, p. 356*) From the ANOVA table in Example 9.2.4, one finds the p-values for the three two-factor interactions AB, BC, and AC to be $0.295, 0.500$, and 0.500; respectively. Since none of them is significant, there is no indication of a problem with the first-order model.

The adequacy of the model can also be checked using residual analysis, and a convenient way to obtain the residuals is with the reverse Yates' algorithm.

Reverse Yates' Algorithm

To apply the reverse Yates' algorithm to a 2^k design, one takes column (k) from (the forward) Yates' algorithm and writes it down in reverse order. Any effects corresponding to terms that are not in the model of interest are replaced with zeros. In the case of model (11.1.1), all the interaction effects are replaced with zeros. These values are then added and subtracted in pairs as in the forward algorithm. The results in column (k) are 2^k times the predicted values $\widehat{y}$ at the 2^k treatment combination levels. Pairing the response column from the forward algorithm (in reverse order) with column (k) from the reverse algorithm divided by 2^k and subtracting produces the residuals.

Example 11.1.3   Table 11.1 shows the results of applying the reverse Yates' algorithm when the data from the 2^3 design in Example 9.2.4 is used to fit the first-order model (11.1.1). Note that the value associated with the grand mean (the I effect) is zero because the responses in the experiment were coded so that they summed to zero, not because it was replaced with a zero.

Table 11.1. The Reverse Yates' Algorithm Applied to the Data in Example 9.2.4

Effect Measured	$(k)^a$	(1)	(2)	(3)	Predicted[b]	Response[c]	Residual[d]
ABC	0	0	−20	−36	−4.5	−5	−0.5
BC	0	−20	−16	92	11.5	13	1.5
AC	0	48	−20	−132	−16.5	−15	1.5
C	−20	−64	112	−4	−0.5	−3	−2.5
AB	0	0	−20	4	0.5	−1	−1.5
B	48	−20	−112	132	16.5	17	0.5
A	−64	48	−20	−92	−11.5	−11	0.5
I	0	64	16	36	4.5	5	0.5

[a] From the forward Yates' algorithm (Table 9.16, p. 356)
[b] Column (3) divided by 2^3
[c] From Table 9.16 (in reverse order)
[d] Response minus Predicted

The normal probability plot of the residuals given in Figure 11.1 does not indicate any problem with the first-order model.

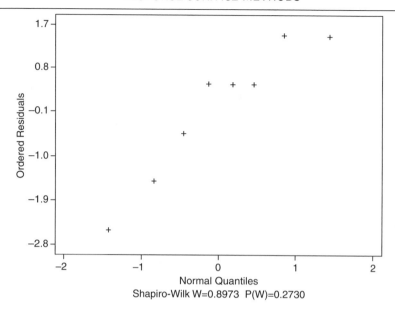

Figure 11.1. A Normal Probability Plot of the Residuals from Example 11.1.3.

Center Points

In order to obtain a pure estimate of the error (i.e., an estimate that measures the error regardless of the correctness of the assumed model) and also another measure of the adequacy of the model, it is usually desirable to add to the 2^k (or 2^{k-p}) design several replicates (say c) of a center point (every factor at level zero in its coded metric). The sample variance of the center points (call it s_c^2) is a pure estimator of the error. Another measure of the adequacy of the model is obtained as follows. Let

$$\beta_{ii} = \text{the coefficient of the model term } x_i^2$$

$$\gamma = \sum_{i=1}^{k} \beta_{ii}$$

$$\bar{y}_f = \text{the average response at points in the factorial design}$$

$$\bar{y}_c = \text{the average response at the center point.}$$

$$s_c^2 = \text{the variance of the responses at the center point.}$$

If γ is not zero, the first-order model is not adequate. An estimate of γ is

$$\boxed{\hat{\gamma} = \bar{y}_f - \bar{y}_c.} \tag{11.1.2}$$

It is not difficult to show that if the true response function is second-order, then

$$\widehat{\gamma} = \overline{Y}_f - \overline{Y}_c$$

is an unbiased estimator of γ (i.e., on average it is correct). Consider a 2^2 design. In that case (assuming the true function is second order) we are fitting the model

$$Y_{ij} = \beta_0 + \beta_1 x_1 + \beta_2 x_2 + \beta_{12} x_1 x_2 + \beta_{11} x_1^2 + \beta_{22} x_2^2 + \epsilon_{ij}$$

and with the x_i's expressed in their coded metrics (i.e., at levels $+1$ or -1)

$$E[Y_{11}] = \beta_0 - \beta_1 - \beta_2 + \beta_{12} + \beta_{11} + \beta_{22}$$
$$E[Y_{12}] = \beta_0 - \beta_1 + \beta_2 - \beta_{12} + \beta_{11} + \beta_{22}$$
$$E[Y_{21}] = \beta_0 + \beta_1 - \beta_2 - \beta_{12} + \beta_{11} + \beta_{22}$$
$$E[Y_{22}] = \beta_0 + \beta_1 + \beta_2 + \beta_{12} + \beta_{11} + \beta_{22}$$
$$E[\overline{Y}_c] = E[Y_c] = \beta_0.$$

Therefore,

$$E[\overline{Y}_f] = \frac{1}{2^2} \sum_{ij} E[Y_{ij}]$$
$$= \beta_0 + \beta_{11} + \beta_{22}$$
$$= \beta_0 + \gamma$$

and

$$E[\widehat{\gamma}] = E[\overline{Y}_f] - E[\overline{Y}_c] = \gamma.$$

For any 2^k or 2^{k-p} design each factor appears as many times at its low level as at its high level. As a result of this symmetry, $E[\overline{Y}_f] = \beta_0 + \gamma$ and $\widehat{\gamma}$ is unbiased.

A Test of H_0: $\gamma = 0$

If $\widehat{\gamma}$ is significantly different from zero, it indicates that the first-order model is not adequate. In general, for a univariate statistic $\widehat{T}$, the sum of squares SS_T is $\widehat{T}^2$, normalized so that $E[\mathrm{SS}_T] = \sigma^2$ under $H_T: T = 0$. With an independent estimate of the error, SS_T can be used in the numerator of an F statistic to test H_T.

Example 11.1.4 In Section 5.4, we made a heuristic argument that one can test the hypothesis H_0: $\mu = 0$ against the alternative H_a: $\mu \neq 0$ by computing

$$t = \frac{\overline{y}}{s/\sqrt{n}}$$

and rejecting H_0 if $|t| > t(\alpha/2; n - 1)$ (see equation (5.4.1) and Table 5.2). An equivalent test can be obtained by applying the above rules. Our estimate of μ is $\overline{y}$, which is a univariate statistic. Under H_0: $\mu = 0$,

$$E[\overline{Y}^2] = \mathrm{Var}(\overline{Y}) = \sigma^2/n$$

and, therefore,

$$SS_{\overline{Y}} = n\overline{Y}^2.$$

Finally,

$$F = \frac{SS_{\overline{y}}}{s^2} = \frac{n\overline{y}^2}{s^2} = t^2$$

and one would reject H_0: $\mu = 0$ if $F > F(\alpha; 1, n-1) = [t(\alpha/2; n-1)]^2$.

One can obtain a test of H_0: $\gamma = 0$ in a similar fashion. Consider a 2^{k-p} design with c center points. Under H_0: $\gamma = 0$,

$$\begin{aligned} E[(\widehat{\gamma})^2] &= \mathrm{Var}(\widehat{\gamma}) \\ &= \mathrm{Var}(\overline{Y}_f) + \mathrm{Var}(\overline{Y}_c) \\ &= \left[\frac{1}{2^{k-p}} + \frac{1}{c}\right]\sigma^2 \end{aligned}$$

and therefore,

$$SS_\gamma = (\widehat{\gamma})^2 / \left[\frac{1}{2^{k-p}} + \frac{1}{c}\right]. \qquad (11.1.3)$$

Now,

$$F_\gamma = \frac{SS_\gamma}{s_c^2}$$

and one would reject H_0: $\gamma = 0$ if $F_\gamma > F(\alpha; 1, n-1)$.

Example 11.1.5   Fuses are welded into the bases of incandescent lamps using an electric welding machine, and previous experimentation had revealed that amperage and time of contact were the two factors that determined the quality of a weld. In order to decrease the percentage of defective welds, an experiment was performed to ascertain the optimum Amperage and Time settings. The initial ranges of the two variables were chosen as 35–55 Amps and 0.6–1 second. A 2^2 design with three replicates of the center point was run, and the results are given in Table 11.2. The values for the variables, coded so that the high and low values are 1 and -1, respectively, are obtained from

$$x_1 = \frac{\text{Amperage} - 45 \text{ Amps}}{10 \text{ Amps}}$$

$$\qquad (11.1.4)$$

$$x_2 = \frac{\text{Time} - 0.8 \text{ seconds}}{0.2 \text{ seconds}}.$$

Table 11.2. Initial 2^2 Design with Three Center Points for Studying the Percentage of Defective Welds as a Function of Amperage and Time of Contact

Trial*	Amperage	Time (seconds)	x_1	x_2	Percent Defective
1	35	0.6	-1	-1	4.9
2	55	0.6	1	-1	4.2
3	35	1	-1	1	5.2
4	55	1	1	1	4.4
5	45	0.8	0	0	4.7
6	45	0.8	0	0	4.8
7	45	0.8	0	0	4.8

* Trials were run in random order

Using the three center points, one computes $s_c^2 = 0.00333$; and using equation (11.1.2), one computes

$$\widehat{\gamma} = \frac{18.7}{4} - \frac{14.3}{3} = -0.092.$$

Applying equation (11.1.3), one obtains

$$\mathrm{SS}_\gamma = (-0.092)^2 / \left[\frac{1}{4} + \frac{1}{3} \right] = 0.0145.$$

An overall measure of lack of fit can be obtained by pooling SS_γ with $\mathrm{SS}_{\text{interactions}}$ and computing an overall F value for lack of fit

$$F_{\mathrm{LF}} = \frac{(\mathrm{SS}_\gamma + \mathrm{SS}_{\text{interactions}})/(1 + \nu_{\text{interactions}})}{s_c^2}. \qquad (11.1.5)$$

The first-order model is rejected as being adequate at level α if

$$\Pr[F(1 + \nu_{\text{interactions}}, c - 1) > F_{\mathrm{LF}}] < \alpha. \qquad (11.1.6)$$

If the first-order model appears to be adequate, the next step should be to confirm that all the factors have significant main effects. Otherwise, either some of the terms can be dropped from the first-order model, or if none of the effects are significant, the model is of no use. The effect of a factor (say A) can be tested by computing the p-value

$$\Pr[F(1, c - 1) > F_A] \qquad (11.1.7)$$

where $F_A = \mathrm{MS}_A / s_c^2$. The addition of center points does not change the estimators $\widehat{\beta}_i$ for $i \geq 1$, but the center points must be included in the computation of $\widehat{\beta}_0$ (the grand mean).

Example 11.1.6   Analyzing the 2^2 part of the design results in Table 11.3; and dividing the values in column (2) by 4; one obtains $\widehat{\beta}_0 = 4.675, \widehat{\beta}_1 = -0.375, \widehat{\beta}_2 = 0.125$, and $\widehat{\beta}_{12} = -0.025$.

Table 11.3. Yates' Algorithm Applied to the Initial 2^2 Design in Example 11.1.5 and the Resulting Sums of Squares

Treatment Combination	Response	(1)	(2)	SS
(1)	4.9	9.1	18.7	–
a	4.2	9.6	−1.5	0.5625
b	5.2	−0.7	0.5	0.0625
ab	4.4	−0.8	−0.1	0.0025

Applying equations (11.1.3) and (11.1.5), one obtains

$$\text{SS}_\gamma = (-0.092)^2 / \left[\frac{1}{4} + \frac{1}{3} \right] = 0.0145$$

$$F_{\text{LF}} = \frac{(0.0145 + 0.0025)/2}{0.00333} = 2.55$$

and the p-value for lack of fit (see inequality (11.1.6)) is 0.282. Thus, the first-order model appears to be adequate. Using formula (11.1.7) one finds that both Amperage (p-value $= 0.006$) and Time (p-value $= 0.049$) are needed in the model. The estimated first-order model (including the center points) is

$$\widehat{y} = 4.71 - 0.375x_1 + 0.125x_2.$$

Path of Steepest Ascent

The path of steepest ascent (the estimated direction to the maximum) is obtained using the $\widehat{\beta}_i$'s. For k factors the response surface is $(k+1)$-dimensional, and the estimated direction to the maximum is along the vector $\widehat{\boldsymbol{\beta}} = (\widehat{\beta}_1, \ldots, \widehat{\beta}_k)$. Normalizing this vector so that it has unit length results in $\mathbf{u} = (u_1, \ldots, u_k)$ where

$$u_i = \widehat{\beta}_i \bigg/ \sqrt{\sum_{i=1}^k \widehat{\beta}_i^2}. \tag{11.1.8}$$

Points along the path of steepest ascent are at $x_1 = \delta u_1, \ldots, x_k = \delta u_k$ for $\delta > 0$. (If one were interested in finding the minimum of the response surface, then the path of steepest descent would be used. It is obtained in a similar fashion using values of $\delta < 0$.) Several experimental trials should be run along the path of steepest ascent, say for $\delta = 2, 4, 6,$ and 8; continuing until the response starts to decrease. The experimental trial resulting in the maximum response is then used as the center point for a new 2^k design, and the entire process is repeated.

Example 11.1.7   For the welding problem (Example 11.1.5), the path of steepest descent is obtained as follows. Using the values in column (2) of Table 11.3 and equation (11.1.8), one computes

$$\widehat{\beta}_1 = -1.5/4 = -0.375$$
$$\widehat{\beta}_2 = 0.5/4 = 0.125$$
$$u_1 = -0.375/0.395 = -0.949$$
$$u_2 = 0.125/0.395 = 0.316.$$

For $\delta = -2, -4, -6$, and -8; the resulting trials and responses are given in Table 11.4. The decoded Amperage and Time values were obtained using equations (11.1.4). Since at trial 11 the response had started to increase, a point near trial 10 with Amperage $= 102$ and Time $= 0.4$ was chosen as the center point for a second 2^2 design. It was decided to study a range of 20 Amps and 0.2 seconds around this center point, resulting in the coded values

$$x_1 = \frac{\text{Amperage} - 102 \text{ Amps}}{10 \text{ Amps}}$$

$$x_2 = \frac{\text{Time} - 0.4 \text{ seconds}}{0.1 \text{ seconds}}. \tag{11.1.9}$$

The second 2^2 design with three center points and the resulting responses are given in Table 11.5. Performing the same analysis as in Example 11.1.5; one obtains (see Problem 11.1.5) $SS_{AB} = 0.0225, SS_\gamma = 1.075, F_{LF} = 164.8$, and p-value $= 0.006$ for lack of fit. Thus, it is clear that the first-order model is not adequate.

Table 11.4. Experimental Trials Performed Along the Path of Steepest Descent (Example 11.1.7)

Trial	δ	x_1	x_2	Amperage	Time	Response
8	-2	1.898	-0.632	64	0.67	3.9
9	-4	3.796	-1.264	83	0.55	3.3
10	-6	5.694	-1.896	102	0.42	2.3
11	-8	7.592	-2.528	121	0.29	5.5

Table 11.5. The 2^2 Design with Three Center Points Performed Around the Minimum Response Found Along the Path of Steepest Descent (Example 11.1.7)

Trial*	x_1	x_2	Amperage	Time	Response
12	0	0	102	0.4	2.4
13	0	0	102	0.4	2.3
14	0	0	102	0.4	2.3
15	−1	−1	92	0.3	2.5
16	1	−1	112	0.3	3.4
17	−1	1	92	0.5	2.7
18	1	1	112	0.5	3.9

* Trials were run in random order

Problems

1. For the Windshield Molding Data in Example 9.2.1 (p. 351) do the following.

   (a) Fit the first-order model (11.1.1).

   (b) Find the residuals for this model using the reverse Yates' algorithm.

   (c) Check for any lack of fit.

2. Suppose that in addition to the Windshield Molding Data in Example 9.2.1 (p. 351) one had values of 785 and 810 for two center points. Perform any possible new tests for lack of fit.

3. Using the data in Problem 2 do the following.

   (a) Find the path of steepest descent.

   (b) For $\delta = -2$, what would the coded factor levels be?

   (c) For $\delta = -2$, what would the actual factor levels be?

4. For the data in `chemreac.txt` do the following.

   (a) Fit the first-order model (11.1.1).

   (b) Find the residuals for this model using the reverse Yates' algorithm.

   (c) Check for any lack of fit.

5. Verify that in Example 11.1.7: $SS_{AB} = 0.0225, SS_{\gamma} = 1.075$, and $F_{\text{LF}} = 164.8$.

11.2 SECOND-ORDER DESIGNS

When the first-order model is not adequate, we need a design that will allow us to fit the second-order model

$$Y = \beta_0 + \sum_{i=1}^{k} \beta_i x_i + \sum_{i=1}^{k} \sum_{j \geq 1} \beta_{ij} x_i x_j + \epsilon. \tag{11.2.1}$$

There are several possibilities, the most obvious of which is a 3^k factorial design (i.e., each of k factors at three levels). That, however, as we will see, turns out not to be the best choice. After looking at the drawbacks to using 3^k designs, we will consider a better alternative: central composite designs.

3^k Factorial Designs

One advantage of 3^k factorial designs is that if one uses three equally spaced levels, it is easy to fit model (11.2.1). With each factor at three levels one can estimate both linear and quadratic effects for each factor, and the interaction effect can be partitioned into four pieces. We will only consider 3^2 designs in any detail since that will be sufficient to see the problems when using 3^k designs to model response surfaces.

Fitting the Second-Order Model
Notation:

$$A_L = \text{the linear effect of factor A}$$
$$A_Q = \text{the quadratic effect of factor A}$$
$$B_L = \text{the linear effect of factor B}$$
$$B_Q = \text{the quadratic effect of factor B}$$
$$A_L B_L = \text{the A linear by B linear interaction effect}$$
$$A_L B_Q = \text{the A linear by B quadratic interaction effect}$$
$$A_Q B_L = \text{the A quadratic by B linear interaction effect}$$
$$A_Q B_Q = \text{the A quadratic by B quadratic interaction effect}$$

The various effect estimates are all contrasts in the responses. A **contrast** is simply a linear combination of items (responses in this case) such that the coefficients sum to zero.

Example 11.2.1   Although we did not refer to them as contrasts, the effect estimates in a 2^k factorial experiment are examples of contrasts. Recall that the effect of factor A was the difference between the average response with A at the high level and the average response with A at the low level. In Example 9.2.1 (p. 351) we found

$$\text{Effect of } A = \frac{600 + 750 + 567 + 647}{4} - \frac{917 + 953 + 735 + 977}{4} = -254.5$$

which is a linear combination of the eight responses with coefficients

$$\frac{1}{4} \; \frac{1}{4} \; \frac{1}{4} \; \frac{1}{4} \; -\frac{1}{4} \; -\frac{1}{4} \; -\frac{1}{4} \; -\frac{1}{4}$$

that sum to zero.

For convenience we will represent the three levels of each factor (low to high) by $0, 1$, and 2. Thus, for a 3^2 design both factors at the low level would be represented by 00, both factors at the medium level would be represented by 11, and so forth. Using this notation the contrast coefficients for the factor and interaction effects in a 3^2 design are given in Table 11.6.

Table 11.6. Contrast Coefficients for a 3^2 Design with Continuous Factors Together with Their Sums of Squares

	Treatment Combinations									
Effect	00	01	02	10	11	12	20	21	22	$\sum c_i^2$
A_L	−1	−1	−1	0	0	0	1	1	1	6
A_Q	1	1	1	−2	−2	−2	1	1	1	18
B_L	−1	0	1	−1	0	1	−1	0	1	6
B_Q	1	−2	1	1	−2	1	1	−2	1	18
$A_L B_L$	1	0	−1	0	0	0	−1	0	1	4
$A_L B_Q$	−1	2	−1	0	0	0	1	−2	1	12
$A_Q B_L$	−1	0	1	2	0	−2	−1	0	1	12
$A_Q B_Q$	1	−2	1	−2	4	−2	1	−2	1	36

Example 11.2.2 For the particle size data in Example 8.3.1 (p. 322), which were collected using a 3^2 factorial design, the effect estimates are (using Table 11.6)

$$\begin{aligned}
A_L &= -1(6.3 + 6.1 + 5.8) + 1(6.1 + 5.8 + 5.5) = -0.8 \\
A_Q &= 1(6.3 + 6.1 + 5.8) - 2(5.9 + 5.6 + 5.3) + 1(6.1 + 5.8 + 5.5) = 2 \\
B_L &= -1(6.3 + 5.9 + 6.1) + 1(5.8 + 5.3 + 5.5) = -1.7 \\
B_Q &= 1(6.3 + 5.9 + 6.1) - 2(6.1 + 5.6 + 5.8) + 1(5.8 + 5.3 + 5.5) = -0.1 \\
A_L B_L &= 6.3 - 5.8 - 6.1 + 5.5 = -0.1 \\
A_L B_Q &= -6.3 + 2(6.1) - 5.8 + 6.1 - 2(5.8) + 5.5 = 0.1 \\
A_Q B_L &= -6.3 + 5.8 + 2(5.9) - 2(5.3) - 6.1 + 5.5 = 0.1 \\
A_Q B_Q &= 6.3 - 2(6.1) + 5.8 - 2(5.9) + 4(5.6) - 2(5.3) + 6.1 - 2(5.8) + 5.5 = -0.1.
\end{aligned}$$

The effect estimates can be used to compute their corresponding sums of squares using

$$SS_{\text{effect}} = \frac{(\text{effect estimate})^2}{n \sum c_i^2}. \tag{11.2.2}$$

Each effect sum of squares has one df. Equation (11.2.2) is a generalization of equation (9.2.1) (see Problem 11.2.1).

Example 11.2.3   For the particle size data in Example 11.2.2 the sum of squares associated with the linear and quadratic effects of factor A are (using equation (11.2.2))

$$SS_{A_L} = \frac{(0.8)^2}{6} = 0.1067$$

and

$$SS_{A_Q} = \frac{(2)^2}{18} = 0.2222.$$

All of the sums of squares are given in Table 11.7 (see Problem 11.2.2). Comparing these sums of squares with those given in the ANOVA table in Example 8.3.2 (p. 326), one finds that (see Problem 11.2.3)

$$SS_A = SS_{A_L} + SS_{A_Q}$$
$$SS_B = SS_{B_L} + SS_{B_Q}$$
$$SS_{AB} = SS_{A_L B_L} + SS_{A_L B_Q} + SS_{A_Q B_L} + SS_{A_Q B_Q}.$$

Thus, we have partitioned the main effects into linear and quadratic portions and the interaction effect into interactions of linear and quadratic effects.

Table 11.7. The Sums of Squares for the 3^2 Design in Example 11.2.2

Source	SS	df
A_L	0.1067	1
A_Q	0.2222	1
B_L	0.4817	1
B_Q	0.0006	1
$A_L B_L$	0.0025	1
$A_L B_Q$	0.0008	1
$A_Q B_L$	0.0008	1
$A_Q B_Q$	0.0003	1

As with the effects from a 2^k factorial design that could be used to fit model (11.1.1), the effects from a 3^k design can be used to fit model (11.2.1). The effects divided by their corresponding $\sum c_i^2$ are the coefficients for the terms in the model. The fitted model (11.2.1) for $k = 2$ is

$$\hat{y} = \bar{y} + \frac{A_L}{6}x_1 + \frac{B_L}{6}x_2 + \frac{A_Q}{18}(3x_1^2 - 2)$$
$$+ \frac{B_Q}{18}(3x_2^2 - 2) + \frac{A_L B_L}{4}x_1 x_2$$

$$(11.2.3)$$

where, as before, x_1 and x_2 are the values for factors A and B, respectively, coded so that they are in the range from -1 to 1.

Example 11.2.4 From the table of sums of squares in Example 11.2.3 it is clear that only the linear and quadratic effects of factor A and the linear effect of factor B are significant (see Problem 11.2.4). Therefore, one would use model (11.2.3) with B_Q and $A_L B_L$ set to zero. This results in

$$\hat{y} = 5.822 + \frac{-0.8}{6}x_1 + \frac{-1.7}{6}x_2 + \frac{2}{18}(3x_1^2 - 2)$$
$$= 5.8222 - 0.1333x_1 - 0.2833x_2 + 0.1111(3x_1^2 - 2)$$
$$= 5.9333 - 0.1333x_1 - 0.2833x_2 + 0.3333x_1^2.$$

This same result could be obtained by fitting the model using the coded factor levels and linear regression (see Problem 11.2.5).

If one wanted the fitted equation in terms of the original factor level units, one could back-transform using

$$\text{Vacuum} = 2x_1 + 22$$
$$\text{Flow Rate} = 5x_2 + 90.$$

This would result in the same equation as when the model is fitted using the original factor units and linear regression (see Problem 11.2.6).

Problems Using a 3^k Design

The main problem with using data from a 3^k factorial design to fit the second-order model (11.2.1) is that the form of the resulting fitted response surface depends heavily on the center point (the point with all factors at coded level 0). Simply changing the value of that one point can result in completely different surfaces.

A good way to graphically represent a three-dimensional response surface is with a **contour plot**. A contour plot consists of a sequence of curves in the plane of the two factor that have constant responses. Each curve is then labeled with the particular response value. Alternatively, one can think of a contour plot as the result of slicing the response surface with a sequence of planes at various heights that are parallel to the plane of the two factors, and projecting the intersections of these planes with the response surface onto the plane of the two factors. If, for example, the response surface was a hemisphere, then the contour plot would be a set of concentric circles.

Example 11.2.5   A contour plot for the particle size model in Example 11.2.4 is given in Figure 11.2.

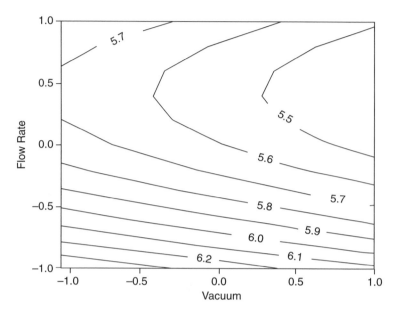

Figure 11.2. A Contour Plot for the Particle Size Model in Example 11.2.4.

Example 11.2.6   Consider the data in Table 11.8, which was collected using a 3^2 factorial design. Fitting model (11.2.1) three times using three different values of y (-5, 0, and 5) results in the three response surfaces contour plots given in Figures 11.3–11.5.

Table 11.8. Data From a 3×2 Factorial Design

		Factor B		
		0	1	2
	0	5	0	-5
Factor A	1	0	y	0
	2	-5	0	5

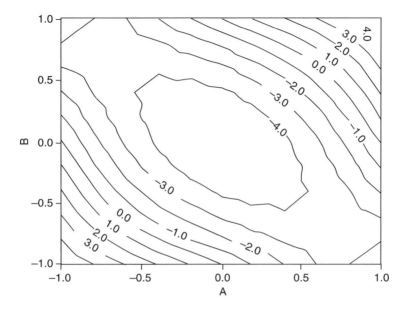

Figure 11.3. Contour Plot for the Fitted Model in Example 11.2.6 When the Center Point is Equal to -5.

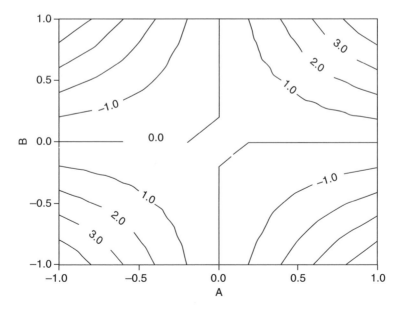

Figure 11.4. Contour Plot for the Fitted Model in Example 11.2.6 When the Center Point is Equal to Zero.

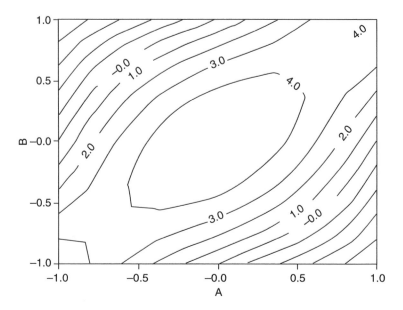

Figure 11.5. Contour Plot for the Fitted Model in Example 11.2.6 When the Center Point is Equal to 5.

It is clear from examining the three contour plots that they represent three entirely different surfaces.

A second problem with using data from a 3^k factorial design to fit model (11.2.1) is that when $k > 2$ the design calls for many more data points than are needed. As we will see in the next section, model (11.2.1) can be fit with far fewer than 3^k data points.

Central Composite Designs

A 2^k design can be easily augmented to allow for fitting the second-order model, Generally, the best way to do this is by adding $2k$ design points (with k factors the design points are k-dimensional) at

$$
\begin{array}{cccccc}
(\pm\alpha, & 0, & 0, & \ldots, & 0) \\
(0, & \pm\alpha, & 0, & \ldots, & 0) \\
& & \vdots & & \\
(0, & \ldots, & 0, & 0, & \pm\alpha)
\end{array}
\tag{11.2.4}
$$

and possibly some additional center points. The additional $2k$ points are called **star points** (or **axial points**), and the resulting design is called a **central composite design**. Central composite designs for $k = 2$ and $k = 3$ are shown in Figure 11.6.

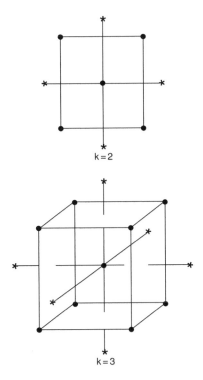

Figure 11.6. Central Composite Designs for $k = 2$ and $k = 3$ Factors.

Note: With $k = 3$ and (say) three center points, a central composite design would require only $2^3 + 2(3) + 3 = 17$ data points versus the 27 points needed for a 3^3 design. The difference between the number of points required for a central composite design with k factors and a 3^k design increases as k increases.

Checking for Lack of Fit

When it is fit using the data from a central composite design, the adequacy of the second-order model (11.2.1) can be checked as follows. Let

$$\overline{Y}_i(j) = \text{the average response when } x_i = j.$$

If the true response function is third order, then

$$\widehat{\gamma}_i = \frac{1}{\alpha^2 - 1} \left[\frac{\overline{Y}_i(\alpha) - \overline{Y}_i(-\alpha)}{2\alpha} - \frac{\overline{Y}_i(1) - \overline{Y}_i(-1)}{2} \right] \qquad (11.2.5)$$

is an unbiased estimator of

$$\gamma_i = \beta_{iii} + \frac{1}{(1 - \alpha^2)} \sum_{j \neq i} \beta_{ijj}. \qquad (11.2.6)$$

Consider the simplest case of a 2^2 design augmented with five star points (see Figure 11.6), and let

$$Y_{ij} = \text{the response with } x_1 = i \text{ and } x_2 = j.$$

Then

$$E[\overline{Y}_1(1)] = \left[\frac{1}{2}\right] E[Y_{1,1} + Y_{1,-1}]$$

$$= \left[\frac{1}{2}\right] (\beta_0 + \beta_1 + \beta_2 + \beta_{12} + \beta_{11} + \beta_{22} + \beta_{112} + \beta_{122} + \beta_{111} + \beta_{222}$$

$$+ \beta_0 + \beta_1 - \beta_2 - \beta_{12} + \beta_{11} + \beta_{22} - \beta_{112} + \beta_{122} + \beta_{111} + \beta_{222})$$

$$= \beta_0 + \beta_1 + \beta_{11} + \beta_{22} + \beta_{122} + \beta_{111}$$

$$E[\overline{Y}_1(-1)] = \left[\frac{1}{2}\right] E[Y_{-1,1} + Y_{-1,-1}]$$

$$= \left[\frac{1}{2}\right] (\beta_0 - \beta_1 + \beta_2 - \beta_{12} + \beta_{11} + \beta_{22} + \beta_{112} - \beta_{122} - \beta_{111} + \beta_{222}$$

$$+ \beta_0 - \beta_1 - \beta_2 + \beta_{12} + \beta_{11} + \beta_{22} - \beta_{112} - \beta_{122} - \beta_{111} - \beta_{222})$$

$$= \beta_0 - \beta_1 + \beta_{11} + \beta_{22} - \beta_{122} - \beta_{111}$$

and

$$\left[\frac{1}{2}\right] E[\overline{Y}_1(1) - \overline{Y}_1(-1)] = \beta_1 + \beta_{122} + \beta_{111}. \tag{11.2.7}$$

Similarly,

$$E[\overline{Y}_1(\alpha)] = Y_{\alpha,0} = \beta_0 + \alpha\beta_1 + \alpha^2\beta_{11} + \alpha^3\beta_{111}$$

$$E[\overline{Y}_1(-\alpha)] = Y_{-\alpha,0} = \beta_0 - \alpha\beta_1 + \alpha^2\beta_{11} - \alpha^3\beta_{111}$$

and

$$\frac{1}{2\alpha} E[\overline{Y}_1(\alpha) - \overline{Y}_1(-\alpha)] = \beta_1 + \alpha^2\beta_{111}. \tag{11.2.8}$$

Combining equations (11.2.7) and (11.2.8), one obtains

$$E[\widehat{\gamma}_1] = \frac{1}{\alpha^2 - 1} E\left[\frac{\overline{Y}_1(\alpha) - \overline{Y}_1(-\alpha)}{2\alpha} - \frac{\overline{Y}_1(1) - \overline{Y}_1(-1)}{2}\right]$$

$$= \frac{1}{\alpha^2 - 1} (\alpha^2\beta_{111} - \beta_{122} - \beta_{111})$$

$$= \beta_{111} + \frac{\beta_{122}}{1 - \alpha^2}$$

$$= \gamma_1.$$

A similar proof holds with k factors and an augmented 2^{k-p} design.

Under H_0: $\gamma_i = 0$,

$$E[(\alpha^2 - 1)\widehat{\gamma}_i^2] = \text{Var}\left[\frac{\overline{Y}_i(\alpha) - \overline{Y}_i(-\alpha)}{2\alpha} - \frac{\overline{Y}_i(1) - \overline{Y}_i(-1)}{2}\right]$$

$$= \frac{1}{4\alpha^2}\text{Var}\left[\overline{Y}_i(\alpha) - \overline{Y}_i(-\alpha)\right] + \frac{1}{4}\text{Var}\left[\overline{Y}_i(1) - \overline{Y}_i(-1)\right]$$

$$= \frac{1}{2\alpha^2}\text{Var}\left[\overline{Y}_i(\alpha)\right] + \frac{1}{2}\text{Var}\left[\overline{Y}_i(1)\right]$$

$$= \left[\frac{1}{2\alpha^2} + \frac{1}{2^{k-p}}\right]\sigma^2$$

and the contribution of $\widehat{\gamma}_i$ to the sum of squares for lack of fit is

$$\text{SS}_{\gamma_i} = (\alpha^2 - 1)^2(\widehat{\gamma}_i)^2 / \left[\frac{1}{2^{k-p}} + \frac{1}{2\alpha^2}\right]. \tag{11.2.9}$$

If additional center points were run with the star points, then two measures of overall curvature are available, and their difference should not be statistically significant if the second-order model is adequate. Specifically, the difference of the two measures of curvature is

$$d = \overline{Y}_f - \overline{Y}_{c_f} - \frac{k}{\alpha^2}(\overline{Y}_s - \overline{Y}_{c_s}) \tag{11.2.10}$$

where

$\overline{Y}_f$ = the average response at the points in the factorial design
$\overline{Y}_{c_f}$ = the average response at the center points run with the factorial design
$\overline{Y}_s$ = the average response at the star points
$\overline{Y}_{c_s}$ = the average response at the center points run with the star points

and the associated sum of squares is

$$\text{SS}_d = d^2 / \left[\frac{1}{2^{k-p}} + \frac{1}{c_f} + \frac{k}{2\alpha^4} + \frac{k^2}{\alpha^4 c_s}\right] \tag{11.2.11}$$

where c_f and c_s are the numbers of center points run with the factorial design and the star points, respectively. Therefore, the overall sum of squares for lack of fit (with $k + 1$ degrees of freedom) is

$$\text{SS}_{\text{LF}} = \text{SS}_d + \sum_{i=1}^{k}\text{SS}_{\gamma_i}. \tag{11.2.12}$$

Also, a pure estimate of error can be obtained by pooling the sample variance of the center points run with the factorial design and the sample variance of the center points run with the star points. Namely,

$$s_p^2 = \frac{(c_f - 1)s_{c_f}^2 + (c_s - 1)s_{c_s}^2}{c_f + c_s - 2}. \tag{11.2.13}$$

Therefore, a test for overall lack of fit can be made using

$$F_{\text{LF}} = \frac{\text{SS}_{\text{LF}}/(k+1)}{s_p^2} \tag{11.2.14}$$

and the computing the p-value

$$\Pr[F(k+1, c_f + c_s - 2) > F_{\text{LF}}]. \tag{11.2.15}$$

Rotatable Designs

If a 2_R^{k-p} design is being augmented, where $R \geq V$, and α is chosen such that

$$\alpha = (2^{k-p})^{1/4} \tag{11.2.16}$$

then the design has the desirable property of being rotatable. That is, the variance of the predicted response $\widehat{Y}$ depends only on the distance from the center point and, hence, is unchanged if the design is rotated about the center point.

Example 11.2.7   Since in Example 11.1.7 the first-order model was found to be inadequate, the factorial design was augmented with star points and two additional center points to form a central composite design. In order to make the design rotatable, α was chosen (using equation (11.2.16)) as $\sqrt{2}$. The results of these trials are given in Table 11.9.

Table 11.9. Star Point and Additional Center Points Added to the 2^2 Design in Table 11.4 to Form a Central Composite Design (Example 11.2.7)

Trial*	x_1	x_2	Amperage	Time	Response
19	$-\sqrt{2}$	0	88	0.4	2.9
20	$\sqrt{2}$	0	116	0.4	4.3
21	0	$-\sqrt{2}$	102	0.26	2.8
22	0	$\sqrt{2}$	102	0.54	3.1
23	0	0	102	0.4	2.2
24	0	0	102	0.4	2.3

* Trials were run in random order.

The first step in the analysis is to check the adequacy of the second-order model. Using equations (11.2.5) and (11.2.9), one computes

$$\widehat{\gamma}_1 = \frac{4.3 - 2.9}{2\sqrt{2}} - \frac{(3.4 + 3.9)/2 - (2.5 + 2.7)/2}{2} = -0.0300$$

$$\text{SS}_{\gamma_1} = (-0.0300)^2 / \left[\frac{1}{4} + \frac{1}{4}\right] = 0.0018$$

$$\widehat{\gamma}_2 = \frac{3.1 - 2.8}{2} - \frac{(2.7 + 3.9)/2 - (2.5 + 3.4)/2}{2} = -0.06889$$

$$SS_{\gamma_2} = (-0.0689)^2 / \left[\frac{1}{4} + \frac{1}{4}\right] = 0.0095$$

and using equations (11.2.10) and (11.2.11), one computes

$$d = (3.125 - 2.35) - (3.275 - 2.25) = -0.25$$

$$SS_d = (-0.25)^2 / \left[\frac{1}{4} + \frac{1}{2} + \frac{1}{4} + \frac{1}{2}\right] = 0.0417.$$

The pure estimate of error (equation (11.2.13)) is

$$s_p^2 = [(2.4 - 2.35)^2 + \cdots + (2.3 - 2.25)^2]/2 = 0.005.$$

Combining these results using equations (11.2.12) and (11.2.14) gives

$$SS_{LF} = 0.0018 + 0.0095 + 0.0417 = 0.053$$

$$F_{LF} = \frac{0.053/3}{0.005} = 3.53$$

and (formula (11.2.15)) p-value $= 0.228$. Thus, there is nothing to indicate the second-order model is not adequate.

Fitting the Second-Order Model

Fitting the second-order model (11.2.1) is best done using a computer and the general theory of least squares. Once the fitted model is obtained, a potential maximum (or minimum) point on the response surface in the region of the design points can be obtained by taking partial derivatives of the fitted model with respect to each x_i, setting the results equal to zero, and solving the system of equations. For $k = 2$ this results in

$$x_1 = \frac{2\widehat{\beta}_1\widehat{\beta}_{22} - \widehat{\beta}_2\widehat{\beta}_{12}}{(\widehat{\beta}_{12})^2 - 4\widehat{\beta}_{11}\widehat{\beta}_{22}} \tag{11.2.17}$$

$$x_2 = \frac{2\widehat{\beta}_2\widehat{\beta}_{11} - \widehat{\beta}_1\widehat{\beta}_{12}}{(\widehat{\beta}_{12})^2 - 4\widehat{\beta}_{11}\widehat{\beta}_{22}} \tag{11.2.18}$$

and for $k > 2$ the solution is most easily obtained numerically using a computer. Additional trials should be run at the potential maximum (minimum) point to establish the actual behavior there.

Example 11.2.8 Using least squares and a computer to fit the second-order model (11.2.1) to the central composite design (trials 12–24) in Tables 11.5 and 11.9, one obtains Table 11.10 and

Table 11.10. Least-Squares Fit of the Second-Order Model

UNWEIGHTED LEAST SQUARES LINEAR REGRESSION OF RESPONSE

PREDICTOR VARIABLES	COEFFICIENT	STD ERROR	STUDENT'S T	P
CONSTANT	2.30000	0.04670	49.26	0.0000
X1	0.50999	0.03692	13.81	0.0000
X2	0.14053	0.03692	3.81	0.0067
X12	0.07500	0.05221	1.44	0.1940
X11	0.61251	0.03959	15.47	0.0000
X22	0.28750	0.03959	7.26	0.0002

$$\widehat{y} = 2.3 + 0.51x_1 + 0.1405x_2 + 0.075x_1x_2 + 0.6125x_1^2 + 0.2875x_2^2.$$

Using equations (11.2.17) and (11.2.18), the minimum value of the estimated response surface is obtained when

$$x_1 = \frac{2(0.51)(0.2875) - (0.1405)(0.075)}{(0.075)^2 - 4(0.6125)(0.2875)} = -0.405$$

$$x_2 = \frac{2(0.1405)(0.6125) - (0.51)(0.075)}{(0.075)^2 - 4(0.6125)(0.2875)} = -0.192$$

and that minimum value is

$$\widehat{y} = 2.3 + 0.51(-0.405) + 0.1405(-0.192) + 0.075(-0.405)(-0.192)$$
$$+0.6125(-0.405)^2 + 0.2875(-0.192)^2$$
$$= 2.18.$$

The predicted minimum percent defective (using equations (11.1.4)) occurs when

$$\text{Amperage} = 10(-0.405) + 102 = 97.95 \text{ Amps}$$
$$\text{Time} = 0.1(-0.192) + 0.4 = 0.3808 \text{ seconds.}$$

Three additional trials were run at 98 Amps and 0.38 seconds, and resulted in 2.2, 2.1, and 2.1 percent defective.

Face-Centered Designs

A face-centered design is a central composite design with $\alpha = 1$. It has the disadvantage that it is not a rotatable design. However, it has the advantage that only three levels of each factor need to be studied, rather than the five levels that would be required for the rotatable design. A face-centered design for $k = 3$ is shown in Figure 11.7.

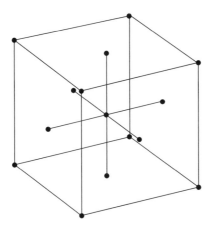

Figure 11.7. A Face-Centered Design for $k = 3$ Factors.

Example 11.2.9   The chemical division of a large company was interested in finding the optimum operating conditions for a residue cleavage process. Initial experimentation had narrowed the possibilities to a reasonably small region. The three factors that had potentially significant effects on the efficiency of hydrogen utilization (the response of interest) are listed in Table 11.11. The response surface

Table 11.11. The Three Factors Studied with a Face-Centered Design

Factor Name	Description
Temperature	Reactor temperature
Ratio	HCL/residue feed ratio
Rate	Residue feed rate

in the region of interest was not first order (see Problem 11.2.9). A face-centered design with two center points (16 trials) was run, and the results are given in Table 11.12. Using a computer to fit model (11.2.1), one obtains Table 11.13.

Table 11.12. Results From the Face-Centered Design

		Temperature								
		−1			0			1		
		Rate			Rate			Rate		
		−1	0	1	−1	0	1	−1	0	1
Ratio	−1	74.4		79.6		83.6		73.5		90.5
	0		89.0		83.4	89.5, 92.0	94.4		88.5	
	1	48.0		58.2		52.2		47.3		54.4

Table 11.13. Least-Squares Analysis of the Data in Table 11.12

```
UNWEIGHTED LEAST SQUARES LINEAR REGRESSION OF RESPONSE


PREDICTOR
VARIABLES     COEFFICIENT    STD ERROR    STUDENT'S T       P
---------     -----------    ---------    -----------    ------

CONSTANT       89.7889        1.43795        62.44       0.0000
RATE            5.05000       0.84602         5.97       0.0019
RATIO         -14.1500        0.84602       -16.73       0.0000
TEMP            0.50000       0.84602         0.59       0.5802
RATE2          -0.96111       1.66837        -0.58       0.5895
RATIO2        -21.9611        1.66837       -13.16       0.0000
TEMP2          -1.11111       1.66837        -0.67       0.5349
RATERATIO      -0.61250       0.94588        -0.65       0.5458
RATETEMP        1.08750       0.94588         1.15       0.3023
RATIOTEMP      -1.81250       0.94588        -1.92       0.1135


R-SQUARED             0.9912    RESID. MEAN SQUARE (MSE)    7.15747
ADJUSTED R-SQUARED    0.9754    STANDARD DEVIATION          2.67535


SOURCE        DF        SS           MS          F        P
---------     ---    ----------   ----------   -----    ------

REGRESSION     9      4037.53      448.614      62.68    0.0001
RESIDUAL       5        35.7874      7.15747
TOTAL         14      4073.31
```

It is clear from the above output that Rate has a significant linear effect (p-value = 0.0019), and Ratio has both a linear (p-value = 0.0000) and a quadratic (p-value = 0.0000) effect. It is also possible that the Ratio/Temperature interaction (p-value = 0.1135) has an effect. However, since this design does not lead to statistically independent estimates of the effects, dropping some terms from the model may effect the significance of those terms remaining. Some trial and error leads to the model

(see the output in Table 11.14)

$$\hat{y} = 88.96 + 5.05x_1 - 14.15x_2 - 22.79x_2^2 - 1.8125x_2x_3. \qquad (11.2.19)$$

Searching numerically for the factor levels that maximize equation (11.2.19), one obtains

$$\text{Rate} = 1 \quad \text{Ratio} = -0.35 \quad \text{Temperature} = 1$$

with a predicted efficiency of 96.8.

Table 11.14. The Reduced Model for Example 11.2.9

UNWEIGHTED LEAST SQUARES LINEAR REGRESSION OF RESPONSE

PREDICTOR VARIABLES	COEFFICIENT	STD ERROR	STUDENT'S T	P
CONSTANT	88.9600	1.08168	82.24	0.0000
RATE	5.05000	0.76486	6.60	0.0001
RATIO	-14.1500	0.76486	-18.50	0.0000
RATIO2	-22.7900	1.32478	-17.20	0.0000
RATIOTEMP	-1.81250	0.85514	-2.12	0.0601

R-SQUARED	0.9856	RESID. MEAN SQUARE (MSE)	5.85017
ADJUSTED R-SQUARED	0.9799	STANDARD DEVIATION	2.41871

SOURCE	DF	SS	MS	F	P
REGRESSION	4	4014.81	1003.70	171.57	0.0000
RESIDUAL	10	58.5017	5.85017		
TOTAL	14	4073.31			

Problems

1. Show that equation (11.2.2) (p. 408) is a generalization of equation (9.2.1) (p. 357).

2. Verify that the sums of squares in Example 11.2.3 (p. 408) are correct.

3. Compare the sums of squares in Example 11.2.3 (p. 408) with those given in the ANOVA table in Example 8.3.2 (p. 326).

4. Using the sum of squares in Example 11.2.3 (p. 408), test the significance of the linear and quadratic effects of factor A and the linear effect of factor B.

5. Show that the fitted model using coded factor levels in Example 11.2.4 (p. 409) can be obtained using linear regression.

6. Show that in Example 11.2.4 (p. 409) back-transforming to the original factor level units in Example 11.2.3 (p. 408) results in the same fitted equation as when the model is fitted using the original factor levels and linear regression (see Example 3.5.5, p. 73).

7. Find the quadratic effects for each of the two factors in Example 11.2.6 (p. 410) for each of the three different values of y studied.

8. Find the fitted response surface for the data in Example 11.2.6 (p. 410) when $y = 0$.

9. Using the 2^3 part of the design and the center points in Example 11.2.9 (p. 419), show that the first-order model is not adequate.

10. Test for lack of fit of the second-order model in Example 11.2.9 (p. 419).

11.3 CHAPTER PROBLEMS

1. Using the data in `reflect.txt`, do the following.
   (a) Using the results from Yates' algorithm, fit the first-order model.
   (b) Check for any lack of fit in the first-order model.
   (c) Find the path of steepest ascent.
   (d) If one wanted to perform additional experiments using a rotatable central composite design, what should α be?

2. For the data set `bright.txt` do the following.
   (a) Fit the first-order model.
   (b) Check for any lack of fit in the first-order model.
   (c) Find the path of steepest ascent.
   (d) If one also had two center points with values of 53.95 and 53.16, perform any possible additional tests for lack of fit.
   (e) If one wanted to perform additional experiments using a rotatable central composite design, what should α be?
   (f) List all the star points for the design in part (e), giving both their coded levels and their actual levels.

3. Consider the data set `defoam.txt`, together with two additional center points whose values are 42.78 and 45.07. Using just what would be the 2^3 part of the design and the three center points, do the following.
   (a) Analyze the data using Yates' algorithm.
   (b) Use the results of part (a) to help fit the first-order model.
   (c) Use the reverse Yates' algorithm to obtain the residuals.
   (d) Construct a normal probability plot of the residuals to check for any lack of fit.
   (e) Test for lack of fit in the first-order model.
   (f) Use a computer to fit the first-order model, and compare this with the result from part (b).
   (g) Find the path of steepest descent.
   (h) For $\delta = -2$ what would the coded factor levels be?

4. Consider again the data set for Problem 3 Using just what would amount to a face-centered design and the three center points, do the following.
   (a) Using a computer, fit the second-order model.
   (b) At what factor levels would you expect to obtain the minimum response?
   (c) Test for lack of fit in the second-order model.
   (d) Compare the second-order model in part (a) with the first-order model in part (b) of Problem 3

12

APPENDICES

12.1 APPENDIX A – DESCRIPTIONS OF DATA SETS

The following is an alphabetical list of all the data sets referenced in the text, and a description of each. When more than one variable is listed, the variables are listed in the order they occur in the data set.

absorb.txt This data set contains data from a silica processing plant. One key quality parameter for silica is the amount of oil it can absorb since silica is often mixed with rubber and oil in various rubber applications (battery separators, tires, shoe soles, etc.) The values reported here are the average oil absorption of the silica produced on one shift. Variables name: `ABSORB`

adhesion.txt This data set contains data from an experiment performed to compare the effect of pH on the adhesive qualities of a lens coating. Ten samples from each formulation were made and tested.

Variable name	Description
`ADHESION`	A measure of the adhesion
`PH`	pH level

adhesion2.txt This data set contains data from an experiment to compare the effect of pH and catalyst on the adhesive qualities of a lens coating. Three levels of pH were studied along with two levels of catalyst. Five measurements were taken for each of the treatment combinations.

Variable name	Description
`CAT`	Type of catalyst (1 or 2)
`PH`	pH level
`ADHESION`	A measure of adhesion

424

alum.txt Over a 2-hour period, twenty-five 200 gm samples were drawn at random from a process that recycles plastic, and the amount of aluminum impurities in ppm was determined for each sample. Variable Name: PPM

applicat.txt The application of powder coating (a type of paint used on appliances and, in limited settings, cars) is done by spraying the material through a "gun" that has an electric charge on it. There are three factors to consider in setting up an application booth. The gun distance from the target item, the charge, and the pressure (flow rate) of the material through the gun. This data set contains data from 18 test runs. The variables are:

Variable name	Description
DISTANCE	Gun distance from target (in.)
CHARGE	Charge (kvolts)
FLOWRATE	Flow Rate (gm/min)
GLOSS	A measure of the resulting gloss

assay.txt Two different chemical processes are being considered for producing a resin needed in the manufacture of paint. Four batches are run with the first process and five batches are run with the second process. Each batch is then assayed to determine the yield. The variables are:

Variable name	Description
PROCESS	Process number
YIELD	Yield

bacteria.txt Purified water is used in one step in the production of a medical device. The water is tested daily for bacteria. This data set is the results from 50 days of testing. The values are the count of a particular strain of bacteria in a 100 ml sample of water. Variable name: LEVEL

bath.txt This data set contains the results of an experiment to investigate the effects of a portion of a production process where the product passes through a water bath. The two factors studied were the time in the bath and the temperature of the bath water. The response of interest was the electrical resistance of the final product. The variables are:

Variable name	Description
TIME	Time in the bath (min.)
TEMP	Temperature of the bath (°F)
ER	Electrical resistance (ohm/in.^{2})

battery.txt This data set contains data from a study on how three processors affect the lifetime of a battery used in notebook computers. Measurements were taken on ten batteries with each of the different processor types. The variables are:

Variable name	Description
TYPE	Type of processor
LIFETIME	Battery lifetime (hours)

break.txt Eighteen samples of steel were stress tested, and their breaking strengths were recorded. Variable Name BSTRENGTH

bright.txt This data set contains the results of a half-fraction experiment to analyze the effects of several variables on the removal of ink from newsprint. The variables are:

Variable name	Description
TYPE	Type of alkali (A or B)
PERCENT	Percent of alkali (25 or 75%)
TIME	Time pulp is soaked (30 or 40 min.)
HARDNESS	Water hardness (150 or 250)
SPEED	Agitation rate (slow or fast)
BRIGHT	Brightness of pulp

calcium.txt This data set contains the calcium levels of 11 test subjects at zero and 3 hours after taking a multi-vitamin containing calcium. The variables are:

Variable name	Description
HRS0	Calcium level at time zero (mgs/dl)
HRS3	Calcium level three hours after a multi-vitamin (mgs/dl)

caliper.txt This data set contains the diameters of 14 metal rods, measured once with each of two calipers. The variables are:

Variable name	Description
PART	Part number
CALIPERA	Caliper type A
CALIPERB	Caliper type B

ccthickn.txt When paint is applied to a car it is applied at a particular film build (i.e., thickness). The thickness of the coating has an effect on the properties of the paint so it is important to monitor the coating process to maintain the correct film build. The final layer of paint on a car is called the clear coat, the film build of this layer was to be 65 microns. This data set contains film build data on the clear-coat layer of 40 cars. Variable name: THICKNESS

cement.txt A contracting firm wished to compare the drying times of various cements used in sidewalks. Type A is a standard concrete, while Types B and C are quick drying compounds (more expensive than the standard compound). The test squares of cement were poured and then tested every 15 minutes until they were dry enough to be walked on. The variables are:

Variable name	Description
TYPE	Type of cement (A, B, or C)
TIME	Time until concrete is dry enough to walk on (hours)

cheese.txt A manufacturer of cheese has a number of production facilities across the country. One product that is produced is used by a major pizza chain, and it is critical that there is consistency in the cheese made by individual plants. This data set contains data from six batches at each of three plants. The factor of interest is the fat content. The variables are:

Variable name	Description
PLANT	Plant designator (A, B, or C)
BATCH	Batch number
FAT	Fat content (%)

chemreac.txt This data set is the result of an experiment designed to improve the yield of a chemical reaction. The variables are:

Variable name	Description
TEMP	Temperature of the reaction (120 or 140°C)
CAT	Catalyst used in the reaction (1 or 2)
TIME	Time the reaction was run (10 or 30 min.)
YIELD	Yield of the reaction

computer.txt An experiment was done to study the repair time (in minutes) for three different brands of computers. Since the repair time might depend on how the computers were configured, the type of computer (business machine, expensive home machine, inexpensive home machine) was considered as a second factor. The variables are:

Variable name	Description
BRAND	Brand of computer
TYPE	Type of computer
	(1 = business, 2 = expensive home, 3 = inexpensive home)
TIME	Repair time (minutes)

cure.txt Time and temperature are key properties in many chemical curing processes. Process engineers are often concerned with determining a "bake window" for a product, that is finding a time/temperature combination that achieves the best results in the smallest amount of time. In one process the engineers studied three temperature levels on their oven at two different time intervals and measured the yield of parts. The variables are:

Variable name	Description
TIME	Curing time
TEMP	Curing temperature
YIELD	Yield of parts

curl.txt A chemist is working on a new coating for transparencies used in ink-jet printers. The current formulation causes the transparencies to curl. There are two components that the chemist has data on that are believed to affect curl. The variables are:

Variable name	Description
CATALYST	Amount of catalyst (the first component)
COMPA	Amount of component A (the second component)
CURL	A measure of the curl

defoam.txt This data set contains the height of a solution containing a "defoamer" in a 50-ml graduated cylinder after being heated to a particular temperature. The goal of a defoamer is to control the amount of foam in a solution (defoamers are often used, for example, in fermentation processes to avoid overflowing kettles). The defoamers studied here had three different pH's, three different concentrations, and were heated to three different temperatures. All three factor levels are given simply as L, M, and H. The variables are:

Variable name	Description
CONC	Concentration of defoamer
PH	pH of defoamer
TEMP	Temperature of defoamer
HEIGHT	Height of the solution

deink.txt This data set contains the results from an experiment on de-inking of newspaper. The factors studied were the amount of alkali in the solution and the hardness of the water used. The response is a measure of brightness of the resulting pulp (that would then be used in new newsprint). The variables are:

Variable name	Description
ALKALI	Amount of alkali in the solution
HARDNESS	Hardness of the water
BRIGHT	A measure of brightness

deink2.txt The R&D group of a chemical manufacturer is working on three formulas to be used to de-ink newspapers (for recycling). To aid in the choice of formulation the chemists conduct a small experiment in which five newspaper specimens (from five different newspapers) are divided into thirds, and one group is treated with each de-inking formulation (assigned at random). Each group is then

graded as to the quality of its color. The variables are:

Variable name	Description
FORMULA	Formula used
NEWSPAPER	Newspaper used
BRIGHT	A measure of brightness

dhaze.txt In the study of coatings (such as for UV protection or anti-reflexiveness) for optical lenses it is important to study the durability of the coating. One such durability test is an abrasion test, simulating day-to-day treatment of lenses (consider how one cleans lenses: on the corner of a T-shirt, etc.). In this study the controlled variable is the surface treatment of the lens, and the response Y is the increase in haze after 150 cycles of abrasion. Minimal increase in haze is desired. The variables are:

Variable name	Description
TREATMENT	Surface treatment
DHAZE	Increase in haze

diagnostic.txt A diagnostic kit (for measuring a particular protein level in blood) can be used to test 16 blood samples. To check for kit-to-kit consistency the manufacturer tested the blood from 16 patients with two different kits. The variables are:

Variable name	Description
KIT1	Measurement using kit 1
KIT2	Measurement using kit 2

diameter.txt A large industrial corporation produces, among other things, balls used in computer mice. These plastic balls have a nominal diameter of 2 cm. Samples of 10 balls were taken from each of two different production lines, and the diameters of the sampled balls were measured. The variables are:

Variable name	Description
LINE	Production line
DIAMETER	Diameter of the ball (cm)

drought.txt In 2002, Colorado experienced a drought and many towns on the front range issued either mandatory or voluntary water restrictions on outdoor watering. This data set contains data from five front range towns on their June 2001 and June 2002 usage in millions of gallons. The variables are:

Variable name	Description
TOWN	Town number
J2001	Water usage for June 2001 (millions of gallons)
J2002	Water usage for June 2002 (millions of gallons)

drums.txt This data set contains the weights of 30 drums of a chemical product. Weights (in pounds) were taken before the drums were filled (empty drum weight) and after the drum was filled (full drum weight). The variables are:

Variable name	Description
NUMBER	Drum number
EMPTY	Empty drum weight (lbs.)
FULL	Full drum weight (lbs.)

dry.txt This data set reports data from an experiment to study the effect of the drying portion of a process on the moisture content of the final product. The plant has two types of dryers available for use, a tunnel dryer where the material passes through the dryer on a conveyor belt and a batch dryer, where the material is placed in the dryer, dried, then removed. The variables are:

Variable name	Description
RATE	Speed of drying (slow or fast)
TYPE	Type of dryer (tunnel or batch)
RESPONSE	Moisture content of final product (%)

epoxy.txt This data set contains data from an experiment to determine the effect of the level of epoxy in an automotive paint formulation. The response considered is the appearance measure longwave (LW). The variables are:

Variable name	Description
EPOXY	Level of epoxy
LW	Longwave measurement

exposure.txt A major automotive paint manufacturer has a "panel farm" on the coast of Florida. The purpose of this "farm" is to expose paints to different environmental conditions (sun, rain, salt, and sand) to test durability. A set of panels from three test formulations was sent to the farm and left out in the elements for six months. After that time they were rated on their appearance using a scale of 1–10 (1 being good). The variables are:

Variable name	Description
FORMULATION	Formulation designator (A, B, C)
RATING	Appearance rating

fbuild.txt A study was performed to quantify how the film build of an automotive paint affects appearance. Typically, higher film builds (i.e., thicker paint) results in better appearance. The variables are:

Variable name	Description
BUILD	Film build
GLOSS	Gloss measurement

fill.txt A process that fills cylinders with small beads is under study. During the process the cylinders are periodically "tamped". Two control factors related to tamping are the distance the cylinder is lifted from the surface (in centimeters) before being tamped and the number of times it is tamped. The file `fill.txt` contains the results of an experiment that varied the tamp distance and the number of tamps and measured the final fill amount (in grams) of material that was used to fill the cylinder. The variables are:

Variable name	Description
DISTANCE	Distance lifted from the surface (cm)
TAMPS	Number of tamps
FILL	Final amount of material (grams)

fillweight.txt A filling machine had just undergone maintenance to replace worn parts. To check to see if the machine was running as desired five tubes of product from each of four filling runs were weighed. The variables are:

Variable name	Description
BATCH	Batch number
WEIGHT	Weight

fish.txt This data file contains the values of a toxin in fish from four sites along a stream that runs near abandoned lead mines. The variables are:

Variable name	Description
SITE	Site location
TOXIN	Amount of toxin (mg/kg wet weight)

fish2.txt An environmental consulting firm collected fish from four sites around a pond feed by a stream. The scientists collected five fish from each site but in testing some samples were destroyed by accident resulting in an uneven number of readings from each site. The variables are:

Variable name	Description
SITE	Site location
TOXIN	Amount of toxin (mg/kg wet weight)

fluoride.txt Fluoride levels were measured in ppm for samples drawn from three separate municipal water sources. The ideal level of fluoride is 1.0 ppm. The variables are:

Variable name	Description
SOURCE	Water source
FLUORIDE	Amount of fluoride (ppm)

gloss.txt The paint on a car consists of multiple layers (an undercoat, a primer, a base coat, and a clear coat). In the development of a clear coat the chemists became concerned that their formulation did not perform the same over all colors (color is a component of the base coat layer). To study the situation they tested their clear coat formulation over four base coats and measured a number of properties, including gloss. The variables are:

Variable name	Description
COLOR	Color in the base coat
GLOSS	A measure of gloss

labcomp.txt This data set is the result of an "inter-lab" study where multiple labs, in this case each located at a different plant, participate in a study to compare measurements within and between the labs. For this study, the surface area of a silica was measured by each lab seven times. All labs had the capability of measuring using a 1-point CTAB procedure, and in addition two labs also measured using a 5-point CTAB procedure. The variables are:

Variable name	Description
LAB	Lab indicator
PT1	Surface area using the 1 point procedure
PT5	Surface area using the 5 point procedure

lw.txt The purpose of this experiment was to choose the best combination of two components in an automotive paint formulation. One factor in the experiment was a chemical denoted as Component 1, which occurred at four levels. The other factor was a second chemical denoted as Component 2, which occurred at three levels. There were two replicate observations for each treatment combination. The variables are:

Variable name	Description
LW	A measure of the quality of the paint
COMP1	Level of component 1
COMP2	Level of component 2

lwsw.txt Longwave (LW) and shortwave (SW) are two appearance measures used in the automotive industry to rate the quality of a paint job. These two measures are generally related. This data set contains values of these appearance measures for 13 cars. The variables are:

Variable name	Description
LW	Longwave measurment
SW	Shortwave measurment

moisture.txt This data set is the result of an experiment to increase the moisture content of a silica product. This experiment was run in a pilot plant at a research location. The variables are:

Variable name	Description
TEMP	Temperature in °F
SPEED	Process speed
SOLIDS	Percent solids
PH	pH of the process

mw.txt This data set contains data on the Mw of a paint as a function of five process factors. (Mw influences certain quality characteristics of the paint.) For proprietary reasons the factors are labeled A–E, with no explanation of what they might represent. Each factor occurs at two levels, designated as "high" and "low", again with no indication of what that stands for.

Variable name	Description
A	Factor A
B	Factor B
C	Factor C
D	Factor D
E	Factor E
MW	Mw

odor.txt Odor, yellowing, and hardness are all performance measures of a monomer used in making optical lenses. This data set records these values for 35 random samples of a monomer. The variables are:

Variable name	Description
ODOR	The odor
YELLOWING	Yellowing
HARDNESS	Hardness

oven.txt The drying ability of several brands of gas and electric lab ovens was measured. For each trial the experimenter recorded the percentage of moisture remaining in a sample of silica after drying for 15 minutes. The variables are:

Variable name	Description
BRAND	Brand of lab oven
TYPE	Type of lab oven (1 = gas; 2 = electric)
MOISTURE	Moisture remaining

ph.txt This data set contains measurements from three batches of material that were made at a chemical plant. For each batch numerous pH readings were taken during the course of production. For this particular material the pH should have been consistent throughout production. The variables are:

Variable name	Description
BATCH	Batch number
PH	pH value

phmeas.txt A lab received a new instrument to measure pH. To compare the new instrument to the old lab instrument eleven samples were measured with both pieces of equipment. The variables are:

Variable name	Description
PHOLD	pH measurement with the old instument
PHNEW	pH measurement with the new instrument

pigment.txt This data set contains the results of an experiment to determine if changing the pigment used in a white paint would improve the yellowing of the paint. Yellowing is measured by Δb, the change in color of the paint over time. Three pigments were to be studied: the current pigment (labeled C) and two new pigments (A and B). A lab batch of paint only contains enough material for three tests so the experimenters used three lab batches of paint, and subdivided each paint batch into three parts so each pigment could be tested with each batch. The variables are:

Variable name	Description
BATCH	Batch number
PIGMENT	Pigment label
DELTAB	Δb value

protein.txt This data set contains the results of a study of a new blood test for a particular disease. The gender, age, duration a patient has had the disease, level of a particular component in the blood (level A), and the value of the protein level obtained by the new test were recorded. The variables are:

Variable name	Description
GENDER	Patient's gender
AGE	Patient's age
DURATION	Duration patient has had the disease
LEVEL A	Batch number
PROTEIN	New test value

purity.txt The data set contains the results from a study to ascertain the effects of two factors (Factors 1 and 2) on the purity of a product from a filtration process. In order to complete all the experiments more quickly, two different technicians were used. The variables are:

Variable name	Description
TECH	Technician number
FAC2	Level of the second factor
FAC1	Level of the first factor
PURITY	Purity

railcar.txt A company ships many products via rail cars. These rail cars are either owned or leased by the company, and keeping track of their whereabouts is critical. The company is interested in minimizing the amount of time that a rail car is held by a customer. This data set contains the number of days a rail car from a particular fleet was held by a customer during a four-month period. Variable name: DAYS

railcar2.txt This data set contains data on rail-car hold times for a second product from the company discussed above. Note the zero hold times. These occur when the rail car is unloaded as soon as it arrives at the customer's site. Variable name: DAYS

railcar3.txt A company suspects that the configuration of the rail cars used to ship their product may have an effect on the moisture level of the product when it reaches its destination. Data from a customer on the product's moisture level when it arrived was collected for two types of rail cars. The variables are:

Variable name	Description
TYPE	Type of rail car
MOISTURE	Moisture level of the product (%)

ratings.txt A company wants to give one of their raw material suppliers a measure of how well a process is running with a new raw material. On a scale of 1– 10 they rate a number of key areas of the process and average those ratings. This average is then given to the supplier as a measure of the success of the raw material. Variable name: RATING

ratings2.txt The company described in the above data set felt that the moisture levels in the raw material were causing them to have problems in their plant. This data set contains both the ratings and the corresponding moisture levels for the raw material. The variables are:

Variable name	Description
RATING	Average rating
MOISTURE	Moisture level of the raw material (%)

reflect.txt This data set contains the results of an experiment to test various components in an anti-reflective coating system (for eyeglass lenses). The response variable is the amount of reflectance (lower values are desired). The variables are:

Variable name	Description
BINDER	Binder type
BASE	Basecoat type
TIME	Etch time (low or high)
RATIO	Ratio of binder to basecoat (1=60/40; 2=85/15)
REFLECT	Reflectance

safety.txt This data set contains the number of recorded safety violations at a company's six plants over a 5-year period. The variables are:

Variable name	Description
YEAR	Year being considered
PLACE	Plant's location
EMPLOYEES	Number of employees at the plant
CASES	Number of recorded safety violations

sales.txt A financial planner for a company wanted to model monthly sales on the previous months capital expenses and power index (an industry specific index). This data set contains the required data for 48 months. The variables are:

Variable name	Description
CAPITAL	Capital expenses
PINDEX	Power index
SALES	Number of sales

sarea.txt This data set contains the surface area of three batches of silica measured by four different lab technicians. The variables are:

Variable name	Description
BATCH	Batch number
TECH	Technician
SAREA	Surface area (m^2/g)

separate.txt This data set contains the results of an experiment to study the effects of three process factors on the electrical resistance of a battery separator (extruded from a rubber, oil, and silica mix). The three factors in the experiment are the type of silica added to the mix, the temperature of a water bath, and the amount of time the material spends in the water bath. Each factor occurs at two levels, designated "High" and "Low". The variables are:

Variable name	Description
SILICA	Type of silica
TEMP	Temperature of bath water
TIME	Time in the bath
Y	Electrical resistance $(ohm/in.^2)$

soap.txt A small company's most popular soap product must be repackaged because the current package type is no longer available. A small consumer study had been performed to determine the best new package for the soap (Box, Foil, or Shrink Wrap). The study had shown that the foil wrapper was the most popular wrapper, however, the director of this project wanted more information. An in-store test was run where all three packages were available and the number of packages of each kind sold in a week was recorded. This was intended to be a balanced experiment, run for five weeks, however after three weeks the company had supplier problems and did not have any more boxes to use in the study. The variables are:

Variable name	Description
BOX	Number of box packages sold
FOIL	Number of foil packages sold
SHRINK	Number of shrink wrap packages sold

stab.txt This data set contains the results of a study to investigate the stability over time of the active ingredient in a chemical product. The product has a nominal 20 mg/ml of the active ingredient. Samples of 5 ml of the chemical are stored at either 23°C (i.e., room temperature) or at 35°C (an extreme condition) in either 0% humidity or 85% relative humidity (again, normal and extreme conditions). At specified time points (1, 4, and 8 weeks) samples are tested and the actual level of active ingredient is recorded. The variables are:

Variable name	Description
TEMP	Storage temperature (room temperature or $35°C$)
HUMIDITY	Relative humidity (0 or 85%)
TIME	Time of storage in weeks (1, 4, or 8)
Y	Level of active ingredient

stretch.txt This data set contains data from an experiment used to study the stretch in hot pizza cheese. Long stretch is a desired property for commercial pizza. Measuring stretch is not a precise science. One uses an implement to lift the cheese and stretch it, measuring the length of the cheese just prior to it breaking. Because of the noise in this measurement, five measurements were taken on each pizza studied. Factors considered were bake temperature and amount of cheese on the pizza. The variables are:

Variable name	Description
TEMP	Baking temperature
CHEESE	Amount of cheese
STRETCH	Amount of stretch

surfarea.txt Samples of a particular type of silica (a chemical product with many applications such as a filler in rubber products) were tested for their surface area (a key property). Variable name: AREA

tablets.txt A process engineer suspected a problem with a batch of chlorine used to make chlorine tablets for home pool chlorinators. The material had already been made into tablets. In order to check the batch the engineer selected ten tablets at random from the suspect batch (Batch 1) and ten tablets from each of two other batches. The lab tested all 30 tablets and recorded the lifetime of each tablet (in hours, where an hour in the lab test corresponds to a day in a home pool chlorinator). The variables are:

Variable name	Description
TIME	Lifetime of the tablet
BATCH	Batch number

temprate.txt This data set contains data from an experiment to study the effect of a water bath on the final product moisture content. The variables are:

Variable name	Description
TEMP	Temperature of water bath (°F)
RATE	Rate product moves through the water bath
RESPONSE	Moisture content of final product (%)

tennis.txt A new fiber is being studied for use in tennis balls. The new fiber is supposed to improve the durability of the tennis ball cover (and hence, increase the lifetime of the tennis ball). Researchers have developed two possible coverings for tennis balls from this new fiber. To gauge their progress they ran a durability test with two commercial products (A and B) and the two candidate coverings (C and D). Five balls of each type were randomly chosen and marked with a stripe to gauge wear. Each ball was put into a machine that simulates a serve. The response variable is the number of serves a ball can withstand before the wear stripe is worn. The variables are:

Variable name	Description
TYPE	Type of covering
WEAR	Number of serves before strip is worn

thinfilm.txt In the development of coatings (such as for paper, optical lenses, computer screens, etc.) a researcher will often study a "thin film"; that is, the coating cured on some substrate such as glass; rather than studying the coating cured on the actual product. This data set contains data from a study of thin films made to determine the effect of the level of a particular component on the strength of a coating. The coatings were applied to two different substrates (glass and foil). The variables are:

Variable name	Description
MATERIAL	Substrate type (glass or foil)
COMP	Level of a component in the coating
MAXLOAD	Strength measurement

timetemp.txt Developers of automotive paint must test not only the appearance of the paint but also the durability. Some durability tests are run on test panels that have been put in a freezer to simulate exposure to cold temperatures. One such test was to be run at $-10°C$. To set up the test protocol a study was done to determine how long it took for a test panel to reach $-10°C$ after being removed from the freezer. This would allow the lab personnel to decide on a temperature for the freezer such that they would have sufficient time to remove the panel and take it to the test equipment. The study was conducted with both OEM (original equipment manufacture) and repaired panels (repaired panels have additional coats of paint on them to simulate touch up work done on a car after the paint job but before it leaves the paint area). The variables are:

Variable name	Description
TIME	Time it took the panel to reach $-10°C$ (seconds)
TEMP	Original temperature ($°C$)
TYPE	Type of panel (1 = repaired; 2 = OEM)

tpaste.txt This data set contains the results of an experiment to study the turbidity of a toothpaste formulation. The three factors studied in the experiment were: NaCl level, reaction temperature, and the addition rate of a particular component in the formulation. The variables are:

Variable name	Description
NACL	NaCl level (2 or 20 ml)
TEMP	Reaction temperature (50 or 80°C)
RATE	Addition rate (30 or 60 seconds)
TURBIDITY	Turbidity

urine.txt As a safety precaution, employees who work in a particular area of a chemical plant are monitored monthly for the level of mercury in their urine. This data set contains the results from four employees over a 1-year time period. The variables are:

Variable name	Description
MONTH	Month of the year
PERSON1	First person's mercury level
PERSON2	Second person's mercury level
PERSON3	Third person's mercury level
PERSON4	Fourth person's mercury level

uvcoatin.txt Scientists developing a new UV coating for eyeglasses felt they were ready to field test the coating (as opposed to running simulated tests in the laboratory). Ten volunteers were to be used in the study in which the scientists wanted to be able to compare the durability of the experimental coating to a current commercial coating. The durability of the coatings would be evaluated by taking an initial haze measurement on each lens and then repeating that measurement after the glasses had been used by the volunteers for 3 months. The change in the haze value (3 month − initial) would be the reported durability number. Smaller values would indicate better durability.

The scientists were concerned that the way the volunteers used their glasses could influence the outcome of the study. For instance, one volunteer was an avid mountain biker, and they expected her lenses to see more extreme conditions than those of the other volunteers. Many factors, such as the care of the lenses, the amount of use, and type of use could all influence the durability measure of the coating. To account for these differences a paired-differences study was designed. Each volunteer in the study would have one lens with the experimental coating and one lens with the commercial coating. Then the differences in the durability of each pair of lenses would be used to ultimately judge the durability of the new coating. The assignment of the coatings to the lenses (i.e., right or left) was done at random. The variables are:

Variable name	Description
A	Difference in haze values for the commercial coating
B	Difference in haze values for the experimental coating
DIFF	Paired differences between commercial and experimental

uvoven.txt A lens coating facility tested the UV absorbance for lenses cured in one of two different ovens.

Variable name	Description
OVEN	Oven number
UV	UV absorbance

viscosity.txt The stability of a paint was tested by subjecting it to increasing times at a high temperature. The viscosity of the paint was tested to determine the point of failure (when the viscosity is too high the paint is said to have gelled and is no longer usable). The time until the material "gelled" is given for 17 samples. Variables name: TIME

vitamin.txt A company was developing a new vitamin aimed at those at risk for ailments associated with low calcium levels. In order to test the effectiveness of three new formulations a group of test subjects had their calcium levels tested before and after taking one of the three vitamins. The variables are:

Variable name	Description
TREATMENT	Formulation
BEFORE	Calcium level before the vitamin
AFTER	Calcium level after the vitamin

wash.txt This data set contains data from a study of a test procedure used for automotive paint. The test consists of applying drops of acid onto a painted steel panel at one minute intervals for thirty minutes. The panel is then rinsed and later rated by a person on how well the paint withstood the acid. It was thought that over time the ratings of the panels changed due to further degradation in the paint from residual acid on the panel. It was also thought that rinsing the panel with a neutralizing wash after the test might solve this degradation problem. A test of 36 panels (18 with the wash, 18 without) was conducted. The panels were rated initially and then again after being stored for two weeks. The variables are:

Variable name	Description
TYPE	Type of treatment (NW = not washed; W = washed)
TIME1	Initial rating
TIME2	Rating after 2 weeks

water.txt A manufacturing process for a medical device uses filtered water that must be strictly monitored for bacteria. This data set contains measurements of the bacteria (in parts per million) from 50 water samples. Variables name: BACTERIA

webtraff.txt An experiment was performed to explore the effect of a marketing campaign on web site traffic. This data set contains the number of weeks into the marketing campaign and the web site traffic. The traffic measure is a daily average for two random 24-hour periods during the particular week. The day that the traffic was sampled was determined before the study began and only weekdays were considered. The value is in 1000's of hits per day. The variables are:

Variable name	Description
WEEK	Number of weeks into the campaign
TRAFFIC	Web traffic (1000's of hits per day)

webvisit.txt This data set contains the number of visits to a web site over a 3-week period. Variable name: VISITS

weight.txt A manufacturer of plastic bags tested a random sample of 43 bags from a production line and measured the weight in pounds they could hold before failing. breaking. Variable name: WEIGHT

whitearea.txt In order to evaluate how well a white raw material has been mixed into a dark product, the percent of "white area" was measured for two different mixing processes.

Variable name	Description
PROCESS	Process number
WHITEAREA	Percent of white area

yellow.txt This data set contains yellowing data (b is a color measure) for 23 samples of monomer used in the casting of optical lenses. The data include an initial measurement and a measurement after one month of storage. It is hoped that the material is stable and that the yellowing value does not change over time.

Variable name	Description
B1	Initial color measurment
B2	Color measurement after one month

yield.txt An industrial engineer wants to develop a model for the yield of a chemical process based on two key variables. The two variables are measurements made on a slurry, which is then further processed into the final product. The two independent variables are the temperature of the slurry as it enters the next process step and the pH of the slurry. The dependent variable of interest is the process yield from the batch of slurry. The variables are:

Variable name	Description
TEMP	Temperature of the slurry
PH	pH of the slurry
YIELD	Process yield (tons)

12.2 APPENDIX B – TABLES

Table B1. Standard Normal Distribution

$$\Phi(z) = \frac{1}{\sqrt{2\pi}} \int_{-\infty}^{z} \exp(-x^2/2)dx$$

z	0.00	0.01	0.02	0.03	0.04	0.05	0.06	0.07	0.08	0.09
0.0	0.5000	0.5040	0.5080	0.5120	0.5160	0.5199	0.5239	0.5279	0.5319	0.5359
0.1	0.5398	0.5438	0.5478	0.5517	0.5557	0.5596	0.5636	0.5675	0.5714	0.5753
0.2	0.5793	0.5832	0.5871	0.5910	0.5948	0.5987	0.6026	0.6064	0.6103	0.6141
0.3	0.6179	0.6217	0.6255	0.6293	0.6331	0.6368	0.6406	0.6443	0.6480	0.6517
0.4	0.6554	0.6591	0.6628	0.6664	0.6700	0.6736	0.6772	0.6808	0.6844	0.6879
0.5	0.6915	0.6950	0.6985	0.7019	0.7054	0.7088	0.7123	0.7157	0.7190	0.7224
0.6	0.7257	0.7291	0.7324	0.7357	0.7389	0.7422	0.7454	0.7486	0.7517	0.7549
0.7	0.7580	0.7611	0.7642	0.7673	0.7704	0.7734	0.7764	0.7794	0.7823	0.7852
0.8	0.7881	0.7910	0.7939	0.7967	0.7995	0.8023	0.8051	0.8078	0.8106	0.8133
0.9	0.8159	0.8186	0.8212	0.8238	0.8264	0.8289	0.8315	0.8340	0.8365	0.8389
1.0	0.8413	0.8438	0.8461	0.8485	0.8508	0.8531	0.8554	0.8577	0.8599	0.8621
1.1	0.8643	0.8665	0.8686	0.8708	0.8729	0.8749	0.8770	0.8790	0.8810	0.8830
1.2	0.8849	0.8869	0.8888	0.8907	0.8925	0.8944	0.8962	0.8980	0.8997	0.9015
1.3	0.9032	0.9049	0.9066	0.9082	0.9099	0.9115	0.9131	0.9147	0.9162	0.9177
1.4	0.9192	0.9207	0.9222	0.9236	0.9251	0.9265	0.9279	0.9292	0.9306	0.9319
1.5	0.9332	0.9345	0.9357	0.9370	0.9382	0.9394	0.9406	0.9418	0.9429	0.9441
1.6	0.9452	0.9463	0.9474	0.9484	0.9495	0.9505	0.9515	0.9525	0.9535	0.9545
1.7	0.9554	0.9564	0.9573	0.9582	0.9591	0.9599	0.9608	0.9616	0.9625	0.9633
1.8	0.9641	0.9649	0.9656	0.9664	0.9671	0.9678	0.9686	0.9693	0.9699	0.9706
1.9	0.9713	0.9719	0.9726	0.9732	0.9738	0.9744	0.9750	0.9756	0.9761	0.9767
2.0	0.9772	0.9778	0.9783	0.9788	0.9793	0.9798	0.9803	0.9808	0.9812	0.9817
2.1	0.9821	0.9826	0.9830	0.9834	0.9838	0.9842	0.9846	0.9850	0.9854	0.9857
2.2	0.9861	0.9864	0.9868	0.9871	0.9875	0.9878	0.9881	0.9884	0.9887	0.9890
2.3	0.9893	0.9896	0.9898	0.9901	0.9904	0.9906	0.9909	0.9911	0.9913	0.9916
2.4	0.9918	0.9920	0.9922	0.9925	0.9927	0.9929	0.9931	0.9932	0.9934	0.9936
2.5	0.9938	0.9940	0.9941	0.9943	0.9945	0.9946	0.9948	0.9949	0.9951	0.9952
2.6	0.9953	0.9955	0.9956	0.9957	0.9959	0.9960	0.9961	0.9962	0.9963	0.9964
2.7	0.9965	0.9966	0.9967	0.9968	0.9969	0.9970	0.9971	0.9972	0.9973	0.9974
2.8	0.9974	0.9975	0.9976	0.9977	0.9977	0.9978	0.9979	0.9979	0.9980	0.9981
2.9	0.9981	0.9982	0.9982	0.9983	0.9984	0.9984	0.9985	0.9985	0.9986	0.9986
3.0	0.9987	0.9987	0.9987	0.9988	0.9988	0.9989	0.9989	0.9989	0.9990	0.9990
3.1	0.9990	0.9991	0.9991	0.9991	0.9992	0.9992	0.9992	0.9992	0.9993	0.9993
3.2	0.9993	0.9993	0.9994	0.9994	0.9994	0.9994	0.9994	0.9995	0.9995	0.9995
3.3	0.9995	0.9995	0.9995	0.9996	0.9996	0.9996	0.9996	0.9996	0.9996	0.9997
3.4	0.9997	0.9997	0.9997	0.9997	0.9997	0.9997	0.9997	0.9997	0.9997	0.9998
3.5	0.9998	0.9998	0.9998	0.9998	0.9998	0.9998	0.9998	0.9998	0.9998	0.9998
3.6	0.9998	0.9998	0.9999	0.9999	0.9999	0.9999	0.9999	0.9999	0.9999	0.9999

For values of $z < 0$ use $\Phi(-z) = 1 - \Phi(z)$.

Table B2. Critical Values of t Distributions, $t(\alpha; \nu)$

ν	0.1	0.05	0.025	0.01	0.005	0.001	0.0005
1	3.078	6.314	12.706	31.821	63.657	318.31	636.62
2	1.886	2.920	4.303	6.965	9.925	22.327	31.598
3	1.638	2.353	3.182	4.541	5.841	10.215	12.924
4	1.533	2.132	2.776	3.747	4.604	7.173	8.610
5	1.476	2.015	2.571	3.365	4.032	5.893	6.869
6	1.440	1.943	2.447	3.143	3.707	5.208	5.959
7	1.415	1.895	2.365	2.998	3.499	4.785	5.408
8	1.397	1.860	2.306	2.896	3.355	4.501	5.041
9	1.383	1.833	2.262	2.821	3.250	4.297	4.781
10	1.372	1.812	2.228	2.764	3.169	4.144	4.587
11	1.363	1.796	2.201	2.718	3.106	4.025	4.437
12	1.356	1.782	2.179	2.681	3.055	3.930	4.318
13	1.350	1.771	2.160	2.650	3.012	3.852	4.221
14	1.345	1.761	2.145	2.624	2.977	3.787	4.140
15	1.341	1.753	2.131	2.602	2.947	3.733	4.073
16	1.337	1.746	2.120	2.583	2.921	3.686	4.015
17	1.333	1.740	2.110	2.567	2.898	3.646	3.965
18	1.330	1.734	2.101	2.552	2.878	3.610	3.922
19	1.328	1.729	2.093	2.539	2.861	3.579	3.883
20	1.325	1.725	2.086	2.528	2.845	3.552	3.849
24	1.318	1.711	2.064	2.492	2.797	3.467	3.745
30	1.310	1.697	2.042	2.457	2.750	3.385	3.646
40	1.303	1.684	2.021	2.423	2.704	3.307	3.551
60	1.296	1.671	2.000	2.390	2.660	3.232	3.460
120	1.289	1.658	1.980	2.358	2.617	3.160	3.373
∞	1.282	1.645	1.960	2.326	2.576	3.090	3.291

Table B3. Sample Sizes for the One-Sample t Test

Sample size to achieve the desired power for $d = |\mu_1 - \mu_0|/\sigma$.
For a two-sided test at significance level 2α, the sample sizes are approximate.

α (2α)	d	Power 0.50	0.60	0.70	0.75	0.80	0.90	0.95	0.99
0.1	0.1	166	237	328	384	452	658	858	1303
(0.2)	0.2	42	60	83	97	114	166	215	327
	0.4	12	16	22	25	30	42	55	83
	0.6	6	8	10	12	14	20	25	38
	0.8	4	5	7	7	8	12	15	22
	1.0	3	4	5	5	6	8	10	14
	1.2	3	3	4	4	5	6	7	10
	1.4	3	3	3	4	4	5	6	8
	1.6	2	3	3	3	3	4	5	7
	1.8	2	3	3	3	3	4	4	6
	2.0	2	2	3	3	3	3	4	5
	3.0	2	2	2	2	2	3	3	3
0.05	0.1	272	362	473	540	620	858	1084	1579
(0.1)	0.2	70	92	120	136	156	216	272	396
	0.4	19	24	31	36	41	55	70	100
	0.6	9	12	15	17	19	26	32	46
	0.8	6	8	9	10	12	15	19	27
	1.0	5	6	7	7	8	11	13	18
	1.2	4	5	5	6	6	8	10	13
	1.4	4	4	5	5	5	7	8	10
	1.6	3	4	4	4	5	6	6	8
	1.8	3	3	4	4	4	5	6	7
	2.0	3	3	3	4	4	4	5	6
	3.0	3	3	3	3	3	3	4	4
0.025	0.1	387	492	620	696	787	1053	1302	1840
(0.05)	0.2	98	125	157	176	199	265	327	462
	0.4	26	33	41	46	52	68	84	117
	0.6	13	16	20	22	24	32	39	54
	0.8	9	10	12	13	15	19	23	31
	1.0	6	7	9	10	10	13	16	21
	1.2	5	6	7	7	8	10	12	15
	1.4	5	5	6	6	7	8	9	12
	1.6	4	5	5	5	6	7	8	10
	1.8	4	4	5	5	5	6	7	8
	2.0	4	4	4	4	5	5	6	7
	3.0	3	3	3	4	4	4	4	5
0.01	0.1	544	669	816	904	1007	1305	1580	2168
(0.02)	0.2	139	170	206	228	254	329	397	544
	0.4	37	45	54	60	66	85	102	139
	0.6	18	22	26	28	31	39	47	63
	0.8	12	14	16	17	19	24	28	37
	1.0	9	10	11	12	13	16	19	25
	1.2	7	8	9	10	10	12	14	18
	1.4	6	7	7	8	8	10	11	14
	1.6	5	6	7	7	7	9	10	12
	1.8	5	5	6	6	6	7	8	10
	2.0	5	5	5	6	6	7	7	9
	3.0	4	4	4	4	4	5	5	6
0.005	0.1	667	804	965	1060	1172	1492	1785	2407
(0.01)	0.2	170	204	244	268	296	376	449	605
	0.4	45	54	64	70	77	97	115	154
	0.6	22	26	31	33	36	45	53	71
	0.8	14	16	19	20	22	27	32	41
	1.0	10	12	13	14	16	19	22	28
	1.2	8	9	11	11	12	14	16	21
	1.4	7	8	9	9	10	12	13	16
	1.6	6	7	8	8	8	10	11	13
	1.8	6	6	7	7	8	8	10	11
	2.0	5	6	6	6	7	8	8	10
	3.0	4	4	5	5	5	6	6	7

Table B4. Sample Sizes for the Two-Sample t Test

Sample size to achieve the desired power for $d = |\mu_1 - \mu_2|/\sigma$.
For a two-sided test at significance level 2α, the sample sizes are approximate.

α (2α)	d	Power							
		0.50	0.60	0.70	0.75	0.80	0.90	0.95	0.99
0.1 (0.2)	0.1	330	473	654	767	903	1316	1715	2605
	0.2	83	119	164	193	227	330	430	652
	0.4	22	31	42	49	58	83	108	164
	0.6	11	14	19	23	26	38	49	74
	0.8	7	9	12	13	15	22	28	42
	1.0	5	6	8	9	10	15	19	27
	1.2	4	5	6	7	8	11	13	19
	1.4	3	4	5	5	6	8	10	15
	1.6	3	3	4	5	5	7	8	12
	1.8	3	3	4	4	4	6	7	10
	2.0	3	3	3	4	4	5	6	8
	3.0	2	2	3	3	3	3	4	5
0.05 (0.1)	0.1	543	723	943	1078	1239	1715	2166	3155
	0.2	137	182	237	271	311	430	543	790
	0.4	36	47	61	69	79	109	137	199
	0.6	17	22	28	32	36	49	62	90
	0.8	10	13	17	19	21	29	36	51
	1.0	7	9	11	13	14	19	24	33
	1.2	6	7	9	9	11	14	17	24
	1.4	5	6	7	8	8	11	13	18
	1.6	4	5	6	6	7	9	10	14
	1.8	4	4	5	5	6	7	9	12
	2.0	4	4	5	5	5	6	8	10
	3.0	3	3	3	3	4	4	5	6
0.025 (0.05)	0.1	771	982	1237	1391	1572	2102	2600	3676
	0.2	195	247	311	349	395	528	652	921
	0.4	50	64	80	89	101	134	165	232
	0.6	24	30	37	41	46	61	75	105
	0.8	15	18	22	24	27	35	43	60
	1.0	10	12	15	16	18	24	28	39
	1.2	8	9	11	12	13	17	21	28
	1.4	7	8	9	10	11	13	16	21
	1.6	6	6	7	8	9	11	13	17
	1.8	5	6	6	7	7	9	11	14
	2.0	5	5	6	6	7	8	9	12
	3.0	3	4	4	4	5	5	6	7
0.01 (0.02)	0.1	1085	1334	1628	1804	2008	2604	3155	4330
	0.2	274	336	410	453	505	654	792	1085
	0.4	71	86	105	116	129	166	200	274
	0.6	33	40	48	53	59	76	91	124
	0.8	20	24	29	31	35	44	53	71
	1.0	14	17	20	21	23	29	35	47
	1.2	11	13	15	16	17	21	25	33
	1.4	9	10	12	13	14	17	19	25
	1.6	8	9	10	10	11	14	16	20
	1.8	7	7	8	9	10	11	13	17
	2.0	6	7	7	8	8	10	11	14
	3.0	4	5	5	5	6	6	7	8
0.005 (0.01)	0.1	1331	1605	1926	2114	2337	2977	3565	4808
	0.2	336	404	484	532	588	748	895	1205
	0.4	87	104	124	136	150	190	227	304
	0.6	41	48	57	63	69	86	103	137
	0.8	25	29	34	37	40	50	60	79
	1.0	17	20	23	25	27	34	40	52
	1.2	13	15	17	19	20	25	29	37
	1.4	11	12	14	15	16	19	22	28
	1.6	9	10	11	12	13	16	18	23
	1.8	8	9	10	10	11	13	15	19
	2.0	7	8	9	9	10	11	13	16
	3.0	5	5	6	6	6	7	8	9

Table B5. Balanced ANOM Critical Values $h(\alpha; k, \nu)$
Level of Significance = 0.25

ν	Number of Means Being Compared, k																		
	2	3	4	5	6	7	8	9	10	11	12	13	14	15	16	17	18	19	20
1	2.41	3.70	4.29	4.71	5.02	5.27	5.48	5.66	5.82	5.96	6.08	6.19	6.29	6.39	6.47	6.55	6.63	6.70	6.76
2	1.60	2.34	2.67	2.90	3.07	3.21	3.33	3.43	3.52	3.60	3.67	3.73	3.79	3.85	3.90	3.94	3.99	4.03	4.06
3	1.42	2.04	2.31	2.49	2.64	2.75	2.85	2.93	3.01	3.07	3.13	3.18	3.23	3.27	3.32	3.35	3.39	3.42	3.45
4	1.34	1.91	2.15	2.32	2.45	2.55	2.64	2.71	2.78	2.84	2.89	2.94	2.98	3.02	3.06	3.09	3.13	3.16	3.19
5	1.30	1.84	2.06	2.22	2.34	2.44	2.52	2.59	2.65	2.71	2.76	2.80	2.84	2.88	2.91	2.95	2.98	3.01	3.03
6	1.27	1.79	2.01	2.16	2.27	2.37	2.44	2.51	2.57	2.62	2.67	2.71	2.75	2.79	2.82	2.85	2.88	2.91	2.93
7	1.25	1.76	1.97	2.11	2.23	2.32	2.39	2.46	2.51	2.56	2.61	2.65	2.69	2.72	2.75	2.78	2.81	2.84	2.86
8	1.24	1.74	1.94	2.08	2.19	2.28	2.35	2.41	2.47	2.52	2.56	2.60	2.64	2.67	2.71	2.73	2.76	2.79	2.81
9	1.23	1.72	1.92	2.06	2.16	2.25	2.32	2.38	2.44	2.49	2.53	2.57	2.60	2.64	2.67	2.70	2.72	2.75	2.77
10	1.22	1.71	1.90	2.04	2.14	2.23	2.30	2.36	2.41	2.46	2.50	2.54	2.57	2.61	2.64	2.67	2.69	2.72	2.74
11	1.21	1.70	1.89	2.02	2.13	2.21	2.28	2.34	2.39	2.44	2.48	2.52	2.55	2.58	2.61	2.64	2.67	2.69	2.71
12	1.21	1.69	1.88	2.01	2.11	2.19	2.26	2.32	2.37	2.42	2.46	2.50	2.53	2.56	2.59	2.62	2.65	2.67	2.69
13	1.20	1.68	1.87	2.00	2.10	2.18	2.25	2.31	2.36	2.40	2.44	2.48	2.52	2.55	2.58	2.60	2.63	2.65	2.67
14	1.20	1.67	1.86	1.99	2.09	2.17	2.24	2.29	2.35	2.39	2.43	2.47	2.50	2.53	2.56	2.59	2.61	2.64	2.66
15	1.20	1.67	1.85	1.98	2.08	2.16	2.23	2.28	2.33	2.38	2.42	2.46	2.49	2.52	2.55	2.57	2.60	2.62	2.64
16	1.19	1.66	1.85	1.98	2.07	2.15	2.22	2.27	2.32	2.37	2.41	2.44	2.48	2.51	2.54	2.56	2.59	2.61	2.63
17	1.19	1.66	1.84	1.97	2.07	2.14	2.21	2.27	2.32	2.36	2.40	2.44	2.47	2.50	2.53	2.55	2.58	2.60	2.62
18	1.19	1.65	1.84	1.96	2.06	2.14	2.20	2.26	2.31	2.35	2.39	2.43	2.46	2.49	2.52	2.54	2.57	2.59	2.61
19	1.19	1.65	1.83	1.96	2.06	2.13	2.20	2.25	2.30	2.35	2.38	2.42	2.45	2.48	2.51	2.54	2.56	2.58	2.60
20	1.18	1.65	1.83	1.95	2.05	2.13	2.19	2.25	2.30	2.34	2.38	2.41	2.45	2.48	2.50	2.53	2.55	2.57	2.60
24	1.18	1.64	1.82	1.94	2.04	2.11	2.17	2.23	2.28	2.32	2.36	2.39	2.42	2.45	2.48	2.51	2.53	2.55	2.57
30	1.17	1.63	1.81	1.93	2.02	2.09	2.16	2.21	2.26	2.30	2.34	2.37	2.40	2.43	2.46	2.48	2.51	2.53	2.55
40	1.17	1.62	1.79	1.91	2.01	2.08	2.14	2.19	2.24	2.28	2.32	2.35	2.38	2.41	2.43	2.46	2.48	2.50	2.52
60	1.16	1.61	1.78	1.90	1.99	2.06	2.12	2.17	2.22	2.26	2.30	2.33	2.36	2.39	2.41	2.43	2.46	2.48	2.50
120	1.16	1.60	1.77	1.89	1.98	2.05	2.11	2.16	2.20	2.24	2.27	2.31	2.34	2.36	2.39	2.41	2.43	2.45	2.47
∞	1.15	1.59	1.76	1.87	1.96	2.03	2.09	2.14	2.18	2.22	2.25	2.29	2.31	2.34	2.36	2.39	2.41	2.43	2.45

Table B5 (continued). Balanced ANOM Critical Values $h(\alpha; k, \nu)$

Level of Significance $= 0.1$

ν									Number of Means Being Compared, k										
	2	3	4	5	6	7	8	9	10	11	12	13	14	15	16	17	18	19	20
1	6.31	9.52	11.0	12.0	12.8	13.4	14.0	14.4	14.8	15.2	15.5	15.7	16.0	16.2	16.5	16.7	16.8	17.0	17.2
2	2.92	4.05	4.56	4.92	5.19	5.41	5.60	5.76	5.91	6.03	6.14	6.25	6.34	6.43	6.51	6.58	6.65	6.72	6.78
3	2.35	3.16	3.50	3.75	3.94	4.09	4.23	4.34	4.44	4.53	4.61	4.68	4.74	4.81	4.86	4.91	4.96	5.01	5.05
4	2.13	2.81	3.09	3.29	3.45	3.58	3.69	3.78	3.86	3.93	4.00	4.06	4.11	4.17	4.21	4.26	4.30	4.34	4.37
5	2.02	2.63	2.88	3.05	3.19	3.30	3.40	3.48	3.55	3.62	3.68	3.73	3.78	3.83	3.87	3.91	3.94	3.98	4.01
6	1.94	2.52	2.74	2.91	3.03	3.13	3.22	3.30	3.36	3.42	3.48	3.53	3.57	3.61	3.65	3.69	3.72	3.75	3.78
7	1.89	2.44	2.65	2.81	2.92	3.02	3.10	3.17	3.24	3.29	3.34	3.39	3.43	3.47	3.51	3.54	3.57	3.60	3.63
8	1.86	2.39	2.59	2.73	2.85	2.94	3.02	3.08	3.14	3.19	3.24	3.29	3.33	3.36	3.40	3.43	3.46	3.49	3.52
9	1.83	2.34	2.54	2.68	2.79	2.87	2.95	3.01	3.07	3.12	3.17	3.21	3.25	3.28	3.32	3.35	3.38	3.40	3.43
10	1.81	2.31	2.50	2.64	2.74	2.83	2.90	2.96	3.02	3.06	3.11	3.15	3.19	3.22	3.25	3.28	3.31	3.34	3.36
11	1.80	2.29	2.47	2.60	2.70	2.79	2.86	2.92	2.97	3.02	3.06	3.10	3.14	3.17	3.20	3.23	3.26	3.28	3.31
12	1.78	2.27	2.45	2.57	2.67	2.75	2.82	2.88	2.93	2.98	3.02	3.06	3.10	3.13	3.16	3.19	3.21	3.24	3.26
13	1.77	2.25	2.43	2.55	2.65	2.73	2.79	2.85	2.90	2.95	2.99	3.03	3.06	3.09	3.12	3.15	3.18	3.20	3.23
14	1.76	2.23	2.41	2.53	2.63	2.70	2.77	2.83	2.88	2.92	2.96	3.00	3.03	3.06	3.09	3.12	3.15	3.17	3.19
15	1.75	2.22	2.39	2.51	2.61	2.68	2.75	2.80	2.85	2.90	2.94	2.97	3.01	3.04	3.07	3.09	3.12	3.14	3.17
16	1.75	2.21	2.38	2.50	2.59	2.67	2.73	2.79	2.83	2.88	2.92	2.95	2.99	3.02	3.04	3.07	3.10	3.12	3.14
17	1.74	2.20	2.37	2.49	2.58	2.65	2.71	2.77	2.82	2.86	2.90	2.93	2.97	3.00	3.02	3.05	3.08	3.10	3.12
18	1.73	2.19	2.36	2.47	2.56	2.64	2.70	2.75	2.80	2.84	2.88	2.92	2.95	2.98	3.01	3.03	3.06	3.08	3.10
19	1.73	2.18	2.35	2.46	2.55	2.63	2.69	2.74	2.79	2.83	2.87	2.90	2.93	2.96	2.99	3.02	3.04	3.05	3.08
20	1.72	2.18	2.34	2.45	2.54	2.62	2.68	2.73	2.78	2.82	2.86	2.89	2.92	2.95	2.98	3.00	3.03	3.05	3.07
24	1.71	2.15	2.31	2.43	2.51	2.58	2.64	2.69	2.74	2.78	2.81	2.85	2.88	2.91	2.93	2.96	2.98	3.00	3.02
30	1.70	2.13	2.29	2.40	2.48	2.55	2.61	2.66	2.70	2.74	2.77	2.81	2.84	2.86	2.89	2.91	2.93	2.96	2.97
40	1.68	2.11	2.26	2.37	2.45	2.52	2.57	2.62	2.66	2.70	2.73	2.77	2.79	2.82	2.84	2.87	2.89	2.91	2.93
60	1.67	2.09	2.24	2.34	2.42	2.49	2.54	2.59	2.63	2.66	2.70	2.73	2.75	2.78	2.80	2.82	2.84	2.86	2.88
120	1.66	2.07	2.22	2.32	2.39	2.45	2.51	2.55	2.59	2.62	2.66	2.69	2.71	2.74	2.76	2.78	2.80	2.82	2.84
∞	1.65	2.05	2.19	2.29	2.36	2.42	2.47	2.52	2.55	2.59	2.62	2.65	2.67	2.69	2.72	2.74	2.75	2.77	2.79

Table B5 (continued). Balanced ANOM Critical Values $h(\alpha; k, \nu)$
Level of Significance = 0.05

ν	Number of Means Being Compared, k																		
	2	3	4	5	6	7	8	9	10	11	12	13	14	15	16	17	18	19	20
1	12.7	19.1	22.0	24.1	25.7	26.9	28.0	28.9	29.7	30.4	31.0	31.6	32.1	32.5	33.0	33.4	33.8	34.1	34.4
2	4.30	5.89	6.60	7.10	7.49	7.81	8.07	8.30	8.50	8.68	8.84	8.99	9.12	9.24	9.36	9.46	9.56	9.65	9.74
3	3.18	4.18	4.60	4.91	5.15	5.34	5.50	5.65	5.77	5.88	5.98	6.08	6.16	6.24	6.31	6.38	6.44	6.50	6.55
4	2.78	3.56	3.89	4.12	4.30	4.45	4.58	4.69	4.79	4.87	4.95	5.02	5.09	5.15	5.21	5.26	5.31	5.35	5.40
5	2.57	3.25	3.53	3.72	3.88	4.00	4.11	4.21	4.29	4.36	4.43	4.49	4.55	4.60	4.65	4.69	4.73	4.77	4.81
6	2.45	3.07	3.31	3.49	3.62	3.73	3.83	3.91	3.99	4.05	4.11	4.17	4.22	4.26	4.31	4.35	4.39	4.42	4.45
7	2.36	2.95	3.17	3.33	3.45	3.56	3.64	3.72	3.79	3.85	3.90	3.95	4.00	4.04	4.08	4.12	4.15	4.19	4.22
8	2.31	2.86	3.07	3.21	3.33	3.43	3.51	3.58	3.64	3.70	3.75	3.80	3.84	3.88	3.92	3.95	3.99	4.02	4.05
9	2.26	2.79	2.99	3.13	3.24	3.33	3.41	3.48	3.54	3.59	3.64	3.68	3.72	3.76	3.80	3.83	3.86	3.89	3.92
10	2.23	2.74	2.93	3.07	3.17	3.26	3.33	3.40	3.45	3.51	3.55	3.59	3.63	3.67	3.70	3.73	3.76	3.79	3.82
11	2.20	2.70	2.88	3.01	3.12	3.20	3.27	3.33	3.39	3.44	3.48	3.52	3.56	3.60	3.63	3.66	3.69	3.71	3.74
12	2.18	2.67	2.85	2.97	3.07	3.15	3.22	3.28	3.33	3.38	3.42	3.46	3.50	3.53	3.57	3.59	3.62	3.65	3.67
13	2.16	2.64	2.81	2.94	3.03	3.11	3.18	3.24	3.29	3.33	3.38	3.42	3.45	3.48	3.51	3.54	3.57	3.59	3.62
14	2.14	2.62	2.79	2.91	3.00	3.08	3.14	3.20	3.25	3.30	3.34	3.37	3.41	3.44	3.47	3.50	3.52	3.55	3.57
15	2.13	2.60	2.76	2.88	2.97	3.05	3.11	3.17	3.22	3.26	3.30	3.34	3.37	3.40	3.43	3.46	3.49	3.51	3.53
16	2.12	2.58	2.74	2.86	2.95	3.02	3.09	3.14	3.19	3.23	3.27	3.31	3.34	3.37	3.40	3.43	3.45	3.48	3.50
17	2.11	2.57	2.73	2.84	2.93	3.00	3.06	3.12	3.16	3.21	3.25	3.28	3.31	3.34	3.37	3.40	3.42	3.45	3.47
18	2.10	2.55	2.71	2.82	2.91	2.98	3.04	3.10	3.14	3.18	3.22	3.26	3.29	3.32	3.35	3.37	3.40	3.42	3.44
19	2.09	2.54	2.70	2.81	2.89	2.96	3.02	3.08	3.12	3.16	3.20	3.24	3.27	3.30	3.32	3.35	3.37	3.40	3.42
20	2.09	2.53	2.68	2.79	2.88	2.95	3.01	3.06	3.11	3.15	3.18	3.22	3.25	3.28	3.30	3.33	3.35	3.38	3.40
24	2.06	2.50	2.65	2.75	2.83	2.90	2.96	3.01	3.05	3.09	3.13	3.16	3.19	3.22	3.24	3.27	3.29	3.31	3.33
30	2.04	2.47	2.61	2.71	2.79	2.85	2.91	2.96	3.00	3.04	3.07	3.10	3.13	3.16	3.18	3.20	3.23	3.25	3.26
40	2.02	2.43	2.57	2.67	2.75	2.81	2.86	2.91	2.95	2.98	3.01	3.04	3.07	3.10	3.12	3.14	3.16	3.18	3.20
60	2.00	2.40	2.54	2.63	2.70	2.76	2.81	2.86	2.90	2.93	2.96	2.99	3.01	3.04	3.06	3.08	3.10	3.12	3.14
120	1.98	2.37	2.50	2.59	2.66	2.72	2.77	2.81	2.85	2.88	2.91	2.93	2.96	2.98	3.00	3.02	3.04	3.06	3.08
∞	1.96	2.34	2.47	2.56	2.62	2.68	2.72	2.76	2.80	2.83	2.86	2.88	2.90	2.93	2.95	2.97	2.98	3.00	3.02

Table B5 (continued). Balanced ANOM Critical Values $h(\alpha; k, \nu)$

Level of Significance $= 0.01$

ν	2	3	4	5	6	7	8	9	10	11	12	13	14	15	16	17	18	19	20
									Number of Means Being Compared, k										
1	63.7	95.7	110.	121.	129.	135.	140.	145.	149.	152.	155.	158.	160.	163.	165.	167.	169.	171.	172.
2	9.92	13.4	15.0	16.1	17.0	17.7	18.3	18.8	19.3	19.7	20.0	20.4	20.7	20.9	21.2	21.4	21.6	21.9	22.1
3	5.84	7.51	8.22	8.73	9.13	9.46	9.74	9.98	10.2	10.4	10.6	10.7	10.9	11.0	11.1	11.2	11.3	11.4	11.5
4	4.60	5.74	6.20	6.54	6.81	7.03	7.22	7.38	7.52	7.65	7.77	7.88	7.97	8.07	8.15	8.23	8.30	8.37	8.44
5	4.03	4.93	5.29	5.55	5.75	5.92	6.07	6.19	6.30	6.40	6.50	6.58	6.66	6.73	6.79	6.86	6.91	6.97	7.02
6	3.71	4.48	4.77	4.98	5.16	5.30	5.42	5.52	5.62	5.70	5.78	5.85	5.91	5.97	6.03	6.08	6.13	6.18	6.22
7	3.50	4.19	4.44	4.63	4.78	4.90	5.01	5.10	5.18	5.25	5.32	5.38	5.44	5.49	5.54	5.59	5.63	5.67	5.71
8	3.36	3.98	4.21	4.38	4.52	4.63	4.72	4.80	4.88	4.94	5.01	5.06	5.11	5.16	5.20	5.24	5.28	5.32	5.36
9	3.25	3.84	4.05	4.20	4.33	4.43	4.51	4.59	4.66	4.72	4.78	4.83	4.87	4.92	4.96	5.00	5.03	5.07	5.10
10	3.17	3.73	3.92	4.07	4.18	4.28	4.36	4.43	4.49	4.55	4.60	4.65	4.69	4.73	4.77	4.81	4.84	4.87	4.90
11	3.11	3.64	3.83	3.96	4.07	4.16	4.23	4.30	4.36	4.41	4.46	4.51	4.55	4.59	4.62	4.66	4.69	4.72	4.75
12	3.05	3.57	3.75	3.87	3.98	4.06	4.13	4.20	4.25	4.31	4.35	4.39	4.43	4.47	4.50	4.54	4.57	4.59	4.62
13	3.01	3.51	3.68	3.80	3.90	3.98	4.05	4.11	4.17	4.22	4.26	4.30	4.34	4.37	4.41	4.44	4.47	4.49	4.52
14	2.98	3.46	3.63	3.74	3.84	3.92	3.98	4.04	4.09	4.14	4.18	4.22	4.26	4.29	4.32	4.35	4.38	4.41	4.43
15	2.95	3.42	3.58	3.69	3.78	3.86	3.93	3.98	4.03	4.08	4.12	4.16	4.19	4.22	4.26	4.28	4.31	4.34	4.36
16	2.92	3.38	3.54	3.65	3.74	3.81	3.88	3.93	3.98	4.02	4.06	4.10	4.13	4.17	4.20	4.22	4.25	4.27	4.30
17	2.90	3.35	3.50	3.61	3.70	3.77	3.83	3.89	3.93	3.98	4.02	4.05	4.08	4.12	4.14	4.17	4.20	4.22	4.24
18	2.88	3.33	3.47	3.58	3.66	3.73	3.79	3.85	3.89	3.94	3.97	4.01	4.04	4.07	4.10	4.12	4.15	4.17	4.20
19	2.86	3.30	3.45	3.55	3.63	3.70	3.76	3.81	3.86	3.90	3.94	3.97	4.00	4.03	4.06	4.08	4.11	4.13	4.15
20	2.85	3.28	3.42	3.53	3.61	3.67	3.73	3.78	3.83	3.87	3.90	3.94	3.97	4.00	4.02	4.05	4.07	4.09	4.12
24	2.80	3.21	3.35	3.44	3.52	3.58	3.64	3.69	3.73	3.77	3.80	3.83	3.86	3.89	3.91	3.94	3.96	3.98	4.00
30	2.75	3.15	3.28	3.37	3.44	3.50	3.55	3.59	3.63	3.67	3.70	3.73	3.76	3.78	3.81	3.83	3.85	3.87	3.89
40	2.70	3.09	3.21	3.29	3.36	3.42	3.46	3.51	3.54	3.58	3.61	3.63	3.66	3.68	3.70	3.72	3.74	3.76	3.78
60	2.66	3.03	3.14	3.22	3.28	3.34	3.38	3.42	3.45	3.49	3.51	3.54	3.56	3.58	3.61	3.62	3.64	3.66	3.68
120	2.62	2.97	3.08	3.15	3.21	3.26	3.30	3.34	3.37	3.40	3.42	3.45	3.47	3.49	3.51	3.53	3.54	3.56	3.58
∞	2.58	2.91	3.01	3.08	3.14	3.19	3.22	3.26	3.29	3.32	3.34	3.36	3.38	3.40	3.42	3.44	3.45	3.47	3.48

Table B5 (continued). Balanced ANOM Critical Values $h(\alpha; k, \nu)$

Level of Significance $= 0.001$

ν	Number of Means Being Compared, k																		
	2	3	4	5	6	7	8	9	10	11	12	13	14	15	16	17	18	19	20
1	637.	957.	1103.	1207.	1285.	1349.	1401.	1446.	1485.	1520.	1551.	1579.	1605.	1628.	1650.	1670.	1689.	1707.	1724.
2	31.6	42.7	47.7	51.2	54.0	56.2	58.1	59.7	61.1	62.4	63.5	64.6	65.5	66.4	67.2	67.9	68.6	69.3	69.9
3	12.9	16.5	18.0	19.1	20.0	20.7	21.3	21.8	22.3	22.7	23.0	23.4	23.7	24.0	24.2	24.5	24.7	24.9	25.1
4	8.61	10.6	11.4	12.0	12.5	12.8	13.2	13.5	13.7	13.9	14.2	14.3	14.5	14.7	14.8	15.0	15.1	15.2	15.3
5	6.87	8.25	8.79	9.19	9.51	9.77	10.0	10.2	10.4	10.5	10.7	10.8	10.9	11.0	11.1	11.2	11.3	11.4	11.5
6	5.96	7.04	7.45	7.75	7.99	8.20	8.37	8.52	8.66	8.78	8.89	8.99	9.08	9.17	9.25	9.33	9.40	9.47	9.53
7	5.41	6.31	6.65	6.89	7.09	7.25	7.40	7.52	7.63	7.73	7.82	7.90	7.98	8.05	8.12	8.18	8.24	8.30	8.35
8	5.04	5.83	6.12	6.33	6.49	6.63	6.75	6.86	6.96	7.04	7.12	7.19	7.26	7.32	7.38	7.43	7.48	7.53	7.58
9	4.78	5.49	5.74	5.93	6.07	6.20	6.30	6.40	6.48	6.56	6.63	6.69	6.75	6.80	6.85	6.90	6.95	6.99	7.03
10	4.59	5.24	5.46	5.63	5.76	5.87	5.97	6.05	6.13	6.20	6.26	6.32	6.37	6.42	6.47	6.51	6.55	6.59	6.62
11	4.44	5.05	5.25	5.40	5.52	5.63	5.71	5.79	5.86	5.92	5.98	6.03	6.08	6.13	6.17	6.21	6.25	6.28	6.31
12	4.32	4.89	5.08	5.22	5.34	5.43	5.51	5.58	5.65	5.71	5.76	5.81	5.85	5.89	5.93	5.97	6.01	6.04	6.07
13	4.22	4.77	4.94	5.08	5.18	5.27	5.35	5.42	5.48	5.53	5.58	5.63	5.67	5.71	5.74	5.78	5.81	5.84	5.87
14	4.14	4.66	4.83	4.96	5.06	5.14	5.21	5.28	5.33	5.38	5.43	5.48	5.52	5.55	5.59	5.62	5.65	5.68	5.71
15	4.07	4.57	4.74	4.86	4.95	5.03	5.10	5.16	5.21	5.26	5.31	5.35	5.39	5.42	5.46	5.49	5.52	5.54	5.57
16	4.01	4.50	4.66	4.77	4.86	4.94	5.00	5.06	5.11	5.16	5.20	5.24	5.28	5.31	5.34	5.37	5.40	5.43	5.45
17	3.97	4.44	4.59	4.70	4.78	4.86	4.92	4.98	5.03	5.07	5.11	5.15	5.19	5.22	5.25	5.28	5.30	5.33	5.35
18	3.92	4.38	4.53	4.63	4.72	4.79	4.85	4.90	4.95	4.99	5.03	5.07	5.10	5.14	5.17	5.19	5.22	5.24	5.27
19	3.88	4.33	4.47	4.57	4.66	4.73	4.79	4.84	4.88	4.93	4.96	5.00	5.03	5.06	5.09	5.12	5.14	5.17	5.19
20	3.85	4.29	4.42	4.52	4.60	4.67	4.73	4.78	4.83	4.87	4.90	4.94	4.97	5.00	5.03	5.05	5.08	5.10	5.12
24	3.75	4.16	4.28	4.37	4.44	4.51	4.56	4.61	4.65	4.68	4.72	4.75	4.78	4.81	4.83	4.85	4.88	4.90	4.92
30	3.65	4.03	4.14	4.23	4.29	4.35	4.40	4.44	4.48	4.51	4.54	4.57	4.60	4.62	4.65	4.67	4.69	4.71	4.72
40	3.55	3.91	4.01	4.09	4.15	4.20	4.25	4.28	4.32	4.35	4.38	4.40	4.43	4.45	4.47	4.49	4.51	4.52	4.54
60	3.46	3.79	3.89	3.96	4.01	4.06	4.10	4.14	4.17	4.20	4.22	4.24	4.27	4.29	4.30	4.32	4.34	4.35	4.37
120	3.37	3.68	3.77	3.84	3.89	3.93	3.96	4.00	4.02	4.05	4.07	4.09	4.11	4.13	4.15	4.17	4.18	4.19	4.21
∞	3.29	3.58	3.66	3.72	3.76	3.80	3.84	3.86	3.89	3.91	3.93	3.95	3.97	3.99	4.00	4.02	4.03	4.04	4.06

Table B6. Unbalanced ANOM Critical Values $m(\alpha; k, \nu)$
Level of Significance = 0.1

							Number of Means Being Compared, k												
ν	2	3	4	5	6	7	8	9	10	11	12	13	14	15	16	17	18	19	20
1	8.96	10.5	11.6	12.5	13.2	13.7	14.2	14.6	15.0	15.3	15.6	15.8	16.1	16.3	16.5	16.7	16.9	17.1	17.2
2	3.83	4.38	4.77	5.06	5.30	5.50	5.67	5.82	5.96	6.08	6.18	6.28	6.37	6.45	6.53	6.60	6.67	6.74	6.80
3	2.99	3.37	3.64	3.84	4.01	4.15	4.27	4.38	4.47	4.55	4.63	4.70	4.76	4.82	4.88	4.93	4.98	5.02	5.07
4	2.66	2.98	3.20	3.37	3.51	3.62	3.72	3.81	3.89	3.96	4.02	4.08	4.13	4.18	4.23	4.27	4.31	4.35	4.38
5	2.49	2.77	2.96	3.12	3.24	3.34	3.43	3.51	3.58	3.64	3.69	3.75	3.79	3.84	3.88	3.92	3.95	3.99	4.02
6	2.38	2.64	2.82	2.96	3.07	3.17	3.25	3.32	3.38	3.44	3.49	3.54	3.58	3.62	3.66	3.70	3.73	3.76	3.79
7	2.31	2.56	2.73	2.86	2.96	3.05	3.13	3.19	3.25	3.31	3.35	3.40	3.44	3.48	3.51	3.55	3.58	3.61	3.63
8	2.26	2.49	2.66	2.78	2.88	2.96	3.04	3.10	3.16	3.21	3.26	3.30	3.34	3.37	3.41	3.44	3.47	3.50	3.52
9	2.22	2.45	2.60	2.72	2.82	2.90	2.97	3.03	3.09	3.13	3.18	3.22	3.26	3.29	3.32	3.35	3.38	3.41	3.44
10	2.19	2.41	2.56	2.68	2.77	2.85	2.92	2.98	3.03	3.08	3.12	3.16	3.20	3.23	3.26	3.29	3.32	3.34	3.37
11	2.17	2.38	2.53	2.64	2.73	2.81	2.88	2.93	2.98	3.03	3.07	3.11	3.15	3.18	3.21	3.24	3.26	3.29	3.31
12	2.15	2.36	2.50	2.61	2.70	2.78	2.84	2.90	2.95	2.99	3.03	3.07	3.10	3.14	3.17	3.19	3.22	3.24	3.27
13	2.13	2.34	2.48	2.59	2.67	2.75	2.81	2.87	2.91	2.96	3.00	3.04	3.07	3.10	3.13	3.16	3.18	3.21	3.23
14	2.12	2.32	2.46	2.57	2.65	2.72	2.79	2.84	2.89	2.93	2.97	3.01	3.04	3.07	3.10	3.13	3.15	3.17	3.20
15	2.11	2.31	2.44	2.55	2.63	2.70	2.76	2.82	2.87	2.91	2.95	2.98	3.01	3.04	3.07	3.10	3.12	3.15	3.17
16	2.10	2.29	2.43	2.53	2.62	2.69	2.75	2.80	2.85	2.89	2.93	2.96	2.99	3.02	3.05	3.08	3.10	3.12	3.15
17	2.09	2.28	2.42	2.52	2.60	2.67	2.73	2.78	2.83	2.87	2.91	2.94	2.97	3.00	3.03	3.06	3.08	3.10	3.12
18	2.08	2.27	2.41	2.51	2.59	2.66	2.72	2.77	2.81	2.85	2.89	2.92	2.96	2.99	3.01	3.04	3.06	3.08	3.10
19	2.07	2.26	2.40	2.50	2.58	2.64	2.70	2.75	2.80	2.84	2.88	2.91	2.94	2.97	3.00	3.02	3.04	3.07	3.09
20	2.07	2.26	2.39	2.49	2.57	2.63	2.69	2.74	2.79	2.83	2.86	2.90	2.93	2.96	2.98	3.01	3.03	3.05	3.07
24	2.05	2.23	2.36	2.46	2.53	2.60	2.66	2.70	2.75	2.79	2.82	2.85	2.88	2.91	2.94	2.96	2.98	3.01	3.03
30	2.03	2.21	2.33	2.43	2.50	2.57	2.62	2.67	2.71	2.75	2.78	2.81	2.84	2.87	2.89	2.92	2.94	2.96	2.98
40	2.01	2.18	2.30	2.40	2.47	2.53	2.58	2.63	2.67	2.71	2.74	2.77	2.80	2.82	2.85	2.87	2.89	2.91	2.93
60	1.99	2.16	2.28	2.37	2.44	2.50	2.55	2.59	2.63	2.67	2.70	2.73	2.76	2.78	2.80	2.83	2.85	2.87	2.88
120	1.97	2.14	2.25	2.34	2.41	2.47	2.52	2.56	2.60	2.63	2.66	2.69	2.72	2.74	2.76	2.78	2.80	2.82	2.84
∞	1.95	2.11	2.23	2.31	2.38	2.43	2.48	2.52	2.56	2.59	2.62	2.65	2.67	2.70	2.72	2.74	2.76	2.77	2.79

Table B6 (continued).   Unbalanced ANOM Critical Values $m(\alpha; k, \nu)$
Level of Significance $= 0.05$

ν		Number of Means Being Compared, k																	
	2	3	4	5	6	7	8	9	10	11	12	13	14	15	16	17	18	19	20
1	18.0	21.1	23.4	25.0	26.4	27.5	28.5	29.3	30.0	30.6	31.3	31.8	32.3	32.7	33.1	33.5	33.9	34.2	34.5
2	5.57	6.34	6.89	7.31	7.65	7.93	8.17	8.38	8.57	8.74	8.89	9.03	9.16	9.28	9.39	9.49	9.59	9.68	9.77
3	3.96	4.43	4.76	5.02	5.23	5.41	5.56	5.69	5.81	5.92	6.01	6.10	6.18	6.26	6.33	6.39	6.45	6.51	6.57
4	3.38	3.74	4.00	4.20	4.37	4.50	4.62	4.72	4.82	4.90	4.97	5.04	5.11	5.17	5.22	5.27	5.32	5.37	5.41
5	3.09	3.40	3.62	3.79	3.93	4.04	4.14	4.23	4.31	4.38	4.45	4.51	4.56	4.61	4.66	4.70	4.74	4.78	4.82
6	2.92	3.19	3.39	3.54	3.66	3.77	3.86	3.94	4.01	4.07	4.13	4.18	4.23	4.28	4.32	4.36	4.39	4.43	4.46
7	2.80	3.06	3.24	3.38	3.49	3.59	3.67	3.74	3.80	3.86	3.92	3.96	4.01	4.05	4.09	4.13	4.16	4.19	4.22
8	2.72	2.96	3.13	3.26	3.36	3.45	3.53	3.60	3.66	3.71	3.76	3.81	3.85	3.89	3.93	3.96	3.99	4.02	4.05
9	2.66	2.89	3.05	3.17	3.27	3.36	3.43	3.49	3.55	3.60	3.65	3.69	3.73	3.77	3.80	3.84	3.87	3.90	3.92
10	2.61	2.83	2.98	3.10	3.20	3.28	3.35	3.41	3.47	3.52	3.56	3.60	3.64	3.68	3.71	3.74	3.77	3.80	3.82
11	2.57	2.78	2.93	3.05	3.14	3.22	3.29	3.35	3.40	3.45	3.49	3.53	3.57	3.60	3.63	3.66	3.69	3.72	3.74
12	2.54	2.75	2.89	3.00	3.09	3.17	3.24	3.29	3.34	3.39	3.43	3.47	3.51	3.54	3.57	3.60	3.63	3.65	3.68
13	2.51	2.72	2.86	2.97	3.06	3.13	3.19	3.25	3.30	3.34	3.39	3.42	3.46	3.49	3.52	3.55	3.57	3.60	3.62
14	2.49	2.69	2.83	2.94	3.02	3.09	3.16	3.21	3.26	3.30	3.34	3.38	3.41	3.45	3.48	3.50	3.53	3.55	3.58
15	2.47	2.67	2.81	2.91	2.99	3.06	3.13	3.18	3.23	3.27	3.31	3.35	3.38	3.41	3.44	3.46	3.49	3.51	3.54
16	2.46	2.65	2.78	2.89	2.97	3.04	3.10	3.15	3.20	3.24	3.28	3.31	3.35	3.38	3.40	3.43	3.46	3.48	3.50
17	2.44	2.63	2.77	2.87	2.95	3.02	3.08	3.13	3.17	3.21	3.25	3.29	3.32	3.35	3.38	3.40	3.43	3.45	3.47
18	2.43	2.62	2.75	2.85	2.93	3.00	3.05	3.11	3.15	3.19	3.23	3.26	3.29	3.32	3.35	3.38	3.40	3.42	3.44
19	2.42	2.61	2.73	2.83	2.91	2.98	3.04	3.09	3.13	3.17	3.21	3.24	3.27	3.30	3.33	3.35	3.38	3.40	3.42
20	2.41	2.59	2.72	2.82	2.90	2.96	3.02	3.07	3.11	3.15	3.19	3.22	3.25	3.28	3.31	3.33	3.36	3.38	3.40
24	2.38	2.56	2.68	2.77	2.85	2.91	2.97	3.02	3.06	3.10	3.13	3.16	3.19	3.22	3.25	3.27	3.29	3.31	3.33
30	2.35	2.52	2.64	2.73	2.80	2.87	2.92	2.96	3.00	3.04	3.07	3.11	3.13	3.16	3.18	3.21	3.23	3.25	3.27
40	2.32	2.49	2.60	2.69	2.76	2.82	2.87	2.91	2.95	2.99	3.02	3.05	3.08	3.10	3.12	3.14	3.17	3.18	3.20
60	2.29	2.45	2.56	2.65	2.72	2.77	2.82	2.86	2.90	2.93	2.96	2.99	3.02	3.04	3.06	3.08	3.10	3.12	3.14
120	2.26	2.42	2.53	2.61	2.67	2.73	2.77	2.81	2.85	2.88	2.91	2.94	2.96	2.98	3.01	3.02	3.04	3.06	3.08
∞	2.24	2.39	2.49	2.57	2.63	2.68	2.73	2.77	2.80	2.83	2.86	2.88	2.91	2.93	2.95	2.97	2.98	3.00	3.02

Table B6 (continued). Unbalanced ANOM Critical Values $m(\alpha; k, \nu)$
Level of Significance = 0.01

ν	Number of Means Being Compared, k																		
	2	3	4	5	6	7	8	9	10	11	12	13	14	15	16	17	18	19	20
1	90.0	106.	117.	125.	132.	138.	142.	146.	150.	153.	156.	159.	161.	164.	166.	168.	170.	171.	173.
2	12.7	14.4	15.7	16.6	17.4	18.0	18.5	19.0	19.4	19.8	20.1	20.5	20.8	21.0	21.3	21.5	21.7	21.9	22.1
3	7.13	7.91	8.48	8.92	9.28	9.58	9.84	10.1	10.3	10.4	10.6	10.8	10.9	11.0	11.2	11.3	11.4	11.5	11.6
4	5.46	5.99	6.36	6.66	6.90	7.10	7.27	7.43	7.57	7.69	7.80	7.91	8.00	8.09	8.17	8.25	8.32	8.39	8.45
5	4.70	5.11	5.40	5.63	5.81	5.97	6.11	6.23	6.33	6.43	6.52	6.60	6.67	6.74	6.81	6.87	6.93	6.98	7.03
6	4.27	4.61	4.86	5.05	5.20	5.33	5.45	5.55	5.64	5.72	5.80	5.86	5.93	5.99	6.04	6.09	6.14	6.18	6.23
7	4.00	4.30	4.51	4.68	4.81	4.93	5.03	5.12	5.20	5.27	5.34	5.39	5.45	5.50	5.55	5.60	5.64	5.68	5.72
8	3.81	4.08	4.27	4.42	4.55	4.65	4.74	4.82	4.89	4.96	5.02	5.07	5.12	5.17	5.21	5.25	5.29	5.33	5.36
9	3.67	3.92	4.10	4.24	4.35	4.45	4.53	4.61	4.67	4.73	4.79	4.84	4.88	4.92	4.96	5.00	5.04	5.07	5.10
10	3.57	3.80	3.97	4.10	4.20	4.29	4.37	4.44	4.50	4.56	4.61	4.66	4.70	4.74	4.78	4.81	4.84	4.88	4.91
11	3.48	3.71	3.87	3.99	4.09	4.17	4.25	4.31	4.37	4.42	4.47	4.51	4.55	4.59	4.63	4.66	4.69	4.72	4.75
12	3.42	3.63	3.78	3.90	4.00	4.08	4.15	4.21	4.26	4.31	4.36	4.40	4.44	4.48	4.51	4.54	4.57	4.60	4.63
13	3.36	3.57	3.71	3.83	3.92	4.00	4.06	4.12	4.18	4.22	4.27	4.31	4.34	4.38	4.41	4.44	4.47	4.50	4.52
14	3.32	3.52	3.66	3.77	3.85	3.93	3.99	4.05	4.10	4.15	4.19	4.23	4.26	4.30	4.33	4.36	4.39	4.41	4.44
15	3.28	3.47	3.61	3.71	3.80	3.87	3.93	3.99	4.04	4.08	4.12	4.16	4.20	4.23	4.26	4.29	4.31	4.34	4.36
16	3.25	3.43	3.57	3.67	3.75	3.82	3.88	3.94	3.99	4.03	4.07	4.11	4.14	4.17	4.20	4.23	4.25	4.28	4.30
17	3.22	3.40	3.53	3.63	3.71	3.78	3.84	3.89	3.94	3.98	4.02	4.06	4.09	4.12	4.15	4.17	4.20	4.22	4.25
18	3.19	3.37	3.50	3.60	3.68	3.74	3.80	3.85	3.90	3.94	3.98	4.01	4.04	4.07	4.10	4.13	4.15	4.18	4.20
19	3.17	3.35	3.47	3.57	3.65	3.71	3.77	3.82	3.86	3.90	3.94	3.97	4.01	4.03	4.06	4.09	4.11	4.13	4.16
20	3.15	3.32	3.45	3.54	3.62	3.68	3.74	3.79	3.83	3.87	3.91	3.94	3.97	4.00	4.03	4.05	4.07	4.10	4.12
24	3.09	3.25	3.37	3.46	3.53	3.59	3.64	3.69	3.73	3.77	3.80	3.83	3.86	3.89	3.91	3.94	3.96	3.98	4.00
30	3.03	3.18	3.29	3.38	3.45	3.50	3.55	3.60	3.64	3.67	3.70	3.73	3.76	3.78	3.81	3.83	3.85	3.87	3.89
40	2.97	3.12	3.22	3.30	3.37	3.42	3.47	3.51	3.54	3.58	3.61	3.63	3.66	3.68	3.71	3.73	3.74	3.76	3.78
60	2.91	3.05	3.15	3.23	3.29	3.34	3.38	3.42	3.46	3.49	3.51	3.54	3.56	3.59	3.61	3.63	3.64	3.66	3.68
120	2.86	2.99	3.09	3.16	3.21	3.26	3.30	3.34	3.37	3.40	3.43	3.45	3.47	3.49	3.51	3.53	3.55	3.56	3.58
∞	2.81	2.93	3.02	3.09	3.14	3.19	3.23	3.26	3.29	3.32	3.34	3.36	3.38	3.40	3.42	3.44	3.45	3.47	3.48

Table B6 (continued). Unbalanced ANOM Critical Values $m(\alpha; k, \nu)$

Level of Significance = 0.001

ν	2	3	4	5	6	7	8	9	10	11	12	13	14	15	16	17	18	19	20
1	900.	1058.	1169.	1253.	1320.	1375.	1423.	1464.	1501.	1533.	1562.	1589.	1614.	1636.	1657.	1677.	1695.	1712.	1728.
2	40.4	45.8	49.7	52.6	55.0	57.1	58.8	60.3	61.6	62.8	63.9	64.9	65.8	66.6	67.4	68.2	68.8	69.5	70.1
3	15.7	17.3	18.6	19.5	20.3	20.9	21.5	22.0	22.4	22.8	23.1	23.5	23.8	24.1	24.3	24.6	24.8	25.0	25.2
4	10.1	11.0	11.7	12.2	12.6	13.0	13.3	13.5	13.8	14.0	14.2	14.4	14.6	14.7	14.9	15.0	15.1	15.3	15.4
5	7.88	8.50	8.95	9.31	9.60	9.85	10.1	10.2	10.4	10.6	10.7	10.8	11.0	11.1	11.2	11.3	11.4	11.4	11.5
6	6.74	7.21	7.56	7.83	8.06	8.25	8.41	8.56	8.69	8.80	8.91	9.01	9.10	9.19	9.27	9.34	9.41	9.48	9.54
7	6.05	6.44	6.73	6.95	7.13	7.29	7.43	7.55	7.65	7.75	7.84	7.92	8.00	8.07	8.13	8.20	8.25	8.31	8.36
8	5.60	5.93	6.18	6.37	6.53	6.66	6.78	6.88	6.97	7.06	7.13	7.20	7.27	7.33	7.39	7.44	7.49	7.54	7.58
9	5.28	5.58	5.79	5.96	6.10	6.22	6.32	6.41	6.50	6.57	6.64	6.70	6.76	6.81	6.86	6.91	6.95	7.00	7.04
10	5.04	5.31	5.51	5.66	5.78	5.89	5.99	6.07	6.14	6.21	6.27	6.33	6.38	6.43	6.47	6.52	6.56	6.59	6.63
11	4.86	5.11	5.29	5.43	5.54	5.64	5.73	5.80	5.87	5.93	5.99	6.04	6.09	6.13	6.17	6.21	6.25	6.29	6.32
12	4.71	4.94	5.11	5.24	5.35	5.44	5.52	5.59	5.66	5.71	5.77	5.81	5.86	5.90	5.94	5.98	6.01	6.04	6.07
13	4.59	4.81	4.97	5.10	5.20	5.28	5.36	5.42	5.48	5.54	5.59	5.63	5.67	5.71	5.75	5.78	5.82	5.85	5.87
14	4.50	4.71	4.86	4.97	5.07	5.15	5.22	5.28	5.34	5.39	5.44	5.48	5.52	5.56	5.59	5.62	5.65	5.68	5.71
15	4.41	4.61	4.76	4.87	4.96	5.04	5.11	5.17	5.22	5.27	5.31	5.35	5.39	5.43	5.46	5.49	5.52	5.55	5.57
16	4.34	4.54	4.68	4.78	4.87	4.95	5.01	5.07	5.12	5.16	5.21	5.25	5.28	5.32	5.35	5.38	5.41	5.43	5.46
17	4.28	4.47	4.60	4.71	4.79	4.86	4.93	4.98	5.03	5.08	5.12	5.15	5.19	5.22	5.25	5.28	5.31	5.33	5.36
18	4.23	4.41	4.54	4.64	4.72	4.79	4.85	4.91	4.95	5.00	5.04	5.07	5.11	5.14	5.17	5.20	5.22	5.25	5.27
19	4.19	4.36	4.49	4.58	4.66	4.73	4.79	4.84	4.89	4.93	4.97	5.00	5.04	5.07	5.09	5.12	5.15	5.17	5.19
20	4.14	4.32	4.44	4.53	4.61	4.68	4.73	4.78	4.83	4.87	4.91	4.94	4.97	5.00	5.03	5.06	5.08	5.10	5.13
24	4.02	4.18	4.29	4.38	4.45	4.51	4.56	4.61	4.65	4.69	4.72	4.75	4.78	4.81	4.83	4.86	4.88	4.90	4.92
30	3.90	4.05	4.15	4.23	4.30	4.35	4.40	4.44	4.48	4.51	4.54	4.57	4.60	4.62	4.65	4.67	4.69	4.71	4.73
40	3.79	3.92	4.02	4.09	4.15	4.20	4.25	4.29	4.32	4.35	4.38	4.40	4.43	4.45	4.47	4.49	4.51	4.53	4.54
60	3.68	3.81	3.89	3.96	4.02	4.06	4.10	4.14	4.17	4.20	4.22	4.24	4.27	4.29	4.31	4.32	4.34	4.36	4.37
120	3.58	3.69	3.78	3.84	3.89	3.93	3.97	4.00	4.03	4.05	4.07	4.10	4.11	4.13	4.15	4.17	4.18	4.19	4.21
∞	3.48	3.59	3.66	3.72	3.76	3.80	3.84	3.87	3.89	3.91	3.93	3.95	3.97	3.99	4.00	4.02	4.03	4.04	4.06

Number of Means Being Compared, k

Table B7. Critical Values of F Distributions, $F(\alpha; \nu_1, \nu_2)$

$$\alpha = 0.1$$

ν_2	ν_1 1	2	3	4	5	6	7	8	9	10	12	14	16	18	20	24	30	40	∞
1	39.863	49.500	53.593	55.833	57.240	58.204	58.906	59.439	59.858	60.195	60.705	61.073	61.350	61.566	61.740	62.002	62.265	62.529	63.328
2	8.526	9.000	9.162	9.243	9.293	9.326	9.349	9.367	9.381	9.392	9.408	9.420	9.429	9.436	9.441	9.450	9.458	9.466	9.491
3	5.538	5.462	5.391	5.343	5.309	5.285	5.266	5.252	5.240	5.230	5.216	5.205	5.196	5.190	5.184	5.176	5.168	5.160	5.134
4	4.545	4.325	4.191	4.107	4.051	4.010	3.979	3.955	3.936	3.920	3.896	3.878	3.864	3.853	3.844	3.831	3.817	3.804	3.761
5	4.060	3.780	3.619	3.520	3.453	3.404	3.368	3.339	3.316	3.297	3.268	3.247	3.230	3.217	3.207	3.191	3.174	3.157	3.105
6	3.776	3.463	3.289	3.181	3.107	3.055	3.014	2.983	2.958	2.937	2.905	2.881	2.863	2.848	2.836	2.818	2.800	2.781	2.722
7	3.589	3.257	3.074	2.961	2.883	2.827	2.785	2.752	2.725	2.703	2.668	2.643	2.623	2.607	2.595	2.575	2.555	2.535	2.471
8	3.458	3.113	2.924	2.806	2.726	2.668	2.624	2.589	2.561	2.538	2.502	2.475	2.454	2.438	2.425	2.404	2.383	2.361	2.293
9	3.360	3.006	2.813	2.693	2.611	2.551	2.505	2.469	2.440	2.416	2.379	2.351	2.329	2.312	2.298	2.277	2.255	2.232	2.159
10	3.285	2.924	2.728	2.605	2.522	2.461	2.414	2.377	2.347	2.323	2.284	2.255	2.233	2.215	2.201	2.178	2.155	2.132	2.055
11	3.225	2.860	2.660	2.536	2.451	2.389	2.342	2.304	2.273	2.248	2.209	2.179	2.156	2.138	2.123	2.100	2.076	2.052	1.972
12	3.177	2.807	2.606	2.480	2.394	2.331	2.283	2.245	2.214	2.188	2.147	2.117	2.094	2.075	2.060	2.036	2.012	1.986	1.904
13	3.136	2.763	2.560	2.434	2.347	2.283	2.234	2.195	2.164	2.138	2.097	2.066	2.042	2.023	2.007	1.983	1.958	1.931	1.846
14	3.102	2.726	2.522	2.395	2.307	2.243	2.193	2.154	2.122	2.095	2.054	2.022	1.998	1.979	1.962	1.938	1.912	1.885	1.797
15	3.073	2.695	2.490	2.361	2.273	2.208	2.158	2.119	2.086	2.059	2.017	1.985	1.961	1.941	1.924	1.899	1.873	1.845	1.755
16	3.048	2.668	2.462	2.333	2.244	2.178	2.128	2.088	2.055	2.028	1.985	1.953	1.928	1.908	1.891	1.866	1.839	1.811	1.718
17	3.026	2.645	2.437	2.308	2.218	2.152	2.102	2.061	2.028	2.001	1.958	1.925	1.900	1.879	1.862	1.836	1.809	1.781	1.686
18	3.007	2.624	2.416	2.286	2.196	2.130	2.079	2.038	2.005	1.977	1.933	1.900	1.875	1.854	1.837	1.810	1.783	1.754	1.657
19	2.990	2.606	2.397	2.266	2.176	2.109	2.058	2.017	1.984	1.956	1.912	1.878	1.852	1.831	1.814	1.787	1.759	1.730	1.631
20	2.975	2.589	2.380	2.249	2.158	2.091	2.040	1.999	1.965	1.937	1.892	1.859	1.833	1.811	1.794	1.767	1.738	1.708	1.607
24	2.927	2.538	2.327	2.195	2.103	2.035	1.983	1.941	1.906	1.877	1.832	1.797	1.770	1.748	1.730	1.702	1.672	1.641	1.533
30	2.881	2.489	2.276	2.142	2.049	1.980	1.927	1.884	1.849	1.819	1.773	1.737	1.709	1.686	1.667	1.638	1.606	1.573	1.456
40	2.835	2.440	2.226	2.091	1.997	1.927	1.873	1.829	1.793	1.763	1.715	1.678	1.649	1.625	1.605	1.574	1.541	1.506	1.377
60	2.791	2.393	2.177	2.041	1.946	1.875	1.819	1.775	1.738	1.707	1.657	1.619	1.589	1.564	1.543	1.511	1.476	1.437	1.291
120	2.748	2.347	2.130	1.992	1.896	1.824	1.767	1.722	1.684	1.652	1.601	1.562	1.530	1.504	1.482	1.447	1.409	1.368	1.193
∞	2.706	2.303	2.084	1.945	1.847	1.774	1.717	1.670	1.632	1.599	1.546	1.505	1.471	1.444	1.421	1.383	1.342	1.295	1.000

Table B7 (continued). Critical Values of F Distributions, $F(\alpha; \nu_1, \nu_2)$

$$\alpha = 0.05$$

ν_2	ν_1 1	2	3	4	5	6	7	8	9	10	12	14	16	18	20	24	30	40	∞
1	161.45	199.50	215.71	224.58	230.16	233.99	236.77	238.88	240.54	241.88	243.91	245.36	246.46	247.32	248.01	249.05	250.10	251.14	254.31
2	18.513	19.000	19.164	19.247	19.296	19.330	19.353	19.371	19.385	19.396	19.412	19.424	19.433	19.440	19.446	19.454	19.462	19.471	19.496
3	10.128	9.552	9.277	9.117	9.013	8.941	8.887	8.845	8.812	8.786	8.745	8.715	8.692	8.675	8.660	8.639	8.617	8.594	8.526
4	7.709	6.944	6.591	6.388	6.256	6.163	6.094	6.041	5.999	5.964	5.912	5.873	5.844	5.821	5.803	5.774	5.746	5.717	5.628
5	6.608	5.786	5.409	5.192	5.050	4.950	4.876	4.818	4.772	4.735	4.678	4.636	4.604	4.579	4.558	4.527	4.496	4.464	4.365
6	5.987	5.143	4.757	4.534	4.387	4.284	4.207	4.147	4.099	4.060	4.000	3.956	3.922	3.896	3.874	3.841	3.808	3.774	3.669
7	5.591	4.737	4.347	4.120	3.972	3.866	3.787	3.726	3.677	3.637	3.575	3.529	3.494	3.467	3.445	3.410	3.376	3.340	3.230
8	5.318	4.459	4.066	3.838	3.687	3.581	3.500	3.438	3.388	3.347	3.284	3.237	3.202	3.173	3.150	3.115	3.079	3.043	2.928
9	5.117	4.256	3.863	3.633	3.482	3.374	3.293	3.230	3.179	3.137	3.073	3.025	2.989	2.960	2.936	2.900	2.864	2.826	2.707
10	4.965	4.103	3.708	3.478	3.326	3.217	3.135	3.072	3.020	2.978	2.913	2.865	2.828	2.798	2.774	2.737	2.700	2.661	2.538
11	4.844	3.982	3.587	3.357	3.204	3.095	3.012	2.948	2.896	2.854	2.788	2.739	2.701	2.671	2.646	2.609	2.570	2.531	2.404
12	4.747	3.885	3.490	3.259	3.106	2.996	2.913	2.849	2.796	2.753	2.687	2.637	2.599	2.568	2.544	2.505	2.466	2.426	2.296
13	4.667	3.806	3.411	3.179	3.025	2.915	2.832	2.767	2.714	2.671	2.604	2.554	2.515	2.484	2.459	2.420	2.380	2.339	2.206
14	4.600	3.739	3.344	3.112	2.958	2.848	2.764	2.699	2.646	2.602	2.534	2.484	2.445	2.413	2.388	2.349	2.308	2.266	2.131
15	4.543	3.682	3.287	3.056	2.901	2.790	2.707	2.641	2.588	2.544	2.475	2.424	2.385	2.353	2.328	2.288	2.247	2.204	2.066
16	4.494	3.634	3.239	3.007	2.852	2.741	2.657	2.591	2.538	2.494	2.425	2.373	2.333	2.302	2.276	2.235	2.194	2.151	2.010
17	4.451	3.592	3.197	2.965	2.810	2.699	2.614	2.548	2.494	2.450	2.381	2.329	2.289	2.257	2.230	2.190	2.148	2.104	1.960
18	4.414	3.555	3.160	2.928	2.773	2.661	2.577	2.510	2.456	2.412	2.342	2.290	2.250	2.217	2.191	2.150	2.107	2.063	1.917
19	4.381	3.522	3.127	2.895	2.740	2.628	2.544	2.477	2.423	2.378	2.308	2.256	2.215	2.182	2.156	2.114	2.071	2.026	1.878
20	4.351	3.493	3.098	2.866	2.711	2.599	2.514	2.447	2.393	2.348	2.278	2.225	2.184	2.151	2.124	2.082	2.039	1.994	1.843
24	4.260	3.403	3.009	2.776	2.621	2.508	2.423	2.355	2.300	2.255	2.183	2.130	2.088	2.054	2.027	1.984	1.939	1.892	1.733
30	4.171	3.316	2.922	2.690	2.534	2.421	2.334	2.266	2.211	2.165	2.092	2.037	1.995	1.960	1.932	1.887	1.841	1.792	1.622
40	4.085	3.232	2.839	2.606	2.449	2.336	2.249	2.180	2.124	2.077	2.003	1.948	1.904	1.868	1.839	1.793	1.744	1.693	1.509
60	4.001	3.150	2.758	2.525	2.368	2.254	2.167	2.097	2.040	1.993	1.917	1.860	1.815	1.778	1.748	1.700	1.649	1.594	1.389
120	3.920	3.072	2.680	2.447	2.290	2.175	2.087	2.016	1.959	1.910	1.834	1.775	1.728	1.690	1.659	1.608	1.554	1.495	1.254
∞	3.841	2.996	2.605	2.372	2.214	2.099	2.010	1.938	1.880	1.831	1.752	1.692	1.644	1.604	1.571	1.517	1.459	1.394	1.000

Table B7 (continued). Critical Values of F Distributions, $F(\alpha; \nu_1, \nu_2)$

$$\alpha = 0.01$$

ν_2	ν_1 1	2	3	4	5	6	7	8	9	10	12	14	16	18	20	24	30	40	∞
1	4052.2	4999.5	5403.4	5624.6	5763.7	5859.0	5928.4	5981.1	6022.5	6055.9	6106.3	6142.7	6170.1	6191.5	6208.7	6234.6	6260.7	6286.8	6365.9
2	98.503	99.000	99.166	99.249	99.299	99.333	99.356	99.374	99.388	99.399	99.416	99.428	99.437	99.444	99.449	99.458	99.466	99.474	99.499
3	34.116	30.817	29.457	28.710	28.237	27.911	27.672	27.489	27.345	27.229	27.052	26.924	26.827	26.751	26.690	26.598	26.505	26.411	26.125
4	21.198	18.000	16.694	15.977	15.522	15.207	14.976	14.799	14.659	14.546	14.374	14.249	14.154	14.080	14.020	13.929	13.838	13.745	13.463
5	16.258	13.274	12.060	11.392	10.967	10.672	10.456	10.289	10.158	10.051	9.888	9.770	9.680	9.610	9.553	9.466	9.379	9.291	9.020
6	13.745	10.925	9.780	9.148	8.746	8.466	8.260	8.102	7.976	7.874	7.718	7.605	7.519	7.451	7.396	7.313	7.229	7.143	6.880
7	12.246	9.547	8.451	7.847	7.460	7.191	6.993	6.840	6.719	6.620	6.469	6.359	6.275	6.209	6.155	6.074	5.992	5.908	5.656
8	11.259	8.649	7.591	7.006	6.632	6.371	6.178	6.029	5.911	5.814	5.667	5.559	5.477	5.412	5.359	5.279	5.198	5.116	4.859
9	10.561	8.022	6.992	6.422	6.057	5.802	5.613	5.467	5.351	5.257	5.111	5.005	4.924	4.860	4.808	4.729	4.649	4.567	4.311
10	10.044	7.559	6.552	5.994	5.636	5.386	5.200	5.057	4.942	4.849	4.706	4.601	4.520	4.457	4.405	4.327	4.247	4.165	3.909
11	9.646	7.206	6.217	5.668	5.316	5.069	4.886	4.744	4.632	4.539	4.397	4.293	4.213	4.150	4.099	4.021	3.941	3.860	3.602
12	9.330	6.927	5.953	5.412	5.064	4.821	4.640	4.499	4.388	4.296	4.155	4.052	3.972	3.910	3.858	3.780	3.701	3.619	3.361
13	9.074	6.701	5.739	5.205	4.862	4.620	4.441	4.302	4.191	4.100	3.960	3.857	3.778	3.716	3.665	3.587	3.507	3.425	3.165
14	8.862	6.515	5.564	5.035	4.695	4.456	4.278	4.140	4.030	3.939	3.800	3.698	3.619	3.556	3.505	3.427	3.348	3.266	3.004
15	8.683	6.359	5.417	4.893	4.556	4.318	4.142	4.004	3.895	3.805	3.666	3.564	3.485	3.423	3.372	3.294	3.214	3.132	2.868
16	8.531	6.226	5.292	4.773	4.437	4.202	4.026	3.890	3.780	3.691	3.553	3.451	3.372	3.310	3.259	3.181	3.101	3.018	2.753
17	8.400	6.112	5.185	4.669	4.336	4.102	3.927	3.791	3.682	3.593	3.455	3.353	3.275	3.212	3.162	3.084	3.003	2.920	2.653
18	8.285	6.013	5.092	4.579	4.248	4.015	3.841	3.705	3.597	3.508	3.371	3.269	3.190	3.128	3.077	2.999	2.919	2.835	2.566
19	8.185	5.926	5.010	4.500	4.171	3.939	3.765	3.631	3.523	3.434	3.297	3.195	3.117	3.054	3.003	2.925	2.844	2.761	2.489
20	8.096	5.849	4.938	4.431	4.103	3.871	3.699	3.564	3.457	3.368	3.231	3.130	3.051	2.989	2.938	2.859	2.778	2.695	2.421
24	7.823	5.614	4.718	4.218	3.895	3.667	3.496	3.363	3.256	3.168	3.032	2.930	2.852	2.789	2.738	2.659	2.577	2.492	2.211
30	7.562	5.390	4.510	4.018	3.699	3.473	3.305	3.173	3.067	2.979	2.843	2.742	2.663	2.600	2.549	2.469	2.386	2.299	2.006
40	7.314	5.179	4.313	3.828	3.514	3.291	3.124	2.993	2.888	2.801	2.665	2.563	2.484	2.421	2.369	2.288	2.203	2.114	1.805
60	7.077	4.977	4.126	3.649	3.339	3.119	2.953	2.823	2.718	2.632	2.496	2.394	2.315	2.251	2.198	2.115	2.028	1.936	1.601
120	6.851	4.787	3.949	3.480	3.174	2.956	2.792	2.663	2.559	2.472	2.336	2.234	2.154	2.089	2.035	1.950	1.860	1.763	1.381
∞	6.635	4.605	3.782	3.319	3.017	2.802	2.639	2.511	2.407	2.321	2.185	2.082	2.000	1.934	1.878	1.791	1.696	1.592	1.000

Table B8. Sample Sizes for the ANOM, $\alpha = 0.1$

| | $\Delta = 3.00$ | | | | | | | | $\Delta = 2.50$ | | | | | | | | $\Delta = 2.00$ | | | | | | | |
| | Power | | | | | | | | Power | | | | | | | | Power | | | | | | | |
k	.50	.60	.70	.75	.80	.90	.95	.99	.50	.60	.70	.75	.80	.90	.95	.99	.50	.60	.70	.75	.80	.90	.95	.99
3	2	3	3	3	3	4	4	6	3	3	3	4	4	5	6	7	3	4	4	5	5	7	8	11
4	2	3	3	3	3	4	5	6	3	3	4	4	4	5	6	8	4	4	5	5	6	7	9	12
5	2	3	3	3	4	4	5	6	3	3	4	4	5	6	7	9	4	4	5	6	6	8	9	13
6	3	3	3	3	4	4	5	7	3	3	4	4	5	6	7	9	4	5	5	6	7	8	10	13
7	3	3	3	3	4	5	5	7	3	4	4	4	5	6	7	9	4	5	6	6	7	9	10	14
8	3	3	3	4	4	5	5	7	3	4	4	5	5	6	7	9	4	5	6	6	7	9	11	14
9	3	3	3	4	4	5	5	7	3	4	4	5	5	6	7	10	4	5	6	7	7	9	11	14
10	3	3	3	4	4	5	6	7	3	4	4	5	5	6	7	10	4	5	6	7	7	9	11	15
11	3	3	3	4	4	5	6	7	3	4	4	5	5	7	8	10	5	5	6	7	8	10	11	15
12	3	3	3	4	4	5	6	7	3	4	5	5	5	7	8	10	5	6	7	7	8	10	12	15
13	3	3	4	4	4	5	6	7	3	4	5	5	5	7	8	10	5	6	7	7	8	10	12	15
14	3	3	4	4	4	5	6	7	3	4	5	5	6	7	8	10	5	6	7	7	8	10	12	16
15	3	3	4	4	4	5	6	8	3	4	5	5	6	7	8	10	5	6	7	7	8	10	12	16
16	3	3	4	4	4	5	6	8	4	4	5	5	6	7	8	11	5	6	7	8	8	10	12	16
17	3	3	4	4	4	5	6	8	4	4	5	5	6	7	8	11	5	6	7	8	8	11	12	16
18	3	3	4	4	4	5	6	8	4	4	5	5	6	7	8	11	5	6	7	8	9	11	13	16
19	3	3	4	4	4	5	6	8	4	4	5	5	6	7	8	11	5	6	7	8	9	11	13	17
20	3	3	4	4	4	5	6	8	4	4	5	5	6	7	8	11	5	6	7	8	9	11	13	17
24	3	3	4	4	4	5	6	8	4	4	5	6	6	7	9	11	5	6	8	8	9	11	13	17
30	3	3	4	4	5	6	6	8	4	5	5	6	6	8	9	12	6	7	8	9	9	12	14	18
40	3	3	4	4	5	6	7	9	4	5	6	6	7	8	9	12	6	7	8	9	10	12	14	18
60	3	4	4	5	5	6	7	9	5	5	6	7	7	9	10	13	7	8	9	10	11	13	15	20

Table B8 (continued). Sample Sizes for the ANOM, $\alpha = 0.1$

| | $\Delta = 1.75$ | | | | | | | | $\Delta = 1.50$ | | | | | | | | $\Delta = 1.25$ | | | | | | | | $\Delta = 1.00$ | | | | | | | |
| | Power | | | | | | | | Power | | | | | | | | Power | | | | | | | | Power | | | | | | | |
k	.50	.60	.70	.75	.80	.90	.95	.99	.50	.60	.70	.75	.80	.90	.95	.99	.50	.60	.70	.75	.80	.90	.95	.99	.50	.60	.70	.75	.80	.90	.95	.99
3	4	5	5	6	6	8	10	14	5	6	7	7	8	11	13	18	6	7	9	10	11	15	18	26	9	11	13	15	17	23	28	39
4	4	5	6	6	7	9	11	15	5	6	8	8	9	12	15	20	7	8	10	12	13	17	21	28	10	12	16	17	19	26	31	44
5	4	5	6	7	7	10	12	16	6	7	8	9	10	13	16	21	7	9	11	13	14	18	22	30	11	14	17	19	21	28	34	47
6	5	6	7	7	8	10	13	17	6	7	9	10	11	14	17	23	8	10	12	13	15	19	23	32	12	15	18	20	23	30	36	49
7	5	6	7	8	9	11	13	17	6	8	9	10	11	14	17	23	9	11	13	14	16	20	24	33	13	16	19	22	24	31	38	51
8	5	6	7	8	9	11	13	18	7	8	10	11	12	15	18	24	9	11	13	15	16	21	25	34	13	17	20	23	25	32	39	53
9	5	6	8	8	9	12	14	19	7	8	10	11	12	15	19	25	9	11	14	15	17	22	26	35	14	17	21	24	26	34	40	54
10	5	7	8	9	9	12	14	19	7	9	10	11	13	16	19	25	10	12	14	16	18	22	27	36	15	18	22	24	27	34	41	56
11	6	7	8	9	10	12	15	19	7	9	11	12	13	16	19	26	10	12	15	16	18	23	28	37	15	19	23	25	28	35	42	57
12	6	7	8	9	10	12	15	20	7	9	11	12	13	17	20	26	10	13	15	17	19	24	28	37	16	19	23	26	28	36	43	58
13	6	7	8	9	9	13	15	20	8	9	11	12	13	17	20	27	11	13	16	17	19	24	29	38	16	20	24	26	29	37	44	59
14	6	7	8	9	10	13	15	20	8	9	11	12	14	17	20	27	11	13	16	17	19	24	29	39	16	20	24	27	29	38	45	60
15	6	7	9	9	10	13	16	20	8	10	12	13	14	18	21	27	11	14	16	18	20	25	30	40	17	20	25	27	30	38	46	61
16	6	7	9	9	11	13	16	21	8	10	12	13	14	18	21	28	11	14	16	18	20	25	30	40	17	21	25	27	31	39	46	62
17	6	8	9	10	11	13	16	21	8	10	12	13	14	18	21	28	12	14	17	18	20	25	30	40	17	21	25	28	31	39	47	62
18	6	8	9	10	11	14	16	21	8	10	12	13	15	18	22	28	12	14	17	19	20	26	31	41	18	21	26	29	32	40	47	63
19	6	8	9	10	11	14	16	21	9	10	12	13	15	18	22	29	12	14	17	19	21	26	31	41	18	22	26	29	32	40	48	64
20	7	8	9	10	11	14	16	21	9	10	12	14	15	19	22	29	12	15	17	19	21	26	31	41	18	22	27	29	32	41	48	64
24	7	8	10	11	12	14	17	22	9	11	13	14	15	19	23	30	13	15	18	20	22	27	32	43	19	24	28	31	34	43	50	66
30	7	9	10	11	12	15	18	23	10	11	13	15	16	20	24	31	13	16	19	21	23	29	34	44	21	25	30	33	36	45	53	69
40	8	9	11	12	13	16	18	24	10	12	14	16	17	21	25	32	15	17	21	22	25	30	36	46	22	27	32	35	38	47	55	72
60	8	10	12	13	14	17	20	25	11	13	16	17	19	23	27	34	16	19	22	24	27	33	38	49	25	29	35	38	41	51	59	77

Table B8 (continued).   Sample Sizes for the ANOM, $\alpha = 0.05$

$\Delta = 3.00$

k	Power							
	.50	.60	.70	.75	.80	.90	.95	.99
3	3	3	4	4	4	5	5	7
4	3	3	4	4	4	5	6	7
5	3	3	4	4	4	5	6	7
6	3	3	4	4	4	5	6	7
7	3	3	4	4	4	5	6	8
8	3	3	4	4	4	5	6	8
9	3	4	4	4	5	5	6	8
10	3	4	4	4	5	6	6	8
11	3	4	4	4	5	6	6	8
12	3	4	4	4	5	6	6	8
13	3	4	4	4	5	6	7	8
14	3	4	4	4	5	6	7	8
15	3	4	4	5	5	6	7	8
16	3	4	4	5	5	6	7	9
17	3	4	4	5	5	6	7	9
18	3	4	4	5	5	6	7	9
19	3	4	4	5	5	6	7	9
20	3	4	4	5	5	6	7	9
24	3	4	4	5	5	6	7	9
30	4	4	5	5	5	6	7	9
40	4	4	5	5	6	7	8	9
60	4	4	5	5	6	7	8	10

$\Delta = 2.50$

k	Power							
	.50	.60	.70	.75	.80	.90	.95	.99
3	3	4	4	5	5	6	7	9
4	3	4	4	5	5	6	7	9
5	4	4	5	5	5	7	8	10
6	4	4	5	5	6	7	8	10
7	4	4	5	5	6	7	8	10
8	4	4	5	5	6	7	8	11
9	4	4	5	6	6	7	8	11
10	4	5	5	6	6	7	9	11
11	4	5	5	6	6	8	9	11
12	4	5	5	6	6	8	9	11
13	4	5	6	6	6	8	9	11
14	4	5	6	6	6	8	9	12
15	4	5	6	6	7	8	9	12
16	4	5	6	6	7	8	9	12
17	4	5	6	6	7	8	9	12
18	4	5	6	6	7	8	9	12
19	4	5	6	6	7	8	9	12
20	4	5	6	6	7	8	10	12
24	5	5	6	7	7	9	10	13
30	5	5	6	7	7	9	10	13
40	5	6	7	7	8	9	11	13
60	5	6	7	7	8	10	11	14

$\Delta = 2.00$

k	Power							
	.50	.60	.70	.75	.80	.90	.95	.99
3	4	5	6	6	7	8	10	13
4	5	5	6	7	7	9	10	14
5	5	6	6	7	8	9	11	14
6	5	6	7	7	8	10	12	15
7	5	6	7	8	8	10	12	16
8	5	6	7	8	8	10	12	16
9	5	6	7	8	9	11	13	16
10	6	6	8	8	9	11	13	17
11	6	7	8	8	9	11	13	17
12	6	7	8	8	9	11	13	17
13	6	7	8	9	9	12	13	17
14	6	7	8	9	10	12	14	18
15	6	7	8	9	10	12	14	18
16	6	7	8	9	10	12	14	18
17	6	7	8	9	10	12	14	18
18	6	7	9	9	10	12	14	18
19	6	7	9	9	10	12	14	18
20	6	7	9	9	10	12	14	19
24	7	8	9	10	11	13	15	19
30	7	8	9	10	11	13	15	20
40	7	9	10	11	11	14	16	20
60	8	9	11	11	12	15	17	22

Table B8 (continued). Sample Sizes for the ANOM, $\alpha = 0.05$

k	Δ = 1.75								Δ = 1.50								Δ = 1.25								Δ = 1.00							
	Power								Power								Power								Power							
	.50	.60	.70	.75	.80	.90	.95	.99	.50	.60	.70	.75	.80	.90	.95	.99	.50	.60	.70	.75	.80	.90	.95	.99	.50	.60	.70	.75	.80	.90	.95	.99
3	5	6	7	7	8	10	12	16	6	7	9	9	10	13	16	21	8	10	12	13	14	18	22	30	12	14	17	19	21	27	33	45
4	5	6	7	8	9	11	13	17	7	8	10	10	12	15	17	23	9	11	13	14	16	20	24	33	13	16	20	22	24	31	37	50
5	6	7	8	9	9	12	14	18	7	9	10	11	12	16	18	25	10	12	14	16	17	22	26	35	15	18	21	24	26	33	40	54
6	6	7	8	9	10	12	15	19	8	9	10	11	13	16	19	26	10	13	15	16	18	23	27	37	16	19	23	25	28	35	42	56
7	6	7	9	10	10	13	15	20	8	10	11	12	14	17	20	27	11	13	16	17	19	24	29	38	17	20	24	27	29	37	44	58
8	7	8	9	10	11	13	16	20	8	10	12	13	14	18	21	27	12	14	17	18	20	25	30	39	18	21	25	28	30	38	45	60
9	7	8	9	10	11	14	16	21	9	10	12	13	15	18	21	28	12	14	17	19	21	26	30	40	18	22	26	29	31	39	47	62
10	7	8	10	10	11	14	16	21	9	10	13	14	15	18	22	29	12	15	17	19	21	26	31	41	19	23	27	29	32	40	48	63
11	7	8	10	11	12	14	17	22	9	11	13	14	15	19	22	29	13	15	18	20	22	27	32	42	20	23	28	30	33	41	49	64
12	7	8	10	11	12	15	17	22	9	11	13	14	16	19	23	30	13	16	19	20	22	27	32	42	20	24	28	31	34	42	50	66
13	7	9	10	11	12	15	17	22	10	11	13	15	16	20	23	30	14	16	19	21	23	28	33	43	21	24	29	31	34	43	51	67
14	8	9	10	11	12	15	18	23	10	12	14	15	16	20	23	31	14	16	19	21	23	28	33	43	21	25	29	32	35	44	52	67
15	8	9	10	11	12	15	18	23	10	12	14	15	16	20	24	31	14	17	20	21	23	29	34	44	21	25	30	33	36	45	52	68
16	8	9	11	12	13	15	18	23	10	12	14	15	17	21	24	31	14	17	20	22	24	29	34	45	22	26	31	33	36	45	53	69
17	8	9	11	12	13	16	18	23	10	12	14	16	17	21	24	31	15	17	20	22	24	30	35	45	22	26	31	34	37	46	54	70
18	8	9	11	12	13	16	18	24	11	12	14	16	17	21	25	32	15	17	20	22	24	30	35	45	22	27	32	34	37	46	54	71
19	8	9	11	12	13	16	18	24	11	13	15	16	17	21	25	32	15	18	21	22	24	30	35	46	22	27	32	35	38	47	55	71
20	8	10	11	12	13	16	19	24	11	13	15	16	17	21	25	32	15	18	21	23	25	31	36	46	23	28	32	35	38	47	55	72
24	9	10	12	12	14	16	19	25	11	13	15	17	18	22	26	33	16	18	22	24	26	32	37	48	24	29	34	37	40	49	57	74
30	9	10	12	13	14	17	20	25	12	14	16	17	19	23	27	34	17	20	23	25	27	33	38	49	26	30	35	38	42	51	59	77
40	9	11	13	14	15	18	21	26	12	15	17	18	20	24	28	36	18	21	24	26	28	35	40	51	28	32	38	41	44	54	62	80
60	10	12	14	15	16	19	22	28	14	16	18	20	21	26	30	38	20	23	26	28	30	37	43	54	30	35	41	44	47	57	66	84

Table B8 (continued).   Sample Sizes for the ANOM, $\alpha = 0.01$

	$\Delta = 3.00$								$\Delta = 2.50$								$\Delta = 2.00$							
	Power								Power								Power							
k	.50	.60	.70	.75	.80	.90	.95	.99	.50	.60	.70	.75	.80	.90	.95	.99	.50	.60	.70	.75	.80	.90	.95	.99
3	4	4	5	5	5	6	7	9	5	5	6	6	7	8	9	11	6	7	8	9	9	11	13	16
4	4	5	5	5	6	7	7	9	5	6	6	7	7	8	10	12	7	8	9	9	10	12	14	18
5	4	5	5	5	6	7	8	9	5	6	7	7	7	9	10	13	7	8	9	10	11	13	15	18
6	4	5	5	5	6	7	8	9	5	6	7	7	8	9	10	13	7	8	10	10	11	13	15	19
7	4	5	5	6	6	7	8	10	5	6	7	7	8	9	11	13	8	9	10	10	11	13	15	20
8	4	5	5	6	6	7	8	10	5	6	7	7	8	9	11	13	8	9	10	11	12	14	16	20
9	4	5	5	6	6	7	8	10	6	6	7	8	8	10	11	14	8	9	10	11	12	14	16	20
10	4	5	5	6	6	7	8	10	6	6	7	8	8	10	11	14	8	9	10	11	12	14	16	21
11	4	5	5	6	6	7	8	10	6	6	7	8	8	10	11	14	8	9	11	11	12	14	17	21
12	4	5	5	6	6	7	8	10	6	7	7	8	8	10	11	14	8	9	11	11	12	15	17	21
13	4	5	6	6	6	7	8	10	6	7	7	8	8	10	11	14	8	10	11	12	12	15	17	22
14	4	5	6	6	6	7	8	10	6	7	7	8	8	10	11	14	8	10	11	12	13	15	17	22
15	4	5	6	6	6	7	8	10	6	7	7	8	9	10	12	14	9	10	11	12	13	15	17	22
16	4	5	6	6	6	7	8	10	6	7	8	8	9	10	12	15	9	10	11	12	13	15	18	22
17	5	5	6	6	6	7	8	10	6	7	8	8	9	10	12	15	9	10	11	12	13	16	18	22
18	5	5	6	6	6	7	8	11	6	7	8	8	9	10	12	15	9	10	11	12	13	16	18	22
19	5	5	6	6	6	8	9	11	6	7	8	8	9	10	12	15	9	10	12	12	13	16	18	23
20	5	5	6	6	6	8	9	11	6	7	8	8	9	11	12	15	9	10	12	12	13	16	18	23
24	5	5	6	6	7	8	9	11	6	7	8	8	9	11	12	15	9	11	12	13	14	16	19	23
30	5	5	6	6	7	8	9	11	6	7	8	9	9	11	13	16	10	11	12	13	14	17	19	24
40	5	5	6	7	7	8	9	11	7	8	9	9	10	11	13	16	10	11	13	14	15	17	20	25
60	5	6	6	7	7	8	10	12	7	8	9	10	10	12	14	17	11	12	14	15	16	18	21	26

Table B8 (continued). Sample Sizes for the ANOM, $\alpha = 0.01$

k	Δ = 1.75 Power								Δ = 1.50 Power								Δ = 1.25 Power								Δ = 1.00 Power							
	.50	.60	.70	.75	.80	.90	.95	.99	.50	.60	.70	.75	.80	.90	.95	.99	.50	.60	.70	.75	.80	.90	.95	.99	.50	.60	.70	.75	.80	.90	.95	.99
3	8	9	10	11	11	14	16	21	10	11	13	14	15	18	21	27	13	15	17	19	21	25	29	38	19	22	26	28	31	38	45	59
4	8	9	11	12	12	15	18	22	11	12	14	15	16	20	23	30	14	17	19	21	23	28	32	42	21	25	29	31	34	42	49	65
5	9	10	11	12	13	16	18	23	11	13	14	16	17	21	24	31	15	18	21	22	24	29	34	44	23	27	31	34	37	45	53	69
6	9	10	12	13	14	17	19	24	12	14	16	17	18	22	25	33	16	19	22	23	25	31	36	46	24	28	33	36	39	48	55	71
7	9	11	12	13	14	17	20	25	12	14	16	17	19	23	26	34	17	19	23	24	26	32	37	48	26	30	35	37	41	49	57	74
8	10	11	13	13	15	17	20	26	13	14	17	18	19	23	27	35	17	20	23	25	27	33	38	49	27	31	36	39	42	51	59	76
9	10	11	13	14	15	18	21	26	13	15	17	18	20	24	28	35	18	21	24	26	28	34	39	50	28	32	37	40	43	52	61	78
10	10	11	13	14	15	18	21	27	13	15	17	19	20	24	28	36	18	21	25	26	29	35	40	51	28	33	38	41	44	54	62	79
11	10	12	13	14	15	19	21	27	14	16	18	19	21	25	29	36	19	22	25	27	29	35	41	52	29	33	38	41	45	55	63	80
12	10	12	14	15	16	19	22	27	14	16	18	20	21	25	29	37	19	22	26	28	30	36	41	53	30	34	40	43	46	56	64	82
13	11	12	14	15	16	19	22	28	14	16	18	20	21	26	29	37	20	23	26	28	30	36	42	53	30	35	40	43	47	56	65	83
14	11	12	14	15	16	19	22	28	14	16	19	20	22	26	30	38	20	23	26	28	31	37	43	54	30	36	41	44	47	57	66	84
15	11	12	14	15	16	20	22	28	15	17	19	20	22	26	30	38	20	23	27	29	31	37	43	54	32	36	42	45	48	58	67	85
16	11	13	14	15	17	20	23	29	15	17	19	21	22	27	30	38	21	24	27	29	31	38	43	55	32	37	42	45	49	59	68	86
17	11	13	15	16	17	20	23	29	15	17	19	21	22	27	31	39	21	24	27	30	32	38	44	56	32	37	43	46	49	59	68	86
18	11	13	15	16	17	20	23	29	15	17	20	21	23	27	31	39	21	24	28	30	32	39	44	56	33	38	43	46	50	60	69	87
19	11	13	15	16	17	20	23	29	15	17	20	21	23	27	31	39	21	24	28	30	33	39	45	56	33	38	44	47	50	61	69	88
20	12	13	15	16	17	20	23	29	15	17	20	21	23	28	32	40	21	25	28	30	33	39	45	57	34	38	44	47	51	61	70	88
24	12	14	15	16	18	21	24	30	16	18	21	22	24	28	32	41	23	26	29	31	34	40	46	58	35	40	46	49	52	63	72	91
30	12	14	16	17	18	22	25	31	17	19	21	23	25	29	33	42	24	27	31	33	35	42	48	60	36	42	47	51	54	65	74	93
40	13	15	17	18	19	23	26	32	17	20	22	24	26	30	35	43	25	28	32	34	37	44	50	62	38	44	50	53	57	68	77	97
60	14	16	18	19	20	24	27	33	19	21	24	25	27	32	36	45	27	30	34	36	39	46	52	65	41	47	53	56	60	72	81	101

Table B9. Sample Sizes for the ANOVA

Number of Treatments = 2

	α = 0.10				α = 0.05				α = 0.01			
	Power				Power				Power			
Δ	.50	.80	.90	.95	.50	.80	.90	.95	.50	.80	.90	.95
0.4	35	78			49				85			
0.5	23	51	70	88	32	64	86	55	96			
0.6	16	36	49	61	23	45	60	74	39	67	85	
0.7	12	26	36	45	17	34	44	55	29	50	63	75
0.8	100	21	28	35	14	26	34	42	23	39	49	5
0.9	8	16	22	28	11	21	27	34	19	31	39	46
1.0	7	14	18	23	9	17	23	27	15	26	32	38
1.2	5	10	13	16	7	12	16	20	11	18	23	27
1.4	4	8	10	12	6	10	12	15	9	14	17	20
1.6	4	6	8	10	5	8	10	12	7	11	14	16
1.8	3	5	7	8	4	6	8	10	6	10	11	13
2.0	3	4	6	7	4	6	7	8	6	8	10	11
2.5	2	3	4	5	3	4	5	6	4	6	7	8
3.0	2	3	3	4	3	4	4	5	4	5	6	6

Number of Treatments = 3

	α = 0.10				α = 0.05				α = 0.01			
	Power				Power				Power			
Δ	.50	.80	.90	.95	.50	.80	.90	.95	.50	.80	.90	.95
0.4	46	98			63							
0.5	30	63	85		41	79			68			
0.6	21	44	59	74	29	55	72	87	48	79	99	
0.7	16	33	44	54	22	41	53	65	35	59	73	86
0.8	12	25	34	42	17	32	41	50	28	45	57	67
0.9	10	20	27	33	14	25	33	40	22	36	45	53
1.0	8	17	22	27	11	21	27	32	18	30	37	43
1.2	6	12	16	19	8	15	19	23	13	21	26	31
1.4	5	9	12	15	7	11	14	17	10	16	20	23
1.6	4	7	10	12	5	9	11	14	9	13	16	18
1.8	4	6	8	9	5	8	9	11	7	11	13	15
2.0	3	5	7	8	4	6	8	9	6	9	11	12
2.5	3	4	5	6	3	5	6	7	5	7	8	9
3.0	2	3	4	4	3	4	5	5	4	5	6	7

Table B9 (continued). Sample Sizes for the ANOVA

Number of Treatments = 4

	$\alpha = 0.10$				$\alpha = 0.05$				$\alpha = 0.01$			
	Power				Power				Power			
Δ	.50	.80	.90	.95	.50	.80	.90	.95	.50	.80	.90	.95
0.4	54				73							
0.5	35	72	96		48	89			76			
0.6	25	50	67	82	33	62	80	97	54	88		
0.7	18	37	49	61	25	46	59	72	40	65	80	94
0.8	14	29	38	47	20	36	46	55	31	50	62	73
0.9	12	23	30	37	16	28	36	44	25	40	49	58
1.0	10	19	25	30	13	23	30	36	21	33	40	47
1.2	7	14	18	22	10	17	21	25	15	23	29	33
1.4	6	10	13	16	7	13	16	19	11	18	22	25
1.6	5	8	11	13	6	10	13	15	9	14	17	20
1.8	4	7	9	10	5	8	10	12	8	12	14	16
2.0	4	6	7	9	4	7	9	10	7	10	12	13
2.5	3	4	5	6	3	5	6	7	5	7	8	9
3.0	2	3	4	5	3	4	5	5	4	5	6	7

Number of Treatments = 5

	$\alpha = 0.10$				$\alpha = 0.05$				$\alpha = 0.01$			
	Power				Power				Power			
Δ	.50	.80	.90	.95	.50	.80	.90	.95	.50	.80	.90	.95
0.4	60				82							
0.5	39	79			53	97			84			
0.6	27	55	73	89	37	68	87		59	95		
0.7	20	41	54	66	28	50	64	77	44	70	86	
0.8	16	32	42	51	22	39	50	59	34	54	67	78
0.9	13	25	33	40	17	31	39	47	27	43	53	62
1.0	11	21	27	33	14	25	32	39	22	35	43	50
1.2	8	15	19	23	10	18	23	27	16	25	31	36
1.4	6	11	14	17	8	14	17	20	12	19	23	27
1.6	5	9	11	14	7	11	14	16	10	15	18	21
1.8	4	7	9	11	5	9	11	13	8	12	15	17
2.0	4	6	8	9	5	7	9	11	7	10	12	14
2.5	3	4	5	6	4	5	6	7	5	7	9	10
3.0	3	4	4	5	3	4	5	6	4	6	7	7

Table B9 (continued). Sample Sizes for the ANOVA

Number of Treatments = 6

	$\alpha = 0.10$				$\alpha = 0.05$				$\alpha = 0.01$			
	Power				Power				Power			
Δ	.50	.80	.90	.95	.50	.80	.90	.95	.50	.80	.90	.95
0.4	65				89							
0.5	42	85			57				90			
0.6	30	59	78	95	40	73	93		63			
0.7	22	44	58	70	30	54	69	82	47	75	92	
0.8	17	34	44	54	23	42	53	63	36	58	71	82
0.9	14	27	35	43	19	33	42	50	29	46	56	65
1.0	12	22	29	35	15	27	34	41	24	38	46	53
1.2	8	16	20	25	11	19	24	29	17	27	32	38
1.4	7	12	15	19	9	15	18	22	13	20	24	28
1.6	5	9	12	15	7	11	14	17	10	16	19	22
1.8	5	8	10	12	6	9	12	14	9	13	15	18
2.0	4	7	8	10	5	8	10	11	7	11	13	15
2.5	3	5	6	7	4	6	7	8	5	8	9	10
3.0	3	4	4	5	3	4	5	6	4	6	7	8

Number of Treatments = 7

	$\alpha = 0.10$				$\alpha = 0.05$				$\alpha = 0.01$			
	Power				Power				Power			
Δ	.50	.80	.90	.95	.50	.80	.90	.95	.50	.80	.90	.95
0.4	70				95							
0.5	46	90			61				96			
0.6	32	63	83		43	77	98		67			
0.7	24	47	61	74	32	57	73	87	50	79	96	
0.8	19	36	47	57	25	44	56	67	38	61	74	86
0.9	15	29	37	45	20	35	44	53	31	48	59	68
1.0	12	24	31	37	16	29	36	43	25	39	48	56
1.2	9	17	22	26	12	20	26	30	18	28	34	39
1.4	7	13	16	20	9	15	19	23	14	21	25	29
1.6	6	10	13	15	7	12	15	18	11	16	20	23
1.8	5	8	10	12	6	10	12	14	9	13	16	18
2.0	4	7	9	10	5	8	10	12	8	11	13	15
2.5	3	5	6	7	4	6	7	8	6	8	9	10
3.0	3	4	5	5	3	5	5	6	4	6	7	8

Table B9 (continued). Sample Sizes for the ANOVA

Number of Treatments = 8

Δ	$\alpha = 0.10$				$\alpha = 0.05$				$\alpha = 0.01$			
	Power				Power				Power			
	.50	.80	.90	.95	.50	.80	.90	.95	.50	.80	.90	.95
0.4	75											
0.5	48	95			65							
0.6	34	67	87		46	81			71			
0.7	25	49	64	78	34	60	76	91	52	82		
0.8	20	38	49	60	26	46	59	70	40	63	77	90
0.9	16	30	39	48	21	37	47	55	32	51	62	71
1.0	13	25	32	39	17	30	38	45	26	41	50	58
1.2	9	18	23	27	12	21	27	32	19	29	35	41
1.4	7	13	17	20	10	16	20	24	14	22	26	30
1.6	6	10	13	16	8	13	16	18	11	17	21	24
1.8	5	9	11	13	6	10	13	15	9	14	17	19
2.0	4	7	9	11	5	9	11	12	8	12	14	16
2.5	3	5	6	7	4	6	7	8	6	8	9	11
3.0	3	4	5	5	3	5	6	6	4	6	7	8

Number of Treatments = 9

Δ	$\alpha = 0.10$				$\alpha = 0.05$				$\alpha = 0.01$			
	Power				Power				Power			
	.50	.80	.90	.95	.50	.80	.90	.95	.50	.80	.90	.95
0.4	79											
0.5	51				69							
0.6	36	70	91		48	85			74			
0.7	27	52	67	81	36	63	79	94	55	86		
0.8	21	40	52	62	28	48	61	72	42	66	80	93
0.9	17	32	41	50	22	38	48	58	34	53	64	74
1.0	14	26	33	40	18	31	40	47	28	43	52	60
1.2	10	18	24	28	13	22	28	33	20	30	37	42
1.4	8	14	18	21	10	17	21	25	15	23	27	32
1.6	6	11	14	17	8	13	16	19	12	18	21	25
1.8	5	9	11	13	7	11	13	15	10	14	17	20
2.0	4	7	9	11	6	9	11	13	8	12	14	16
2.5	3	5	6	8	4	6	8	9	6	8	10	11
3.0	3	4	5	6	3	5	6	6	5	6	7	8

Table B10. Maximum F Ratio Critical Values $F_{\max}(\alpha; k, \nu)$

$\alpha = 0.1$

ν	3	4	5	6	7	k 8	9	10	11	12
2	42.48	69.13	98.18	129.1	161.7	195.6	230.7	266.8	303.9	341.9
3	16.77	23.95	30.92	37.73	44.40	50.94	57.38	63.72	69.97	76.14
4	10.38	13.88	17.08	20.06	22.88	25.57	28.14	30.62	33.01	35.33
5	7.68	9.86	11.79	13.54	15.15	16.66	18.08	19.43	20.71	21.95
6	6.23	7.78	9.11	10.30	11.38	12.38	13.31	14.18	15.01	15.79
7	5.32	6.52	7.52	8.41	9.20	9.93	10.60	11.23	11.82	12.37
8	4.71	5.68	6.48	7.18	7.80	8.36	8.88	9.36	9.81	10.23
9	4.26	5.07	5.74	6.31	6.82	7.28	7.70	8.09	8.45	8.78
10	3.93	4.63	5.19	5.68	6.11	6.49	6.84	7.16	7.46	7.74
12	3.45	4.00	4.44	4.81	5.13	5.42	5.68	5.92	6.14	6.35
15	3.00	3.41	3.74	4.02	4.25	4.46	4.65	4.82	4.98	5.13
20	2.57	2.87	3.10	3.29	3.46	3.60	3.73	3.85	3.96	4.06
30	2.14	2.34	2.50	2.62	2.73	2.82	2.90	2.97	3.04	3.10
60	1.71	1.82	1.90	1.96	2.02	2.07	2.11	2.14	2.18	2.21

$\alpha = 0.05$

ν	3	4	5	6	7	k 8	9	10	11	12
2	87.49	142.5	202.4	266.2	333.2	403.1	475.4	549.8	626.2	704.4
3	27.76	39.51	50.88	61.98	72.83	83.48	93.94	104.2	114.4	124.4
4	15.46	20.56	25.21	29.54	33.63	37.52	41.24	44.81	48.27	51.61
5	10.75	13.72	16.34	18.70	20.88	22.91	24.83	26.65	28.38	30.03
6	8.36	10.38	12.11	13.64	15.04	16.32	17.51	18.64	19.70	20.70
7	6.94	8.44	9.70	10.80	11.80	12.70	13.54	14.31	15.05	15.74
8	6.00	7.19	8.17	9.02	9.77	10.46	11.08	11.67	12.21	12.72
9	5.34	6.31	7.11	7.79	8.40	8.94	9.44	9.90	10.33	10.73
10	4.85	5.67	6.34	6.91	7.41	7.86	8.27	8.64	8.99	9.32
12	4.16	4.79	5.30	5.72	6.09	6.42	6.72	6.99	7.24	7.48
15	3.53	4.00	4.37	4.67	4.94	5.17	5.38	5.57	5.75	5.91
20	2.95	3.28	3.53	3.74	3.92	4.08	4.22	4.35	4.46	4.57
30	2.40	2.61	2.77	2.90	3.01	3.11	3.19	3.27	3.34	3.40
60	1.84	1.96	2.04	2.11	2.16	2.21	2.21	2.29	2.32	2.35

$\alpha = 0.01$

ν	3	4	5	6	7	k 8	9	10	11	12
2	447.5	729.2	1036.	1362.	1705.	2063.	2432.	2813.	3204.	3604.
3	84.56	119.8	153.8	187.0	219.3	251.1	282.3	313.0	343.2	373.1
4	36.70	48.43	59.09	69.00	78.33	87.20	95.68	103.8	111.7	119.3
5	22.06	27.90	33.00	37.61	41.85	45.81	49.53	53.06	56.42	59.63
6	15.60	19.16	22.19	24.89	27.32	29.57	31.65	33.61	35.46	37.22
7	12.09	14.55	16.60	18.39	20.00	21.47	22.82	24.08	25.26	26.37
8	9.94	11.77	13.27	14.58	15.73	16.78	17.74	18.63	19.46	20.24
9	8.49	9.93	11.10	12.11	12.99	13.79	14.52	15.19	15.81	16.39
10	7.46	8.64	9.59	10.39	11.10	11.74	12.31	12.84	13.33	13.79
12	6.10	6.95	7.63	8.20	8.69	9.13	9.53	9.89	10.23	10.54
15	4.93	5.52	5.99	6.37	6.71	7.00	7.27	7.51	7.73	7.93
20	3.90	4.29	4.60	4.85	5.06	5.25	5.42	5.57	5.70	5.83
30	2.99	3.23	3.41	3.56	3.68	3.79	3.88	3.97	4.04	4.12
60	2.15	2.26	2.35	2.42	2.47	2.52	2.57	2.61	2.64	2.67

Table B11. ANOMV Critical Values $L(\alpha; k, \nu)$ and $U(\alpha; k, \nu)$
$$\alpha = 0.1$$

	Number of Variances Being Compared, k									
	3		4		5		6		7	
ν	lower	upper	lower	upper	lower	upper	lower	upper	lower	upper
3	0.0277	0.7868	0.0145	0.6805	0.0094	0.5964	0.0066	0.5310	0.0050	0.4791
4	0.0464	0.7346	0.0262	0.6255	0.0177	0.5424	0.0130	0.4794	0.0100	0.4301
5	0.0633	0.6962	0.0373	0.5863	0.0259	0.5046	0.0194	0.4437	0.0153	0.3966
6	0.0780	0.6666	0.0474	0.5568	0.0335	0.4767	0.0254	0.4173	0.0202	0.3719
7	0.0908	0.6430	0.0564	0.5337	0.0404	0.4549	0.0310	0.3971	0.0248	0.3529
8	0.1020	0.6236	0.0644	0.5149	0.0466	0.4373	0.0360	0.3808	0.0290	0.3377
9	0.1118	0.6074	0.0715	0.4992	0.0521	0.4227	0.0405	0.3673	0.0328	0.3253
10	0.1205	0.5935	0.0779	0.4860	0.0571	0.4105	0.0446	0.3560	0.0363	0.3148
11	0.1282	0.5815	0.0837	0.4746	0.0617	0.4000	0.0484	0.3464	0.0395	0.3059
12	0.1352	0.5710	0.0889	0.4646	0.0658	0.3909	0.0518	0.3380	0.0424	0.2981
13	0.1415	0.5617	0.0937	0.4558	0.0696	0.3828	0.0549	0.3306	0.0451	0.2914
14	0.1472	0.5534	0.0980	0.4480	0.0731	0.3756	0.0578	0.3241	0.0476	0.2854
15	0.1524	0.5459	0.1021	0.4409	0.0763	0.3693	0.0605	0.3182	0.0499	0.2800
16	0.1573	0.5391	0.1058	0.4346	0.0793	0.3635	0.0630	0.3129	0.0520	0.2751
17	0.1617	0.5329	0.1092	0.4288	0.0821	0.3582	0.0653	0.3082	0.0540	0.2708
18	0.1659	0.5272	0.1124	0.4235	0.0847	0.3534	0.0675	0.3038	0.0559	0.2668
19	0.1697	0.5220	0.1154	0.4187	0.0871	0.3490	0.0695	0.2998	0.0576	0.2631
20	0.1733	0.5172	0.1182	0.4142	0.0894	0.3450	0.0715	0.2961	0.0593	0.2598
21	0.1767	0.5126	0.1208	0.4100	0.0915	0.3412	0.0733	0.2927	0.0608	0.2566
22	0.1798	0.5085	0.1233	0.4062	0.0935	0.3378	0.0749	0.2896	0.0623	0.2538
23	0.1828	0.5046	0.1256	0.4026	0.0955	0.3345	0.0766	0.2866	0.0637	0.2511
24	0.1856	0.5009	0.1278	0.3992	0.0972	0.3315	0.0781	0.2839	0.0650	0.2486
25	0.1882	0.4975	0.1299	0.3961	0.0990	0.3286	0.0795	0.2813	0.0662	0.2463
26	0.1908	0.4942	0.1319	0.3931	0.1006	0.3260	0.0809	0.2789	0.0674	0.2440
27	0.1931	0.4912	0.1338	0.3903	0.1021	0.3235	0.0822	0.2766	0.0686	0.2419
28	0.1954	0.4883	0.1356	0.3876	0.1036	0.3210	0.0834	0.2745	0.0697	0.2400
29	0.1976	0.4855	0.1373	0.3851	0.1050	0.3188	0.0847	0.2724	0.0707	0.2382
30	0.1996	0.4830	0.1390	0.3827	0.1064	0.3167	0.0858	0.2705	0.0717	0.2364
31	0.2016	0.4805	0.1405	0.3805	0.1077	0.3146	0.0869	0.2687	0.0726	0.2347
32	0.2035	0.4781	0.1420	0.3783	0.1089	0.3127	0.0879	0.2670	0.0735	0.2332
33	0.2053	0.4758	0.1434	0.3763	0.1101	0.3109	0.0889	0.2653	0.0744	0.2316
34	0.2070	0.4737	0.1449	0.3743	0.1112	0.3091	0.0899	0.2637	0.0753	0.2302

Table B11 (continued). ANOMV Critical Values $L(\alpha; k, \nu)$ and $U(\alpha; k, \nu)$
$$\alpha = 0.1$$

	Number of Variances Being Compared, k									
	8		9		10		11		12	
ν	lower	upper	lower	upper	lower	upper	lower	upper	lower	upper
3	0.0039	0.4371	0.0032	0.4029	0.0027	0.3736	0.0022	0.3483	0.0019	0.3262
4	0.0081	0.3908	0.0067	0.3583	0.0056	0.3309	0.0048	0.3079	0.0042	0.2878
5	0.0124	0.3589	0.0104	0.3281	0.0089	0.3028	0.0077	0.2809	0.0068	0.2627
6	0.0167	0.3358	0.0140	0.3064	0.0121	0.2823	0.0105	0.2615	0.0093	0.2438
7	0.0206	0.3182	0.0175	0.2898	0.0151	0.2665	0.0132	0.2465	0.0117	0.2298
8	0.0242	0.3040	0.0206	0.2766	0.0179	0.2539	0.0157	0.2348	0.0140	0.2187
9	0.0275	0.2923	0.0235	0.2657	0.0204	0.2436	0.0180	0.2252	0.0161	0.2096
10	0.0305	0.2826	0.0261	0.2565	0.0228	0.2352	0.0201	0.2172	0.0180	0.2019
11	0.0332	0.2742	0.0285	0.2489	0.0249	0.2279	0.0221	0.2104	0.0198	0.1955
12	0.0357	0.2671	0.0308	0.2422	0.0269	0.2217	0.0239	0.2045	0.0214	0.1899
13	0.0381	0.2609	0.0328	0.2363	0.0288	0.2162	0.0256	0.1994	0.0230	0.1850
14	0.0402	0.2553	0.0348	0.2311	0.0305	0.2113	0.0271	0.1948	0.0244	0.1808
15	0.0422	0.2503	0.0365	0.2265	0.0321	0.2070	0.0286	0.1908	0.0257	0.1769
16	0.0441	0.2458	0.0382	0.2224	0.0336	0.2032	0.0299	0.1871	0.0269	0.1735
17	0.0459	0.2418	0.0397	0.2186	0.0350	0.1997	0.0312	0.1838	0.0281	0.1704
18	0.0475	0.2381	0.0412	0.2152	0.0363	0.1965	0.0324	0.1809	0.0292	0.1676
19	0.0490	0.2347	0.0426	0.2121	0.0375	0.1935	0.0335	0.1781	0.0302	0.1650
20	0.0505	0.2316	0.0438	0.2092	0.0387	0.1909	0.0346	0.1755	0.0312	0.1627
21	0.0518	0.2288	0.0451	0.2065	0.0398	0.1884	0.0356	0.1732	0.0321	0.1604
22	0.0531	0.2261	0.0462	0.2041	0.0408	0.1860	0.0365	0.1711	0.0330	0.1584
23	0.0543	0.2237	0.0473	0.2018	0.0418	0.1839	0.0374	0.1691	0.0338	0.1566
24	0.0555	0.2213	0.0483	0.1997	0.0427	0.1820	0.0383	0.1672	0.0346	0.1548
25	0.0566	0.2192	0.0493	0.1976	0.0436	0.1801	0.0391	0.1655	0.0353	0.1531
26	0.0576	0.2172	0.0502	0.1958	0.0445	0.1783	0.0398	0.1639	0.0360	0.1516
27	0.0586	0.2153	0.0512	0.1940	0.0453	0.1767	0.0406	0.1623	0.0367	0.1501
28	0.0596	0.2134	0.0520	0.1923	0.0461	0.1751	0.0413	0.1609	0.0374	0.1488
29	0.0605	0.2117	0.0528	0.1908	0.0468	0.1737	0.0420	0.1595	0.0380	0.1476
30	0.0614	0.2102	0.0536	0.1893	0.0475	0.1723	0.0426	0.1582	0.0386	0.1463
31	0.0623	0.2086	0.0543	0.1879	0.0482	0.1710	0.0432	0.1570	0.0392	0.1451
32	0.0630	0.2072	0.0551	0.1866	0.0488	0.1698	0.0438	0.1558	0.0397	0.1440
33	0.0638	0.2058	0.0558	0.1853	0.0495	0.1686	0.0444	0.1547	0.0402	0.1430
34	0.0646	0.2045	0.0564	0.1841	0.0501	0.1674	0.0449	0.1537	0.0408	0.1420

Table B11 (continued). ANOMV Critical Values $L(\alpha; k, \nu)$ and $U(\alpha; k, \nu)$

$$\alpha = 0.05$$

	Number of Variances Being Compared, k									
	3		4		5		6		7	
ν	lower	upper	lower	upper	lower	upper	lower	upper	lower	upper
3	0.0167	0.8347	0.0090	0.7291	0.0058	0.6426	0.0041	0.5749	0.0031	0.5200
4	0.0315	0.7821	0.0181	0.6705	0.0123	0.5842	0.0090	0.5175	0.0070	0.4654
5	0.0461	0.7420	0.0276	0.6281	0.0192	0.5427	0.0144	0.4781	0.0114	0.4278
6	0.0595	0.7103	0.0366	0.5957	0.0259	0.5114	0.0197	0.4486	0.0157	0.4000
7	0.0715	0.6847	0.0449	0.5700	0.0322	0.4870	0.0248	0.4258	0.0199	0.3789
8	0.0823	0.6635	0.0525	0.5490	0.0380	0.4673	0.0294	0.4074	0.0238	0.3618
9	0.0919	0.6456	0.0593	0.5316	0.0433	0.4510	0.0337	0.3923	0.0274	0.3477
10	0.1005	0.6302	0.0656	0.5167	0.0482	0.4373	0.0377	0.3796	0.0307	0.3358
11	0.1084	0.6168	0.0713	0.5039	0.0527	0.4254	0.0414	0.3686	0.0338	0.3257
12	0.1155	0.6050	0.0765	0.4926	0.0568	0.4150	0.0447	0.3591	0.0367	0.3170
13	0.1220	0.5945	0.0813	0.4827	0.0605	0.4060	0.0478	0.3508	0.0393	0.3092
14	0.1279	0.5851	0.0857	0.4739	0.0640	0.3979	0.0507	0.3434	0.0418	0.3025
15	0.1334	0.5766	0.0898	0.4660	0.0673	0.3906	0.0534	0.3368	0.0441	0.2964
16	0.1384	0.5690	0.0936	0.4588	0.0703	0.3841	0.0559	0.3309	0.0462	0.2909
17	0.1431	0.5619	0.0971	0.4523	0.0732	0.3781	0.0583	0.3255	0.0482	0.2860
18	0.1475	0.5555	0.1005	0.4463	0.0758	0.3727	0.0605	0.3205	0.0501	0.2815
19	0.1515	0.5496	0.1036	0.4408	0.0783	0.3678	0.0626	0.3160	0.0519	0.2774
20	0.1554	0.5440	0.1065	0.4357	0.0806	0.3632	0.0645	0.3118	0.0536	0.2736
21	0.1589	0.5390	0.1092	0.4310	0.0828	0.3590	0.0664	0.3080	0.0551	0.2701
22	0.1623	0.5342	0.1118	0.4266	0.0849	0.3550	0.0681	0.3044	0.0567	0.2668
23	0.1655	0.5297	0.1142	0.4226	0.0869	0.3513	0.0698	0.3011	0.0581	0.2638
24	0.1685	0.5255	0.1166	0.4187	0.0888	0.3479	0.0713	0.2980	0.0594	0.2609
25	0.1714	0.5216	0.1188	0.4151	0.0905	0.3447	0.0728	0.2951	0.0607	0.2583
26	0.1741	0.5179	0.1208	0.4118	0.0923	0.3416	0.0742	0.2924	0.0619	0.2558
27	0.1766	0.5144	0.1229	0.4086	0.0939	0.3388	0.0756	0.2898	0.0631	0.2535
28	0.1791	0.5111	0.1248	0.4056	0.0954	0.3361	0.0769	0.2874	0.0642	0.2513
29	0.1814	0.5080	0.1266	0.4028	0.0969	0.3336	0.0782	0.2851	0.0653	0.2492
30	0.1837	0.5050	0.1283	0.4001	0.0983	0.3311	0.0794	0.2829	0.0663	0.2472
31	0.1858	0.5022	0.1300	0.3975	0.0997	0.3289	0.0805	0.2808	0.0673	0.2453
32	0.1878	0.4995	0.1316	0.3950	0.1010	0.3267	0.0816	0.2789	0.0683	0.2436
33	0.1898	0.4969	0.1331	0.3927	0.1022	0.3246	0.0827	0.2770	0.0692	0.2419
34	0.1917	0.4944	0.1346	0.3905	0.1034	0.3226	0.0837	0.2752	0.0701	0.2402

Table B11 (continued). ANOMV Critical Values $L(\alpha; k, \nu)$ and $U(\alpha; k, \nu)$
$$\alpha = 0.05$$

	Number of Variances Being Compared, k									
	8		9		10		11		12	
ν	lower	upper	lower	upper	lower	upper	lower	upper	lower	upper
3	0.0025	0.4763	0.0020	0.4385	0.0017	0.4075	0.0014	0.3795	0.0012	0.3559
4	0.0056	0.4231	0.0047	0.3887	0.0039	0.3590	0.0034	0.3343	0.0030	0.3129
5	0.0093	0.3878	0.0078	0.3547	0.0066	0.3269	0.0058	0.3038	0.0051	0.2838
6	0.0130	0.3616	0.0109	0.3300	0.0094	0.3038	0.0082	0.2815	0.0073	0.2629
7	0.0165	0.3416	0.0140	0.3112	0.0121	0.2860	0.0106	0.2648	0.0094	0.2467
8	0.0198	0.3254	0.0169	0.2962	0.0147	0.2721	0.0129	0.2515	0.0115	0.2341
9	0.0229	0.3123	0.0196	0.2839	0.0171	0.2606	0.0151	0.2408	0.0135	0.2239
10	0.0258	0.3015	0.0221	0.2737	0.0193	0.2508	0.0171	0.2317	0.0153	0.2154
11	0.0285	0.2921	0.0245	0.2650	0.0214	0.2427	0.0190	0.2240	0.0170	0.2081
12	0.0309	0.2839	0.0267	0.2574	0.0233	0.2356	0.0207	0.2172	0.0186	0.2018
13	0.0332	0.2768	0.0287	0.2509	0.0252	0.2294	0.0224	0.2115	0.0201	0.1964
14	0.0354	0.2706	0.0306	0.2450	0.0268	0.2240	0.0239	0.2065	0.0215	0.1915
15	0.0374	0.2650	0.0323	0.2397	0.0284	0.2191	0.0253	0.2018	0.0228	0.1872
16	0.0392	0.2599	0.0340	0.2351	0.0299	0.2147	0.0267	0.1977	0.0240	0.1833
17	0.0410	0.2554	0.0355	0.2308	0.0313	0.2108	0.0279	0.1940	0.0252	0.1799
18	0.0426	0.2512	0.0370	0.2270	0.0326	0.2072	0.0291	0.1906	0.0263	0.1767
19	0.0442	0.2475	0.0384	0.2235	0.0339	0.2039	0.0302	0.1876	0.0273	0.1738
20	0.0456	0.2440	0.0397	0.2203	0.0350	0.2009	0.0313	0.1848	0.0283	0.1711
21	0.0470	0.2407	0.0409	0.2173	0.0361	0.1981	0.0323	0.1822	0.0292	0.1687
22	0.0484	0.2377	0.0421	0.2145	0.0372	0.1956	0.0333	0.1798	0.0301	0.1664
23	0.0496	0.2350	0.0432	0.2120	0.0382	0.1932	0.0342	0.1776	0.0309	0.1643
24	0.0508	0.2323	0.0443	0.2095	0.0391	0.1910	0.0350	0.1755	0.0317	0.1624
25	0.0519	0.2299	0.0452	0.2073	0.0401	0.1888	0.0359	0.1735	0.0325	0.1605
26	0.0530	0.2277	0.0462	0.2051	0.0409	0.1869	0.0367	0.1716	0.0332	0.1588
27	0.0540	0.2255	0.0471	0.2032	0.0418	0.1850	0.0374	0.1699	0.0339	0.1572
28	0.0550	0.2234	0.0480	0.2013	0.0425	0.1833	0.0382	0.1683	0.0345	0.1557
29	0.0560	0.2215	0.0489	0.1995	0.0433	0.1817	0.0388	0.1668	0.0352	0.1542
30	0.0569	0.2197	0.0497	0.1979	0.0440	0.1801	0.0395	0.1653	0.0358	0.1529
31	0.0577	0.2180	0.0505	0.1963	0.0447	0.1787	0.0401	0.1640	0.0364	0.1516
32	0.0586	0.2164	0.0512	0.1948	0.0454	0.1772	0.0408	0.1626	0.0369	0.1504
33	0.0594	0.2149	0.0519	0.1934	0.0461	0.1759	0.0414	0.1614	0.0375	0.1492
34	0.0601	0.2134	0.0526	0.1920	0.0467	0.1747	0.0419	0.1603	0.0380	0.1481

Table B11 (continued). ANOMV Critical Values $L(\alpha; k, \nu)$ and $U(\alpha; k, \nu)$
$\alpha = 0.01$

	Number of Variances Being Compared, k									
	3		4		5		6		7	
ν	lower	upper	lower	upper	lower	upper	lower	upper	lower	upper
3	0.0055	0.9064	0.0030	0.8130	0.0020	0.7311	0.0014	0.6601	0.0011	0.6042
4	0.0134	0.8592	0.0079	0.7531	0.0054	0.6656	0.0040	0.5952	0.0031	0.5383
5	0.0230	0.8197	0.0140	0.7065	0.0098	0.6176	0.0074	0.5484	0.0058	0.4926
6	0.0329	0.7868	0.0206	0.6702	0.0146	0.5810	0.0112	0.5130	0.0089	0.4589
7	0.0426	0.7592	0.0271	0.6405	0.0195	0.5521	0.0150	0.4852	0.0121	0.4329
8	0.0518	0.7358	0.0334	0.6161	0.0243	0.5284	0.0189	0.4628	0.0153	0.4121
9	0.0604	0.7156	0.0394	0.5954	0.0289	0.5088	0.0225	0.4442	0.0183	0.3945
10	0.0684	0.6980	0.0451	0.5777	0.0332	0.4919	0.0260	0.4285	0.0212	0.3800
11	0.0758	0.6824	0.0504	0.5623	0.0373	0.4775	0.0293	0.4150	0.0240	0.3674
12	0.0827	0.6687	0.0553	0.5488	0.0411	0.4648	0.0324	0.4034	0.0266	0.3566
13	0.0891	0.6563	0.0599	0.5368	0.0447	0.4537	0.0354	0.3931	0.0291	0.3471
14	0.0950	0.6452	0.0642	0.5261	0.0481	0.4437	0.0381	0.3839	0.0314	0.3387
15	0.1006	0.6350	0.0682	0.5164	0.0512	0.4348	0.0407	0.3758	0.0336	0.3312
16	0.1058	0.6258	0.0721	0.5076	0.0542	0.4267	0.0432	0.3684	0.0357	0.3243
17	0.1107	0.6174	0.0756	0.4997	0.0570	0.4194	0.0455	0.3616	0.0377	0.3183
18	0.1153	0.6096	0.0790	0.4923	0.0597	0.4127	0.0477	0.3555	0.0396	0.3126
19	0.1196	0.6024	0.0822	0.4856	0.0622	0.4065	0.0498	0.3500	0.0413	0.3075
20	0.1236	0.5958	0.0852	0.4793	0.0646	0.4008	0.0518	0.3448	0.0430	0.3028
21	0.1275	0.5896	0.0881	0.4736	0.0669	0.3956	0.0536	0.3401	0.0446	0.2984
22	0.1311	0.5837	0.0908	0.4682	0.0691	0.3907	0.0554	0.3356	0.0462	0.2944
23	0.1346	0.5783	0.0934	0.4631	0.0711	0.3862	0.0571	0.3315	0.0476	0.2906
24	0.1379	0.5731	0.0958	0.4584	0.0731	0.3819	0.0588	0.3276	0.0490	0.2872
25	0.1410	0.5683	0.0982	0.4540	0.0749	0.3779	0.0603	0.3240	0.0503	0.2839
26	0.1439	0.5638	0.1004	0.4498	0.0767	0.3741	0.0618	0.3206	0.0516	0.2808
27	0.1468	0.5595	0.1025	0.4459	0.0784	0.3706	0.0632	0.3175	0.0528	0.2779
28	0.1495	0.5554	0.1046	0.4422	0.0801	0.3673	0.0646	0.3144	0.0540	0.2752
29	0.1521	0.5515	0.1065	0.4387	0.0816	0.3642	0.0659	0.3116	0.0551	0.2725
30	0.1546	0.5479	0.1084	0.4353	0.0832	0.3611	0.0672	0.3090	0.0562	0.2701
31	0.1570	0.5444	0.1103	0.4321	0.0846	0.3583	0.0684	0.3064	0.0572	0.2678
32	0.1593	0.5410	0.1120	0.4291	0.0860	0.3556	0.0696	0.3039	0.0583	0.2656
33	0.1615	0.5378	0.1137	0.4262	0.0874	0.3530	0.0707	0.3016	0.0592	0.2635
34	0.1636	0.5348	0.1153	0.4235	0.0887	0.3506	0.0718	0.2994	0.0602	0.2615

Table B11 (continued). ANOMV Critical Values $L(\alpha; k, \nu)$ and $U(\alpha; k, \nu)$
$$\alpha = 0.01$$

ν	Number of Variances Being Compared, k									
	8		9		10		11		12	
	lower	upper	lower	upper	lower	upper	lower	upper	lower	upper
3	0.0009	0.5551	0.0007	0.5141	0.0006	0.4807	0.0005	0.4486	0.0004	0.4185
4	0.0025	0.4914	0.0021	0.4528	0.0018	0.4193	0.0015	0.3901	0.0013	0.3650
5	0.0048	0.4481	0.0040	0.4106	0.0034	0.3788	0.0030	0.3533	0.0026	0.3291
6	0.0074	0.4161	0.0062	0.3801	0.0054	0.3508	0.0047	0.3255	0.0042	0.3036
7	0.0101	0.3913	0.0086	0.3570	0.0074	0.3284	0.0065	0.3043	0.0058	0.2839
8	0.0127	0.3715	0.0109	0.3386	0.0095	0.3110	0.0083	0.2877	0.0074	0.2679
9	0.0154	0.3552	0.0132	0.3231	0.0115	0.2967	0.0101	0.2742	0.0091	0.2552
10	0.0179	0.3415	0.0153	0.3104	0.0134	0.2846	0.0119	0.2631	0.0106	0.2444
11	0.0202	0.3299	0.0174	0.2995	0.0152	0.2745	0.0135	0.2533	0.0121	0.2354
12	0.0225	0.3199	0.0194	0.2902	0.0170	0.2656	0.0151	0.2450	0.0136	0.2277
13	0.0246	0.3110	0.0213	0.2820	0.0187	0.2581	0.0166	0.2378	0.0149	0.2208
14	0.0266	0.3033	0.0231	0.2747	0.0203	0.2512	0.0180	0.2315	0.0162	0.2148
15	0.0285	0.2963	0.0247	0.2682	0.0218	0.2452	0.0194	0.2259	0.0175	0.2095
16	0.0304	0.2900	0.0263	0.2623	0.0232	0.2398	0.0207	0.2208	0.0186	0.2047
17	0.0321	0.2844	0.0278	0.2571	0.0245	0.2348	0.0219	0.2162	0.0198	0.2003
18	0.0337	0.2792	0.0293	0.2524	0.0258	0.2303	0.0231	0.2120	0.0208	0.1964
19	0.0352	0.2744	0.0306	0.2480	0.0271	0.2263	0.0242	0.2082	0.0218	0.1928
20	0.0367	0.2702	0.0319	0.2439	0.0282	0.2226	0.0252	0.2047	0.0228	0.1896
21	0.0381	0.2662	0.0332	0.2403	0.0293	0.2191	0.0262	0.2015	0.0237	0.1865
22	0.0394	0.2624	0.0343	0.2369	0.0304	0.2159	0.0272	0.1985	0.0246	0.1838
23	0.0407	0.2590	0.0355	0.2337	0.0314	0.2130	0.0281	0.1957	0.0254	0.1811
24	0.0419	0.2557	0.0365	0.2307	0.0324	0.2102	0.0290	0.1932	0.0262	0.1787
25	0.0431	0.2527	0.0376	0.2279	0.0333	0.2077	0.0298	0.1907	0.0270	0.1765
26	0.0442	0.2500	0.0386	0.2253	0.0342	0.2052	0.0306	0.1885	0.0277	0.1744
27	0.0452	0.2473	0.0395	0.2228	0.0350	0.2029	0.0314	0.1864	0.0284	0.1724
28	0.0463	0.2448	0.0404	0.2205	0.0358	0.2008	0.0321	0.1844	0.0291	0.1705
29	0.0473	0.2424	0.0413	0.2184	0.0366	0.1987	0.0329	0.1825	0.0298	0.1688
30	0.0482	0.2401	0.0421	0.2163	0.0374	0.1969	0.0336	0.1807	0.0304	0.1670
31	0.0491	0.2380	0.0429	0.2144	0.0381	0.1950	0.0342	0.1790	0.0310	0.1655
32	0.0500	0.2360	0.0437	0.2125	0.0388	0.1933	0.0349	0.1774	0.0316	0.1639
33	0.0509	0.2341	0.0445	0.2108	0.0395	0.1917	0.0355	0.1759	0.0322	0.1626
34	0.0517	0.2323	0.0452	0.2091	0.0402	0.1901	0.0361	0.1745	0.0327	0.1612

Table B12. Tolerance Factors for Normal Populations

	Values of r When a Proportion p of the Population is to Be Covered and $\bar{y}$ is Based on n Observations						Values of u for Confidence Level γ When s Has Degrees of Freedom df			
			p						γ	
n	0.50	0.75	0.90	0.95	0.99	0.999	df	0.90	0.95	0.99
6	0.7322	1.2463	1.7768	2.1127	2.7640	3.5119	6	1.6499	1.9154	2.6230
7	0.7237	1.2326	1.7587	2.0922	2.7399	3.4853	7	1.5719	1.7972	2.3769
8	0.7175	1.2224	1.7448	2.0765	2.7211	3.4644	8	1.5141	1.7110	2.2043
9	0.7127	1.2144	1.7340	2.0641	2.7066	3.4476	9	1.4694	1.6452	2.0762
10	0.7088	1.2080	1.7253	2.0541	2.6945	3.4338	10	1.4337	1.5931	1.9771
11	0.7056	1.2027	1.7182	2.0459	2.6845	3.4223	11	1.4043	1.5506	1.8980
12	0.7030	1.1984	1.7122	2.0390	2.6760	3.4125	12	1.3797	1.5153	1.8332
13	0.7008	1.1947	1.7071	2.0331	2.6688	3.4040	13	1.3587	1.4854	1.7792
14	0.6989	1.1915	1.7027	2.0280	2.6625	3.3967	14	1.3406	1.4597	1.7332
15	0.6973	1.1887	1.6990	2.0236	2.6571	3.3902	15	1.3248	1.4373	1.6936
16	0.6958	1.1863	1.6956	2.0197	2.6523	3.3845	16	1.3108	1.4176	1.6592
17	0.6945	1.1842	1.6926	2.0163	2.6480	3.3794	17	1.2983	1.4001	1.6288
18	0.6934	1.1823	1.6901	2.0132	2.6441	3.3748	18	1.2871	1.3845	1.6019
19	0.6924	1.1807	1.6877	2.0105	2.6407	3.3707	19	1.2770	1.3704	1.5778
20	0.6915	1.1792	1.6855	2.0080	2.6376	3.3670	20	1.2678	1.3576	1.5560
21	0.6907	1.1778	1.6837	2.0058	2.6348	3.3636	21	1.2594	1.3460	1.5363
22	0.6900	1.1765	1.6819	2.0037	2.6322	3.3605	22	1.2517	1.3353	1.5184
23	0.6893	1.1754	1.6803	2.0018	2.6298	3.3576	23	1.2446	1.3255	1.5020
24	0.6887	1.1743	1.6788	2.0001	2.6276	3.3550	24	1.2380	1.3165	1.4868
25	0.6881	1.1734	1.6775	1.9985	2.6256	3.3526	25	1.2319	1.3081	1.4729
26	0.6875	1.1725	1.6762	1.9971	2.6238	3.3503	26	1.2262	1.3002	1.4600
27	0.6870	1.1717	1.6750	1.9957	2.6221	3.3482	27	1.2209	1.2929	1.4479
28	0.6866	1.1709	1.6740	1.9945	2.6205	3.3462	28	1.2159	1.2861	1.4367
29	0.6862	1.1702	1.6730	1.9933	2.6190	3.3444	29	1.2112	1.2797	1.4263
30	0.6858	1.1695	1.6721	1.9922	2.6176	3.3427	30	1.2068	1.2737	1.4164
31	0.6854	1.1689	1.6712	1.9912	2.6163	3.3411	31	1.2026	1.2680	1.4072
32	0.6851	1.1683	1.6704	1.9902	2.6150	3.3396	32	1.1987	1.2627	1.3985
33	0.6848	1.1678	1.6696	1.9893	2.6138	3.3382	33	1.1950	1.2575	1.3903
34	0.6845	1.1673	1.6689	1.9885	2.6128	3.3368	34	1.1914	1.2528	1.3825
35	0.6842	1.1668	1.6682	1.9877	2.6118	3.3356	35	1.1881	1.2482	1.3751

Table B12 (continued). Tolerance Factors for Normal Populations

	Values of r When a Proportion p of the Population is to Be Covered and $\bar{y}$ is Based on n Observations						Values of u for Confidence Level γ When s Has Degrees of Freedom df			
	p							γ		
n	0.50	0.75	0.90	0.95	0.99	0.999	df	0.90	0.95	0.99
36	0.6839	1.1663	1.6676	1.9869	2.6106	3.3344	36	1.1849	1.2438	1.3681
37	0.6836	1.1659	1.6670	1.9862	2.6098	3.3333	37	1.1818	1.2397	1.3615
38	0.6834	1.1655	1.6664	1.9855	2.6090	3.3322	38	1.1789	1.2358	1.3552
39	0.6832	1.1651	1.6658	1.9848	2.6082	3.3311	39	1.1761	1.2320	1.3491
40	0.6830	1.1647	1.6653	1.9842	2.6074	3.3301	40	1.1734	1.2284	1.3434
45	0.6820	1.1632	1.6631	1.9816	2.6039	3.3259	45	1.1616	1.2124	1.3181
50	0.6813	1.1618	1.6612	1.9794	2.6012	3.3225	50	1.1518	1.1993	1.2973
60	0.6801	1.1600	1.6585	1.9762	2.5970	3.3173	60	1.1364	1.1787	1.2651
70	0.6793	1.1586	1.6566	1.9739	2.5940	3.3135	70	1.1248	1.1631	1.2411
80	0.6787	1.1575	1.6551	1.9722	2.5917	3.3107	80	1.1156	1.1510	1.2224
90	0.6782	1.1568	1.6540	1.9708	2.5900	3.3085	90	1.1082	1.1410	1.2072
100	0.6779	1.1561	1.6531	1.9697	2.5886	3.3067	100	1.1019	1.1328	1.1947
110	0.6776	1.1556	1.6523	1.9689	2.5874	3.3053	110	1.0966	1.1258	1.1841
120	0.6773	1.1551	1.6517	1.9681	2.5865	3.3041	120	1.0920	1.1198	1.1750
130	0.6771	1.1548	1.6512	1.9675	2.5857	3.3030	130	1.0880	1.1145	1.1670
140	0.6769	1.1545	1.6507	1.9670	2.5850	3.3022	140	1.0845	1.1098	1.1601
150	0.6767	1.1542	1.6503	1.9665	2.5844	3.3014	150	1.0814	1.1057	1.1539
160	0.6766	1.1539	1.6500	1.9661	2.5838	3.3007	160	1.0785	1.1020	1.1483
170	0.6765	1.1537	1.6497	1.9657	2.5834	3.3001	170	1.0760	1.0986	1.1433
180	0.6764	1.1536	1.6494	1.9654	2.5829	3.2996	180	1.0736	1.0956	1.1387
190	0.6763	1.1534	1.6492	1.9651	2.5826	3.2991	190	1.0715	1.0928	1.1345
200	0.6762	1.1532	1.6490	1.9649	2.5822	3.2987	200	1.0695	1.0902	1.1307
250	0.6758	1.1527	1.6481	1.9639	2.5810	3.2971	250	1.0617	1.0799	1.1154
300	0.6756	1.1523	1.6476	1.9632	2.5801	3.2960	300	1.0559	1.0724	1.1044
400	0.6753	1.1518	1.6469	1.9624	2.5790	3.2946	400	1.0480	1.0620	1.0892
600	0.6751	1.1513	1.6462	1.9616	2.5780	3.2933	600	1.0388	1.0500	1.0717
1000	0.6748	1.1509	1.6457	1.9609	2.5771	3.2922	1000	1.0297	1.0383	1.0547
2000	0.6747	1.1506	1.6453	1.9605	2.5765	3.2914	2000	1.0208	1.0268	1.0381
5000	0.6746	1.1505	1.6450	1.9602	2.5761	3.2909	5000	1.0130	1.0168	1.0238
∞	0.6745	1.1504	1.6449	1.9600	2.5758	3.2905	∞	1.0000	1.0000	1.0000

12.3 APPENDIX C – FIGURES

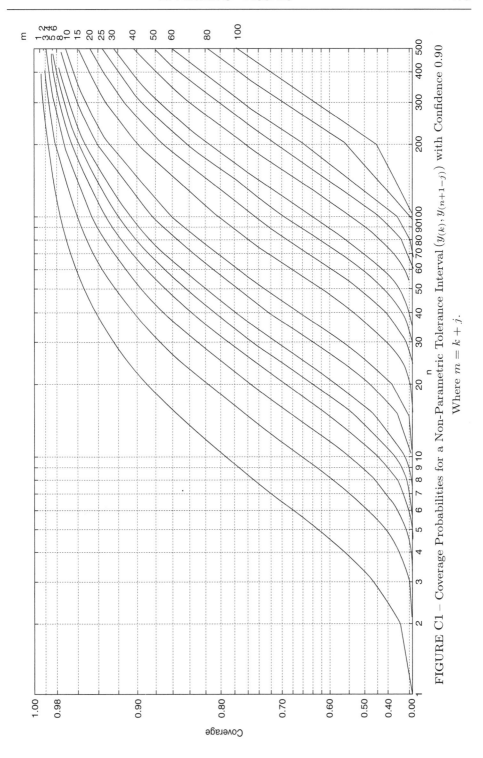

FIGURE C1 – Coverage Probabilities for a Non-Parametric Tolerance Interval $(y_{(k)}, y_{(n+1-j)})$ with Confidence 0.90. Where $m = k + j$.

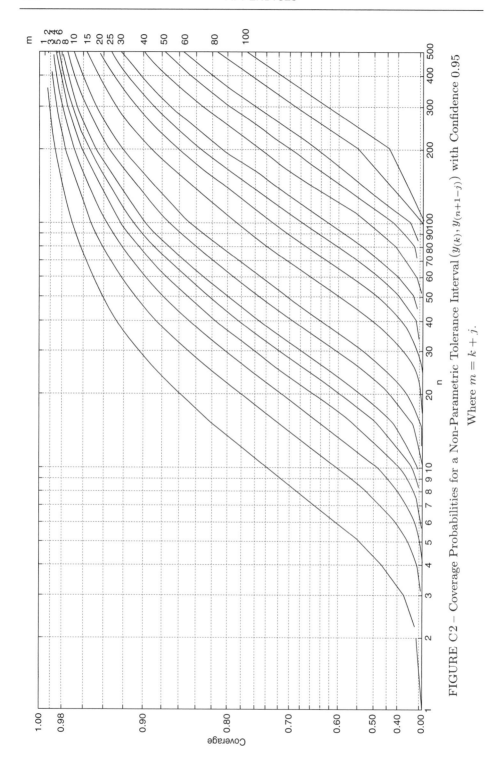

FIGURE C2 – Coverage Probabilities for a Non-Parametric Tolerance Interval $(y_{(k)}, y_{(n+1-j)})$ with Confidence 0.95.

Where $m = k + j$.

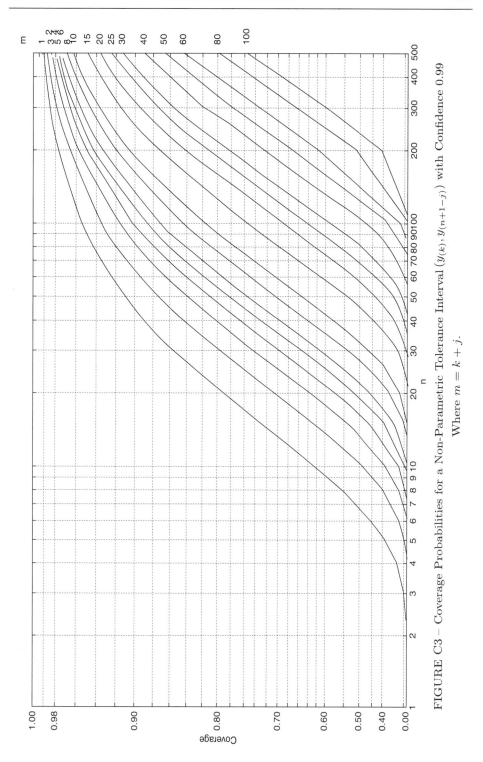

FIGURE C3 – Coverage Probabilities for a Non-Parametric Tolerance Interval $(y_{(k)}, y_{(n+1-j)})$ with Confidence 0.99

Where $m = k + j$.

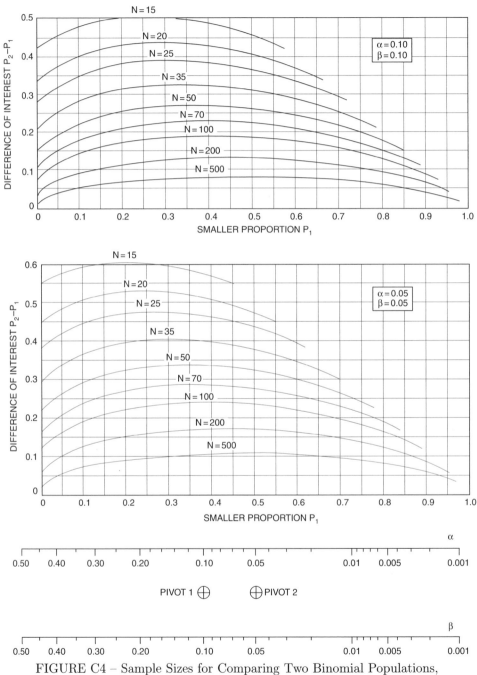

FIGURE C4 – Sample Sizes for Comparing Two Binomial Populations, and Scales for Determining Other α/β Combinations.

12.4 APPENDIX D – SAMPLE PROJECTS

This appendix contains two examples of final project reports based on actual projects. The first report is based on a project done by Josh Branham and Paul Shealy.

The Effect of Launcher Angle and Length of Draw-Back on Water-Balloon Travel

Introduction

Because summer is quickly approaching, water balloons will start to fly around campus. Since the best defense is a good offense, finding the best way to get others wet from the farthest distance is a great way to stay dry. Thus, we were interested in studying the effect of angle and length of draw back on how far water balloons travel when using a three-man water-balloon launcher.

The Experimental Plan

Some initial experimentation showed that reasonable angles were between $0°$ (level) and $45°$, and reasonable draw-back lengths were between 6 and 8 feet. The figure below helps to show how length and angle were measured. The length was measured as the distance between points A and B, and the angle θ was measured relative to when the launcher was in the horizontal (level) position.

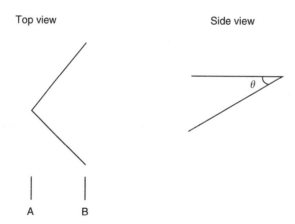

We decided to study three angles and three draw-back lengths using a 3×3 factorial design with four replicates. The three angles used were 0, 30, and $45°$; and the three length were 6, 7, and 8 feet.

The experiment was performed on the Clemson University Tiger Band practice field behind the Brooks Center. Yard lines were already marked on the field so it was only necessary to measure the distance from the closest yard line to the landing point of the balloon. Four people were required to perform the experiment. Two people held the launcher, a third drew back the launcher to launch the balloons, and the fourth measured the distance traveled.

All the balloons were filled to about average fist size to maintain consistency. To achieve randomization the different combinations of angle and draw-back lengths were scattered randomly throughout the 36 launches. Before any data were collected, several practice launches were performed to eliminate the possibility of the launching technique improving as the experiment progressed.

Data Analysis

The data are summarized in the table below.

		Angle (degrees) 0	30	45
		58.7	96.5	178.7
	6	56.0	108.5	202.8
		30.0	78.9	172.5
		36.8	108.8	173.4
		58.0	120.6	234.7
Draw-Back	7	53.0	145.1	193.3
Length (feet)		51.7	99.7	201.2
		41.4	136.3	211.8
		68.0	152.1	232.5
	8	74.0	175.1	232.8
		67.9	127.0	246.5
		55.4	143.6	231.5

Tables of sample means and sample variances based on these data are given in the Appendix.

Since we were planning to analyze the data using the usual model for a two-way layout, namely,

$$Y_{ijk} = \mu + \alpha_i + \beta_j + \alpha\beta_{ij} + \epsilon_{ijk}$$

where $\epsilon_{ijk} \sim N(0, \sigma^2)$, we first checked the assumptions of normality and equal variances. In order to check normality we constructed a normal probability plot of the residuals, which is shown below. For a two-way layout with replicate observations, the residuals are

$$\widehat{\epsilon}_{ijk} = y_{ijk} - \overline{y}_{ij\bullet}.$$

A table of the residuals is given in the Appendix.

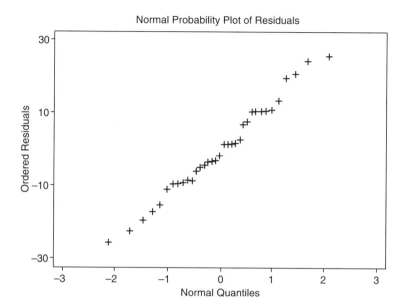

Normal Probability Plot of Residuals

Since the plot is quite linear, there is no problem with the assumption of normality. There is no indication of unequal variance in the normal probability plot, but as a further check we tested for unequal variances using the ANOMV. From the table of sample variances in the Appendix we computed

$$\mathrm{MS}_e = \frac{200.1 + \cdots + 51.0}{9} = 209.15$$

and for $\alpha = 0.1$ the ANOMV decision lines are

$$\mathrm{UDL} = U(0.1; 9, 3)I\ \mathrm{MS}_e = (0.4029)(9)(209.15) = 758.4$$
$$\mathrm{CL} = \mathrm{MS}_e = 209.15$$
$$\mathrm{LDL} = L(0.1; 9; 3)I\ \mathrm{MS}_e = (0.0032)(9)(209.15) = 6.02.$$

The ANOMV chart given below shows no evidence of unequal variances.

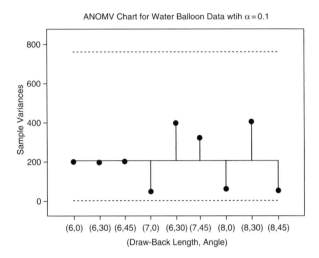

Next we checked for an interaction between the draw-back length and the angle. Using the table of sample means in the Appendix, we first drew an interaction plot, which is shown below.

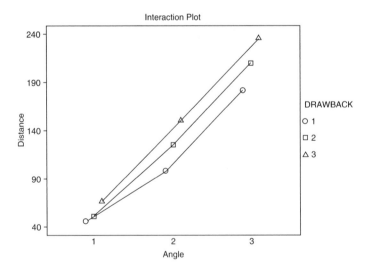

The line segments are fairly parallel, which suggests there is no interaction. As a more precise check, we tested for interaction using the ANOVA. From the ANOVA table given below it is clear that there is no interaction between draw-back length and angle since the p-value for interaction (0.1806) is quite large.

ANALYSIS OF VARIANCE TABLE FOR DISTANCE

SOURCE	DF	SS	MS	F	P
DRAWBACK (A)	2	10620.5	5310.25	25.39	0.0000
ANGLE (B)	2	144715	72357.5	345.97	0.0000
A*B	4	1416.49	354.122	1.69	0.1806
RESIDUAL	27	5646.92	209.145		
TOTAL	35	162399			

Since there is no interaction between the two factors, we can study them separately using the ANOM. From the ANOVA table we know that both factors are extremely significant (p-values that are zero to four significant figures), so we chose to construct ANOM charts using $\alpha = 0.001$. The two factors each have the same number of levels (3), and therefore, the ANOM decision lines will be the same for both factors, namely,

$$\bar{y}_{\bullet\bullet\bullet} \pm h(0.001; 3, 27 \rightarrow 24)\sqrt{209.145}\sqrt{\frac{2}{35}}$$

$$129.3 \pm 4.16(3.46)$$

$$\pm 14.4$$

$$(114.9, 143.7).$$

The two ANOM charts are given below.

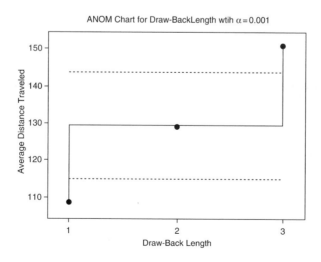

ANOM Chart for Draw-BackLength wtih $\alpha = 0.001$

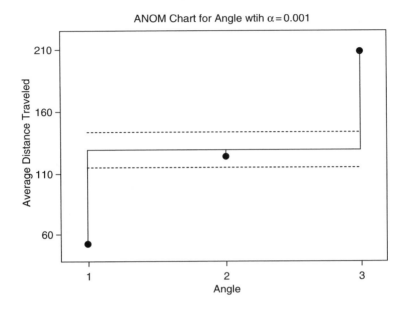

It is clear from the two ANOM charts that both draw-back length and angle have significant ($\alpha = 0.001$) effects due to large draw-back length and large angle producing significantly greater distances and small draw-back length and small angle producing significantly smaller distances.

It is also clear from the ANOM charts that within the range of draw-back lengths and angles studied, the increase in distance traveled is a linear function of both draw-back length and angle. Fitting the multiple linear regression model

$$Y = \beta_0 + \beta_1 x_1 + \beta_2 x_2 + \epsilon$$

where x_1 is the draw-back length and x_2 is the angle, results in

$$\widehat{y} = -100.1 + 21.0 x_1 + 3.29 x_2$$

(see the table below). Thus, for every foot the draw-back length is increased, the distance traveled increases an average of 21 feet, and for every degree the angle is increased the distance traveled increases an average of 3.29 feet.

UNWEIGHTED LEAST SQUARES LINEAR REGRESSION OF DISTANCE

PREDICTOR VARIABLES	COEFFICIENT	STD ERROR	STUDENT'S t	P
CONSTANT	-100.121	31.6551	-3.16	0.0033
DRAWBACK	21.0333	4.43882	4.74	0.0000
ANGLE	3.28750	0.19373	16.97	0.0000

R-SQUARED	0.9039	RESID. MEAN SQUARE (MSE)	472.875
ADJUSTED R-SQUARED	0.8981	STANDARD DEVIATION	21.7457

SOURCE	DF	SS	MS	F	P
REGRESSION	2	146794	73397.0	155.21	0.0000
RESIDUAL	33	15604.9	472.875		
TOTAL	35	162399			

The predicted distance for a draw-back length of 8 feet and an angle of 45° is

$$\widehat{y}_{33} = \widehat{\mu} + \widehat{\alpha}_3 + \widehat{\beta}_3$$
$$= 129.3 + (209.3 - 129.3) + (150.5 - 129.3)$$
$$= 230.5$$

and a 95% prediction interval for the distance a balloon will travel if launched using a draw-back distance of 8 feet and an angle of 45° is

$$\widehat{y}_{33} \pm t(0.025; 27 \to 24)\sqrt{\text{MS}_e}\sqrt{1 + \frac{1}{N}}$$
$$235.8 \pm 2.064\sqrt{209.1}\sqrt{1 + \frac{1}{36}}$$
$$\pm 30.4$$
$$(205.4, 266.2).$$

Conclusions

Using a 3×3 factorial design, we studied the effects of angle and draw-back length on the distance water balloons would travel. We found that the two factors did not interact, and both of them had significant ($\alpha = 0.001$) effects. Both factors had linear effects over the levels studied. Each additional degree of angle resulted in an average of 3.29 feet additional distance, and each additional foot of draw back resulted in an average of 21 feet additional distance. The combination that produced the maximum distance was an angle of 45° with a draw-back length of 8 feet. With that combination a 95% prediction interval for the distance in feet that a balloon will travel is $(205.4, 266.2)$.

Appendix

Tables of sample means and sample variances are given below.

Means

		Factor B			
		1	2	3	
	1	45.4	98.2	181.9	108.5
Factor A	2	51.0	125.4	210.3	128.9
	3	66.3	149.5	235.8	150.5
		54.2	124.4	209.3	129.3

Variances

		Factor B		
		1	2	3
	1	200.1	197.9	202.5
Factor A	2	48.5	396.8	323.1
	3	61.2	401.1	51.0

Using the original data and the table of sample means, we computed the following table of residuals.

		Angle (degrees)		
		0	30	45
		13.3	−1.7	−3.2
	6	10.6	10.3	20.9
		−15.4	−19.3	−9.4
		−8.6	10.6	−8.5
		7.00	−4.8	24.4
Draw-Back	7	2.0	19.7	−17.0
Length (feet)		0.7	−25.7	−9.1
		−9.6	10.9	1.5
		1.7	2.6	−3.3
	8	7.7	25.6	−3.0
		1.6	−22.5	10.7
		−10.9	−5.9	−4.3

The report below is based on a project done by Shannon Driggers.

The Effect of Washer and Dryer Settings on the Size of Your Clothes

Introduction

Every fall, thousands of freshmen students leave for college and venture out on their own, for the first time living away from home. One challenge they will all face is learning how to wash their clothes on their own. Trial and error experiences will teach them over time which settings on the washer and dryer keep their clothes from shrinking. In an effort to bypass these traumatic experiences, I decided to study the effect of different washer and dryer settings on cotton shirts.

The Experimental Procedure

On most standard washers and dryers, there are three temperature settings: cold, warm, and hot. To account for every washing and drying combination possible, nine different treatment combinations were used. Since it is important to have more than one observation in each cell (to calculate variances), 18 T-shirts were used. In order to help ensure independent errors half of the shirts were purchased from one store and half from another. All shirts were the same brand (Fruit-of-the Loom, Active Comfort T-shirts) and the same size (boys medium 6–8). Even though they were all supposed to be the same size, their dimensions varied from shirt to shirt. Therefore, the percent change in the length was used as the response of interest. Prior to any experimentation, the length of every shirt was measured and recorded. For consistency, shirts were measured from top to bottom, along the middle crease, as shown below.

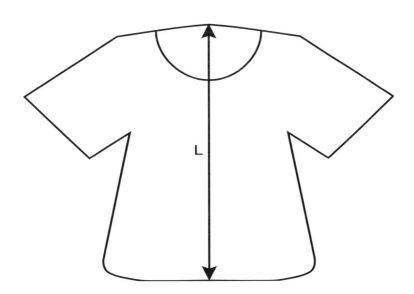

Next, the shirts were divided into nine groups of two (they were mixed up (randomized) so that the store where they were purchased was not a factor), and labeled on the tags so that it would be easier to remember which shirt went to which setting. All 18 shirts were washed and dried separately in a random order (without any detergent or other clothes present in the machines). Measurements were once again recorded for the length of every shirt, and then compared to the initial measurements to calculate a percent change.

Data Analysis

The before and after measurements for each shirt are given in Table A1 in the Appendix. The percent changes in length are displayed in the table below.

		Washer Setting		
		Cold	Warm	Hot
	Cold	8.806	8.635	5.202
		8.381	11.925	3.725
Dryer Setting	Warm	6.024	9.053	4.200
		2.887	4.094	6.015
	Hot	0.398	1.406	0.000
		1.006	0.585	0.202

Tables A2 and A3 in the Appendix contain the cell means and variances for this data.

Since there are two factors in this experiment and more than one observation in each cell, the model

$$Y_{ijk} = \mu + \alpha_i + \beta_j + \alpha\beta_{ij} + \epsilon_{ijk} \qquad (12.4.1)$$

will be used to describe the data, where

$\mu = $ the overall mean

$\alpha_i = $ the effect on the ith level of factor A

$\beta_j = $ the effect of the jth level of factor B

$\alpha\beta_{ij} = $ the interaction between the ith level of A and the jth level of B

$\epsilon_{ijk} = $ the random error.

There are three levels for each factor and two observations per cell. Therefore, $I = 3, J = 3, n = 2$, and $N = 18$. In order to use the above model, the variances in the cells should be equal, and the residuals should be at least approximately normally distributed ($\epsilon_{ijk} \sim N(0, \sigma^2)$).

To test for equal variances one can compute the F_{max} statistic by dividing the largest sample variance by the smallest sample variance to obtain

$$F_{\text{max}} = \frac{s^2_{\text{max}}}{s^2_{\text{min}}} = \frac{12.296}{0.020} = 614.8.$$

The critical value $F_{\max}(0.05; 9, 1)$ is not tabled, but since $614.8 > F_{\max}(0.05; 9, 2) = 475.4$, there is some concern as to whether the variances are equal. The fact that the variances appear unequal is due to the very small variance in the $(3, 3)$ cell, which is in turn due to the suspiciously small value of zero.

There are several possibilities as to how to deal with the unequal variances. One could simply ignore the data in the $(3, 3)$ cell and analyze the remaining data by treating each treatment combination (cell) as a level of a single factor. Another possibility would be to modify the one data value causing the unequal variances and see if changing it has any effect on the conclusions. If the two analyses (modified data and original data with possibly unequal variances) result in the same conclusions, then the fact that the variances appear unequal does not affect the conclusions. Still another possibility would be to try to model the relationship between the cell means and variances, and use that model to transform the data so that the equal variance assumption is satisfied. We will start by analyzing the original data using model (12.4.1) and then see if modifying the zero value (which causes the variances to appear unequal) changes the results.

Analysis One

For model (12.4.1) the residuals are

$$\widehat{\epsilon}_{ijk} = y_{ijk} - \overline{y}_{ij\bullet}$$

and are given in Table A4 in the Appendix. Checking the assumption of normality with a normal probability plot of the residuals (shown below), one finds that the plot is almost a straight line, and therefore the assumption of normality is reasonable.

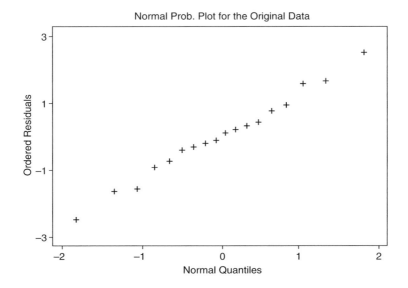

Normal Prob. Plot for the Original Data

The next step is to check to see if there is an interaction between Washer Setting and Dryer Setting. The interaction plot is given below.

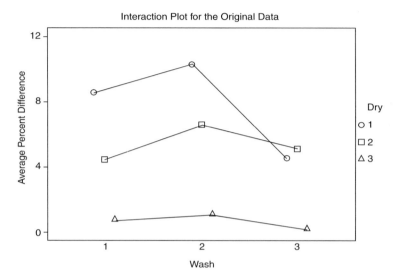

Interaction Plot for the Original Data

The lines are not all close to being parallel, but a more precise check can be done using the analysis of variance (ANOVA). From the ANOVA table given below one sees that (p-value $= 0.2443$) there is no significant interaction.

ANALYSIS OF VARIANCE TABLE FOR PDIFF

SOURCE	DF	SS	MS	F	P
DRY (A)	2	160.300	80.1501	27.74	0.0001
WASH (B)	2	22.2807	11.1403	3.86	0.0617
A*B	4	19.0694	4.76736	1.65	0.2443
RESIDUAL	9	26.0006	2.88895		
TOTAL	17	227.651			

Since there is no significant interaction, we can proceed directly to the analysis of means (ANOM). Using α levels suggested by the ANOVA p-values ($\alpha = 0.1$ for Washer Setting and $\alpha = 0.001$ for Dryer Setting) the decision lines are (for Washer Setting)

$$\bar{y}_{\bullet\bullet\bullet} \pm h(0.1; 3, 9)\sqrt{2.889}\sqrt{\frac{2}{18}}$$
$$4.586 \pm 2.34(0.567)$$
$$\pm 1.33$$
$$(3.26, 5.91)$$

and (for Dryer Setting)

$$\overline{y}_{\bullet\bullet\bullet} \pm h(0.001; 3, 9)\sqrt{3.796}\sqrt{\frac{2}{18}}$$
$$4.586 \pm 5.49(0.567)$$
$$\pm 3.11$$
$$(1.48, 7.70)$$

and the ANOM charts are given below.

ANOM Chart for Washer Settings (Original Data) with α=0.1

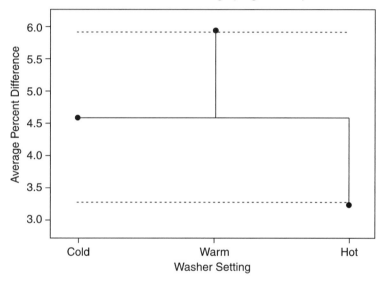

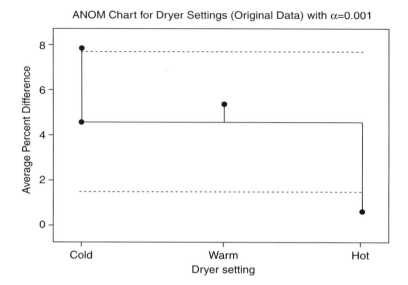

ANOM Chart for Dryer Settings (Original Data) with $\alpha=0.001$

From the ANOM charts, one would conclude that Washer Setting is significant at the $\alpha = 0.1$ level due to the warm setting resulting in significantly more shrinkage and the hot setting resulting in significantly less shrinkage, and Dryer Setting is significant at the $\alpha = 0.001$ level due to the cold setting resulting in significantly more shrinkage and the hot setting resulting in significantly less shrinkage.

Changing the 0.00 in cell $(3, 3)$ to 0.80 results in a sample variance of $s_{33}^2 = 0.179$ for that cell, and an $F_{\max}$ value of

$$F_{\max} = \frac{12.296}{0.090} = 136.6.$$

Since $136.6 < F_{\max}(0.1; 9, 2) = 230.7 < F_{\max}(0.1; 9, 1)$, there is now no evidence of unequal variances. Since only one value has been changed, the normal probability plot and the interaction plot look essentially the same as their previous counterparts. The new ANOVA table is

ANALYSIS OF VARIANCE TABLE FOR PDIFF

SOURCE	DF	SS	MS	F	P
DRY (A)	2	153.993	76.9966	26.49	0.0002
WASH (B)	2	20.1732	10.0866	3.47	0.0764
A*B	4	20.5925	5.14813	1.77	0.2186
RESIDUAL	9	26.1593	2.90659		
TOTAL	17	220.918			

The new row means, column means, and grand mean are given in Table A5 in the Appendix. Using the new MS_e and the new grand mean to compute ANOM decision lines, one obtains (for Washer Setting)

$$\bar{y}_{\bullet\bullet\bullet} \pm h(0.1; 3, 9)\sqrt{2.907}\sqrt{\frac{2}{18}}$$
$$4.630 \pm 2.34(0.568)$$
$$\pm 1.33$$
$$(3.30, 5.96)$$

and (for Dryer Setting)

$$\bar{y}_{\bullet\bullet\bullet} \pm h(0.001; 3, 9)\sqrt{2.907}\sqrt{\frac{2}{18}}$$
$$4.630 \pm 5.49(0.568)$$
$$\pm 3.12$$
$$(1.51, 7.75).$$

From the ANOM charts given below, one sees that Washer Setting is not significant at the $\alpha = 0.1$ level, and Dryer Setting is significant at the $\alpha = 0.001$ level due to the cold setting producing significantly more shrinkage and the hot setting producing significantly less shrinkage. Thus, the unequal variances do not substantially affect the conclusions.

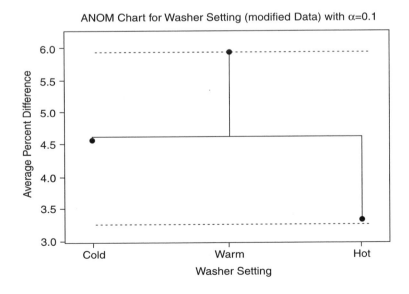

ANOM Chart for Washer Setting (modified Data) with α=0.1

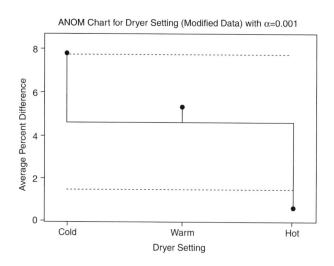

ANOM Chart for Dryer Setting (Modified Data) with α=0.001

Analysis Two

Re-analyzing the original data without using the data in cell $(3,3)$ and using the model

$$Y_{ijk} = \mu + \alpha_{ij} + \epsilon_{ijk}$$

one obtains the following table of sample means and sample variances.

Washer/Dryer Settings

	(Cold, Cold)	(Cold, Warm)	(Cold, Hot)	(Warm, Cold)
$\bar{y}_{ij\bullet}$	8.594	4.455	0.702	10.280
s_{ij}^2	0.090	4.922	0.185	5.413

	(Warm, Warm)	(Warm, Hot)	(Hot, Cold)	(Hot, Warm)
$\bar{y}_{ij\bullet}$	6.573	0.995	4.464	5.107
s_{ij}^2	12.296	0.337	1.091	1.647

From these values one obtains

$$\bar{y}_{\bullet\bullet\bullet} = \frac{8.594 + \cdots + 5.107}{8} = 5.15$$

$$\mathrm{MS}_e = \frac{0.090 + \cdots + 1.647}{8} = 3.25.$$

Checking for equal variances, one would compute

$$F_{\max} = \frac{12.296}{0.090} = 136.6$$

and since $136.6 < F_{\max}(0.1; 8; 2) = 195.6 < F_{\max}(0.1; 8; 1)$, there is no evidence of unequal variances. The residuals for this model are the same as those for model (12.4.1) except the residuals from cell $(3,3)$ are missing. Therefore, we can conclude from the previous normal probability plot that the assumption of normality is reasonable. Decision limits for the ANOM are (using $\alpha = 0.05$)

$$\bar{y}_{\bullet\bullet\bullet} \pm h(0.05; 8, 8)\sqrt{3.25}\sqrt{\frac{7}{16}}$$

$$5.15 \pm 3.51(1.19)$$

$$\pm 4.19$$

$$(0.96, 9.34)$$

and from the ANOM chart given below one sees that there are significant differences at the $\alpha = 0.05$ level due to the (Warm, Cold) combination producing significantly more shrinkage and the (Cold, Hot) combination producing significantly less shrinkage.

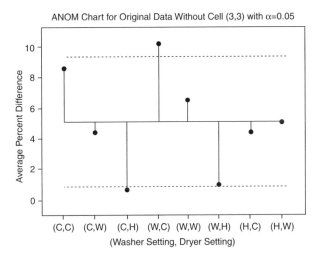

ANOM Chart for Original Data Without Cell (3,3) with α=0.05

(Washer Setting, Dryer Setting)

Analysis Three

If the standard deviations σ_{ij} are related to the means μ_{ij} through the model

$$\sigma_{ij} = c\mu_{ij}^{\lambda}$$

then

$$\ln(\sigma_{ij}) = \ln(c) + \lambda \ln(\mu_{ij})$$

and the parameter λ can be estimated by replacing the unknown σ_{ij}'s and μ_{ij}'s with s_{ij}'s and $\bar{y}_{ij\bullet}$'s, and fitting the simple linear model

$$\ln(s_{ij}) \doteq \ln(c) + \lambda \ln(\bar{y}_{ij\bullet}).$$

Transforming the y_{ijk}'s to $w_{ijk} = y_{ijk}^{1-\widehat{\lambda}}$ should then stabilize the variances. A scatter plot of the $\ln(s_{ij})$'s versus the $\bar{y}_{ij\bullet}$'s is shown below, and it appears reasonable to fit a line to these points.

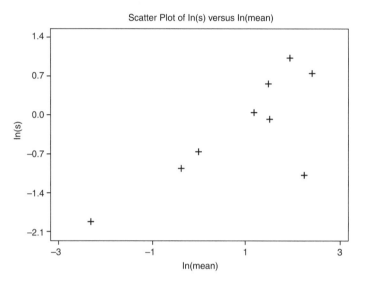

Scatter Plot of ln(s) versus ln(mean)

The resulting fitted line is (see Table A6)

$$\widehat{\ln(s_{ij})} = -0.624 + 0.512\ln(\bar{y}_{ij\bullet})$$

and $\widehat{\lambda} = 0.512 \doteq 0.5$. Therefore, one should try transforming the y_{ijk}'s by taking the square root. The resulting transformed percent differences are given in the table below.

		Washer Setting		
		Cold	Warm	Hot
	Cold	2.968	2.938	2.281
		2.895	3.453	1.930
Dryer Setting	Warm	2.454	3.009	2.049
		1.699	2.023	2.453
	Hot	0.631	1.186	0.000
		1.003	0.765	0.449

Tables A7 and A8 in the Appendix contain the cell means and variances for this data. Testing for equal variances, one obtains

$$F_{\max} = \frac{4.86}{0.003} = 162$$

and since $162 < F_{\max}(0.1; 9, 2) = 230.7 < F_{\max}(0.1; 9, 1)$, there is no evidence of unequal variances. The residuals for the transformed data using model (12.4.1) are given in Table A9. A normal probability plot of the residuals and an interaction plot for the transformed data are given below.

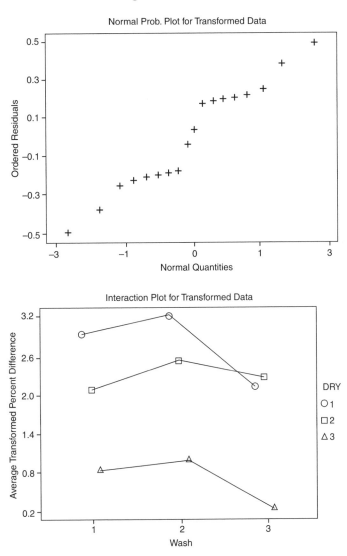

The normal probability plot does not indicate any non-normality serious enough to worry about. Based on a comparison with the previous interaction plots, it seems unlikely that there is any interaction, and this is confirmed by the ANOVA table given below.

ANALYSIS OF VARIANCE TABLE FOR W

SOURCE	DF	SS	MS	F	P
DRY (A)	2	14.1922	7.09610	48.85	0.0000
WASH (B)	2	1.49466	0.74733	5.14	0.0324
A*B	4	0.62129	0.15532	1.07	0.4258
RESIDUAL	9	1.30731	0.14526		
TOTAL	17	17.6155			

Using α levels suggested by the ANOVA p-values ($\alpha = 0.05$ for Washer Setting and $\alpha = 0.001$ for Dryer Setting) the ANOM decision lines are (for Washer Setting)

$$\overline{y}_{\bullet\bullet\bullet} \pm h(0.05; 3, 9)\sqrt{0.1453}\sqrt{\frac{2}{18}}$$
$$1.899 \pm 2.79(0.127)$$
$$\pm 0.354$$
$$(1.545, 2.253)$$

and (for Dryer Setting)

$$\overline{y}_{\bullet\bullet\bullet} \pm h(0.001; 3, 9)\sqrt{3.796}\sqrt{\frac{2}{18}}$$
$$1.899 \pm 5.49(0.127)$$
$$\pm 0.697$$
$$(1.202, 2.596)$$

and the ANOM charts are given below.

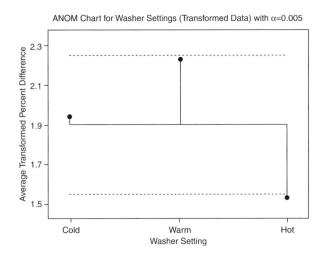

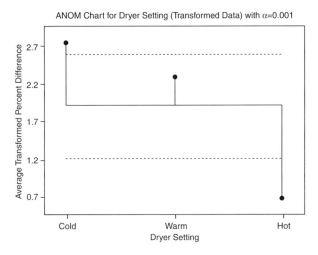

From the ANOM charts, one would conclude that Washer Setting is significant at the $\alpha = 0.05$ level due to the hot setting producing significantly less shrinkage, and that Dryer Setting is significant at the $\alpha = 0.001$ level due to the cold setting resulting in significantly more shrinkage and the hot setting resulting in significantly less shrinkage.

Conclusions

A 3×3 factorial experiment with two replicates was run to help determine which washer and dryer settings would cause cotton shirts to shrink the least. The response measured was the percent shrinkage. Three analyses were performed because of concern with unequal variances. The first analysis was based on a two-way layout using the original data and the original data with one modified value to remove any concern over unequal variances. This analysis suggested that there was no interaction between Washer Setting and Dryer Setting. Dryer Setting had a more significant effect ($\alpha = 0.001$) than Washer Setting ($\alpha = 0.1$), with cold and hot dryer settings producing significantly more and less shrinkage, respectively; and the hot washer setting producing significantly less shrinkage. Thus, the combination leading to significantly less shrinkage was both washer and dryer settings of hot.

The second analysis ignored the values in the (Hot, Hot) cell, which had caused the concern with unequal variances. The remaining cells were treated as levels of a single factor. Obviously, using this analysis, no conclusion could be drawn about the (Hot, Hot) treatment combination. With this analysis it was found that there was a significant ($\alpha = 0.05$) effect due to the (Warm, Cold) Washer/Dryer combination producing significantly more shrinkage.

The third analysis found no interaction and that both Washer Setting ($\alpha = 0.05$) and Dryer Setting ($\alpha = 0.001$) had significant effects. The hot Washer Setting resulted in significantly less shrinkage, the cold Dryer Setting resulted in significantly more shrinkage, and the hot Dryer Setting resulted in significantly less shrinkage.

Since these conclusions are all contrary to the usual washing/drying recommendation, it seems that the (Hot, Hot) settings should be investigated further to confirm that combination's superiority.

Appendix

Table A1 – Before and After
Length Measurements

Washer	Dryer	Before	After	% Difference
1	1	51.1	46.6	8.806
1	1	52.5	48.1	8.381
1	2	49.8	46.8	6.024
1	2	48.5	47.1	2.887
1	3	50.2	50.0	0.398
1	3	49.7	49.2	1.006
2	1	49.8	45.5	8.634
2	1	47.8	42.1	11.925
2	2	47.5	43.2	9.053
2	2	51.3	49.2	4.094
2	3	49.8	49.1	1.406
2	3	51.3	51.0	0.585
3	1	51.9	49.2	5.202
3	1	51.0	49.1	3.725
3	2	50.0	47.9	4.200
3	2	53.2	50.0	6.015
3	3	51.0	51.0	0.000
3	3	49.6	49.5	0.202

Table A2 – Cell Means, Row Means, Column Means, and the Grand Mean

		Washer Setting			
		Cold	Warm	Hot	
Dryer Setting	Cold	8.594	10.280	4.464	7.779
	Warm	4.455	6.573	5.107	5.379
	Hot	0.702	0.995	0.101	0.599
		4.584	5.949	3.224	4.586

Table A3 – Cell Variances

		Washer Setting		
		Cold	Warm	Hot
Dryer Setting	Cold	0.090	5.412	1.091
	Warm	4.922	12.296	1.647
	Hot	0.185	0.337	0.020

Table A4 – Residuals for Model (12.4.1)
Using Original Data

		Washer Setting		
		Cold	Warm	Hot
	Cold	0.213	−1.645	0.738
		−0.213	1.645	−0.738
Dryer Setting	Warm	1.569	2.480	−0.908
		−1.569	−2.480	0.908
	Hot	−0.304	0.410	−0.101
		0.304	−0.410	0.101

Table A5 – Cell Means, Row Means, Column Means,
and the Grand Mean for the Modified Data

		Washer Setting			
		Cold	Warm	Hot	
	Cold	8.594	10.280	4.464	7.779
Dryer Setting	Warm	4.455	6.573	5.107	5.379
	Hot	0.702	0.995	0.501	0.733
		4.584	5.949	3.357	4.630

Table A6 – Regression Analysis of $(\ln(s_{ij}), \ln(\overline{y}_{ij\bullet}))$ pairs

UNWEIGHTED LEAST SQUARES LINEAR REGRESSION OF LNS

PREDICTOR VARIABLES	COEFFICIENT	STD ERROR	STUDENT'S t	P
CONSTANT	-0.62429	0.30771	-2.03	0.0821
LNMEAN	0.51226	0.18062	2.84	0.0252

R-SQUARED	0.5347	RESID. MEAN SQUARE (MSE)	0.60052
ADJUSTED R-SQUARED	0.4682	STANDARD DEVIATION	0.77494

SOURCE	DF	SS	MS	F	P
REGRESSION	1	4.83058	4.83058	8.04	0.0252
RESIDUAL	7	4.20367	0.60052		
TOTAL	8	9.03425			

Table A7 – Cell Means, Row Means, Column Means,
and the Grand Mean for Transformed Data

		Washer Setting			
		Cold	Warm	Hot	
	Cold	2.931	3.196	2.106	2.744
Dryer Setting	Warm	2.077	2.516	2.251	2.281
	Hot	0.817	0.975	0.225	0.672
		1.942	2.229	1.527	1.899

Table A8 – Cell Variances for
Transformed Data

		Washer Setting		
		Cold	Warm	Hot
	Cold	0.003	0.133	0.062
Dryer Setting	Warm	0.285	0.486	0.081
	Hot	0.069	0.089	0.101

13

REFERENCES

Bayes, T. (1763). "An Essay Towards Solving a Problem in the Doctrine of Chances". *Philosophical Transactions of the Royal Society of London* 53, pp. 370-418.

Bernoulli, J. (1713). *Ars Conjectandi.* Thurnisius, Basilea.

Bortkiewicz, L. von (1898). *Das Gesetz der Kleinen Zahlen.* Teubner, Leipzig.

De Moivre, A. (1711). "De Mensura Sortis". *Philosophical Transactions of the Royal Society, No. 329,* 27, pp. 213-264.

De Moivre, A. (1733). "Approximatio ad Summam Ferminorum Binomii $(a + b)^n$ in Seriem expansi". *Supplementum II to Miscellanae Analytica,* pp. 1-7.

Frechét, M. (1927). "Sur la Loi de Probabilité de l'Écart Maximum". *Annales de la Société Polonaise de Mathematique, Cracovie* 6, pp. 93-116.

Galton, F. (1879). "The Geometric Mean in Vital and Social Statistics". *Proceedings of the Royal Society of London* 29, pp. 365-367.

Gauss, C. F. (1809). *Theoria Motus Corporum Coelestium.* Perthes & Besser, Hamburg, Germany.

Gauss, C. F. (1816). "Bestimmung der Genauigkeit der Beobachtungen". *Zeitshrift Astronomi* 1, pp. 185-197.

Hahn, G. J. and Meeker, W. Q. (1991). *Statistical Intervals: A Practical Guide for Practitioners.* John Wiley & Sons, New York, NY.

Laplace, P. S. (1774). "Determiner le milieu que l'on doit prendre entre trois observations données d'un même phénomené". *Mémoires de Mathematique et Physique presentées à l'Académie Royale des Sciences par divers Savans* 6, pp. 621-625.

Laplace, P. S. (1812). *Théorie Analytique des Probabilités*, 1st edn., Paris.

McAlister, D. (1879). "The Law of the Geometric Mean". *Proceedings of the Royal Society of London* 29, pp. 367-375.

Murphy, R. B. (1948). "Non-Parametric Tolerance Limits". *Annals of Mathematical Statistics* 19, pp. 581-589.

Nelson, L. S. (1977). "Tolerance Factors for Normal Distributions". *Journal of Quality Technology* 9, pp. 198-199.

Nelson, L. S. (1980). "Sample Sizes for Comparing Two Proportions". *Journal of Quality Technology* 12, pp. 114-115.

Nelson, L. S. (1985). "Sample Size Tables for Analysis of Variance". *Journal of Quality Technology* 17, pp. 167-169.

Nelson, L. S. (1987). "Upper 10%, 5% and 1% Points of the Maximum F Ratio". *Journal of Quality Technology* 19, pp. 165-167.

Nelson, P. R. (1983). "A Comparison of Sample Sizes for the Analysis of Means and the Analysis of Variance". *Journal of Quality Technology* 15, pp. 33-39.

Nelson, P. R. (1989). "Multiple Comparisons of Means Using Simultaneous Confidence Intervals". *Journal of Quality Technology* 21, pp. 232-241.

Nelson, P. R. (1993). "Additional Uses for the Analysis of Means and Extended Tables of Critical Values". *Technometrics* 35, pp. 61-71.

Pascal, B. (1679). *Varia opera Mathematica D. Petri de Fermat.* Tolossae.

Poisson, S. D. (1830). "Mémoire sur la proportion des naissances des filles et des garons". *Mémoires de l'Académie royale de l'Institut de France* 9, pp. 239-308.

Rosen, P. and Rammler, B. (1933). "The Law Governing the Fineness of Powdered Coal". *Journal of the Institute of Fuels* 6, pp. 29-36.

Rutherford, E. and Geiger, H. (1910). "The Probability Variations in the Distribution of α Particles". *Philosophical Magazine, 6th series* 20, pp. 698-704.

Satterthwaite, F. E. (1946). "An Approximate Distribution of Estimates of Variance Components". *Biometrics Bulletin* 2, pp. 110-114.

Shapiro, S. S. and Wilk, M. B. (1965). "An Analysis of Variance Test for Normality (Complete Samples)". *Biometrika* 52, pp. 591-611.

Weibull, W. (1939a). "A Statistical Theory of the Strength of Materials". Report No. 151, Ingeniörs Vetenskaps Akademiens Handligar, Stockholm.

Weibull, W. (1939b). "The Phenomena of Rupture in Solids". Report No. 153, Ingeniörs Vetenskaps Akademiens Handligar, Stockholm.

Weibull, W. (1951). "A Statistical Distribution of Wide Applicability". *Journal of Applied Mechanics* 18, pp. 293-297.

Wludyka, P. S. and Nelson, P. R. (1997). "An Analysis-of-Means-Type Test for Variances from Normal Populations". *Technometrics* 39, pp. 274-285.

INDEX

511